Automation, Control and Intelligent Systems

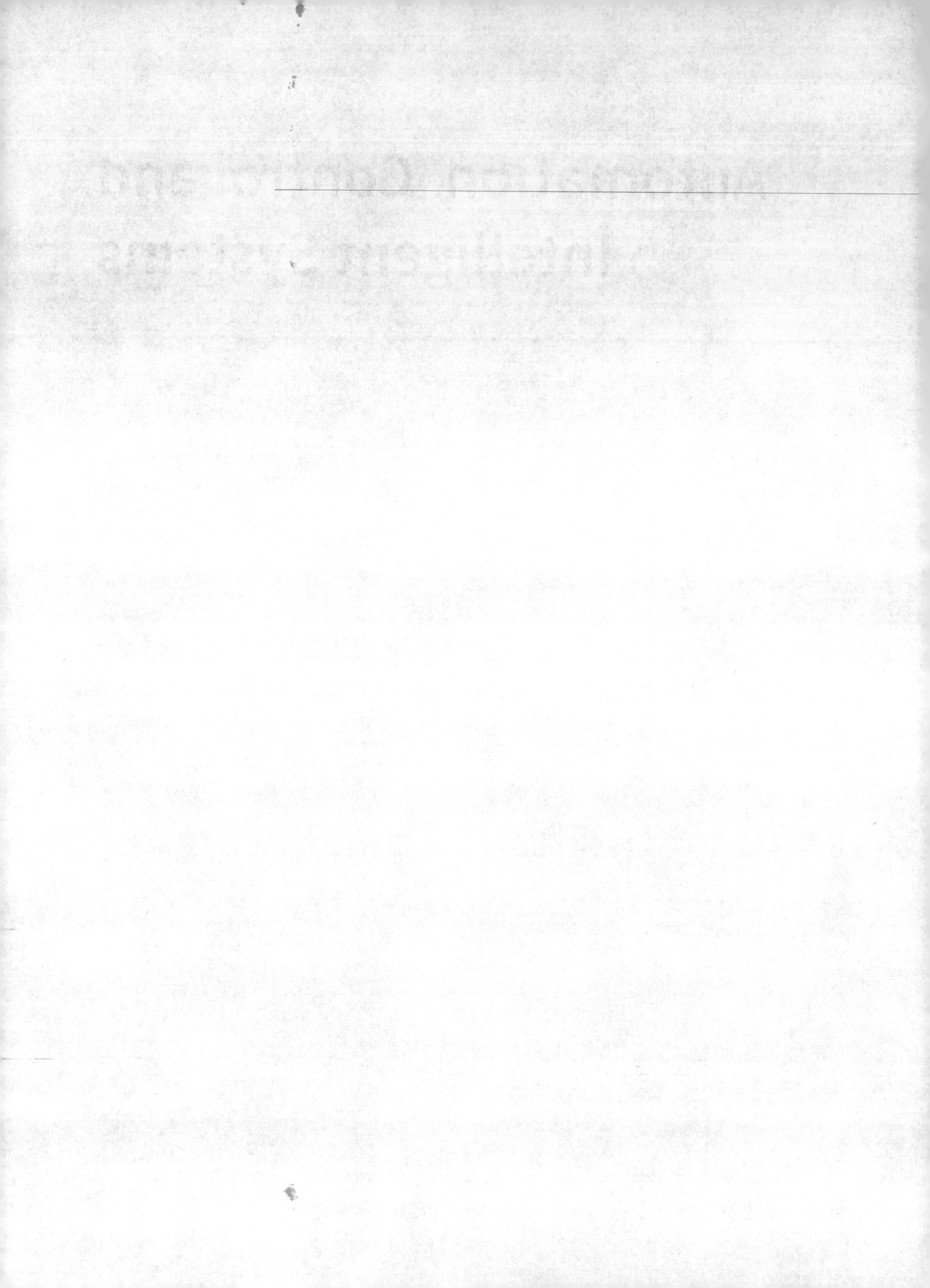

Automation, Control and Intelligent Systems

Edited by
Alfred Silva

Willford Press
www.willfordpress.com

Published by Willford Press,
118-35 Queens Blvd., Suite 400,
Forest Hills, NY 11375, USA

ISBN: 978-1-68285-348-1

Cataloging-in-Publication Data

Automation, control and intelligent systems / edited by Alfred Silva.
p. cm.
Includes bibliographical references and index.
ISBN 978-1-68285-348-1
1. Automation. 2. Automatic control. 3. Computational intelligence. 4. Artificial intelligence.
5. Intelligent control systems. 6. Computer engineering. I. Silva, Alfred.

T59.5 .A98 2017
670.427--dc23

For information on all Willford Press publications visit our website at www.willfordpress.com

Willford Press

Printed in the United States of America.

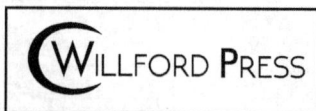

Contents

Preface

In my initial years as a student, I used to run to the library at every possible instance to grab a book and learn something new. Books were my primary source of knowledge and I would not have come such a long way without all that I learnt from them. Thus, when I was approached to edit this book; I became understandably nostalgic. It was an absolute honor to be considered worthy of guiding the current generation as well as those to come. I put all my knowledge and hard work into making this book most beneficial for its readers.

Automation is the process whereby machinery is used for operating various kinds of equipment. This book on automation, control and intelligent systems discusses topics related to the design and manufacture of control systems. Innovation in this field can lead to the creation of mechatronic devices that can perform a variety of tasks and duties. This book discusses the fundamentals as well as modern approaches of automation and control. It elucidates new techniques and their applications in a multidisciplinary approach. The topics covered in the book offer the readers new insights in the field of automation and control systems. This book aims to equip students and experts with the advanced topics and upcoming concepts in this area.

I wish to thank my publisher for supporting me at every step. I would also like to thank all the authors who have contributed their researches in this book. I hope this book will be a valuable contribution to the progress of the field.

Editor

Flatness control of a crane

H. Souilem[1], H. Mekki[2] and N. Derbel[1]

Control, Energy and Management Lab (CEM Lab)
[1]National School of Engineers of Sfax BP.W, 3038, Sfax-Tunisia
[2]National School of Engineers of Sousse

Email address:
haifa.souilem@gmail.com (H. Souilem); hassen.mekki@eniso.rnu.tn (H. Mekki); nabil.derbel@ieee.org (N. Derbel)

Abstract: The aim of this work is to propose a flatness control of a crane detailing adopted mechanisms and approaches in order to be able to control this system and to solve problems encountered during its functioning. The control objective is the sway-free transportation of the crane's load taking the commands of the crane operator into account. Based on the mathematical model linearizing and stabilizing control laws for the slewing and luffing motion are derived using the input/output linearization approach. The method allows for transportation of the payload to a selected point and ensures minimisation of its swings when the motion is finished. To achieve this goal a mathematical model of the control system of the displacement of the payload has been constructed. A theory of control which ensures swing-free stop of the payload is presented. Selected results of numerical simulations are shown. At the end of this work, a comparative study between the real moving and the desired one has been presented.

Keywords: Crane, Flatness Control, Path Planning, Path Tracking

1. Introduction

Flatness-based control techniques have been developed and applied in many industrial processing with a great success in solving planning and tracking problems of reference trajectories such as thermal process control [15], motors control [1], chemical reactor control [16], crane control [10] etc... This theory was introduced in 1992 by M. Fliess and al. [6].

The existence of a variable called a flat output permits to define all other system variables. The dynamics of such process can be then deduced without solving differential equations. Therefore, it is possible to express the state, as well as the input and the output system, as differential functions of the flat output [6] [11].

Conventionally, it is difficult to resolve the path planning problem due to the necessity to solve the differential system equations from the initial conditions to obtain the solution at the final time. In the case of flat systems, this problem can be solved easily without approximation and without solving differential systems equations. Indeed, flatness property ensures the existence of a flat output which allows the parameterization of all system variables as a function of finite number of its derivatives.

The goal of this work is to solve problems encountered during the motion of the load using the technique known by control by fatness whose main objective is to attenuate the undesirable swings of the load [7], [9], [13]. In fact, the differential flatness has been introduced by Fliess and al. [5] in 1995. The states and the in puts of the flat system can be expressed in function of the particular out puts and their successive derivatives. We can find a lot of the literature uses a linear approaches [3], [8], [19] or an approaches of optimal control [12], [20]. Also, several methods are proposed in [9] and [14] in order to decrease the oscillations created by the outsides disruptions. Authors of [2] and [4] use energizing techniques by exploiting the fact that a crane can be identified to a pendulum if we fixes the length of the vertical cable bound to the load. Other techniques can be useful: technique of in put / out put linearization, technique of in put / state linearization, but these techniques present several problems which are so difficult to solve it. We here interested to exploit the concept of the flatness in order to control the system crane: in section 2, we present the dynamic model of the crane. In section 3, the crane is modeled by a flat system. Section 4 deals with flatness and linearization. Section 5 deals with flatness and trajectories generation. Finally, in section 6, we present the flatness and the tracking of trajectory.

2. Formulation of the Problem

The most of the weight handling equipments use ropes and winches in order to displace the load. This system of raising went up on a mechanical structure with one or two articulations.

Among these weight handling equipments, we consider the crane system which its characteristics are the following:

- A load of mass M which its coordinates are:(ξ, ζ)

A weight handling equipment compound a rope, pulleys and winches. The motors manipulating the winches are supposed to be controlled by two couples:T_1 and T_2.

- A mechanical structure entirely articulate on which is fixed the winches of coil of the ropes.
- A rail breaks the movement of the pulley.

The tabular diagram of the crane is given by the following figure: [17]

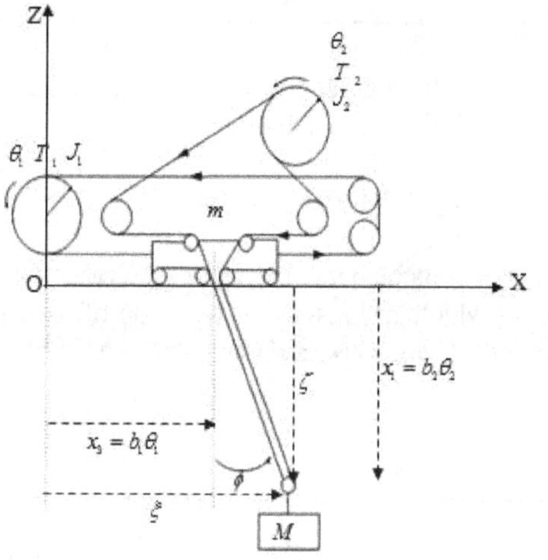

Figure 1. *Crane in dimension 2.*

2.1. Model of the crane

The dynamic model of the crane can be given by: [17]

$$T_1 = [J_1 + (M+m)b_1^2]\ddot{\theta}_1 + Mb_1b_2\,\ddot{\theta}_2 \sin\phi$$
$$+ Mb_1b_2[2\,\dot{\theta}_2\,\dot{\phi}\cos\phi - \theta_2(\dot{\phi})^2 \sin\phi] \tag{1}$$
$$+ Mb_1b_2\,\ddot{\phi}\,\theta_2 \cos\phi$$

$$T_2 = (J_2 + Mb_2^2)\ddot{\theta}_2 - Mb_2^2\theta_2(\dot{\phi})^2$$
$$+ Mb_1b_2\,\ddot{\theta}_1 \sin\phi - Mgb_2 \cos\phi \tag{2}$$

$$b_1\ddot{\theta}_1 \cos\phi + b_2\theta_2\ddot{\phi} + 2b_2\dot{\theta}_2\dot{\phi} + g\sin\phi = 0 \tag{3}$$

The geometric constraints between the coordinates of the wagon and the load:

$$\xi = x_3 + x_1 \sin\phi \tag{4}$$

$$\zeta = x_1 \cos\phi \tag{5}$$

with,

$$x_1 = b_2\theta_2 \text{'} x_3 = b_1\theta_1$$

thus, we gets:

$$\begin{pmatrix} \frac{J_1}{b_1} + (M+m)b_1 & M.b_1 \sin\phi & M.b_1x_1\cos\phi \\ M.b_2 \sin\phi & \frac{J_2}{b_2} + M.b_2 & 0 \\ \cos\phi & 0 & x_1 \end{pmatrix} . \begin{pmatrix} \ddot{x}_3 \\ \ddot{x}_1 \\ \ddot{\phi} \end{pmatrix} =$$

$$\begin{pmatrix} T_1 - Mb_1(2\dot{x}_1\dot{\phi}\cos\phi - x_1\dot{\phi}^2 \sin\phi) \\ T_2 + Mgb_2\cos\phi + M.b_2x_1\dot{\phi}^2 \\ -g\sin\phi - 2\dot{x}_1\dot{\phi} \end{pmatrix} \tag{6}$$

So an explicit representation of the system can be obtained $(x_3, \dot{x}_3, x_1, \dot{x}_1, \emptyset, \dot{\emptyset})$.

3. Flat Systems

Recent research in trajectory tracking control has focused on systems with a property known as differential flatness. A nonlinear system $\dot{x} = f(x,u)$ is differentially flat if an output $y$can be found such that the states x and inputs u can be expressed in terms of y and a finite number of its derivatives [4]. A benefit of flat systems is that flat outputs can follow arbitrary trajectories $y_d(t)$ provided that the trajectory is sufficiently smooth. For this system we can choose the position of the load denoting by (ξ, ζ), and we verify if it presents a flat output, that is to say, verify that all variables and all controls of the system can be expressed in function of this chosen output. According to the model (1), we can see that the third equation can be got by using the derivative in order two of ξ and ζ, more precisely, by using the following expression:

$$\ddot{\xi}\cos\phi - \ddot{\zeta}\sin\phi = -g\sin\phi \tag{7}$$

We obtain:[18]

$$x_3 = \xi - \frac{\ddot{\xi}\zeta}{\ddot{\zeta} - g}, x_1^2 = \zeta^2 + \left(\frac{\ddot{\xi}\zeta}{\ddot{\xi} - g}\right)^2,$$

$$\phi = \arctan\left(\frac{\ddot{\xi}}{\ddot{\xi} - g}\right) \tag{8}$$

$$T_1 = Mb_1(2\dot{x}_1\dot{\phi}\cos\phi - x_1\dot{\phi}^2\sin\phi) + (\frac{J_1}{b_1} + (M+m)b_1)\ddot{x}_3 + Mb_1\sin\phi\,\ddot{x}_1 + Mb_1x_1\cos\phi\,\ddot{\phi}$$
$$= Mb_1[2\frac{d}{dt}(\zeta\sqrt{1+(\frac{\ddot{\xi}}{\ddot{\zeta}-g})^2}).\frac{d}{dt}(\arctan(\frac{\ddot{\xi}}{\ddot{\zeta}-g})).\cos\phi - x_1(\frac{d}{dt}(\arctan(\frac{\ddot{\xi}}{\ddot{\zeta}-g})))^2\sin\phi] + (\frac{J_1}{b_1} + (M+m)b_1).$$
$$\frac{d^2}{dt^2}(\xi - \zeta\frac{\ddot{\xi}}{\ddot{\zeta}-g}) + Mb_1\sin\phi\frac{d^2}{dt^2}(\zeta\sqrt{1+(\frac{\ddot{\xi}}{\ddot{\zeta}-g})^2}) + Mb_1x_1\cos\phi\frac{d^2}{dt^2}(\arctan(\frac{\ddot{\xi}}{\ddot{\zeta}-g})). \quad (9)$$

$$T_2 = -Mgb_2\cos\phi - Mb_2x_1\dot{\phi}^2 + Mb_2\sin\phi\,\ddot{x}_3 + (\frac{J_2}{b_2} + Mb_2)\ddot{x}_1.$$
$$= -Mgb_2\cos\phi - Mb_2x_1(\frac{d}{dt}(\arctan(\frac{\ddot{\xi}}{\ddot{\zeta}-g})))^2 + Mb_2\sin\phi\frac{d^2}{dt^2}(\xi - \zeta\frac{\ddot{\xi}}{\ddot{\zeta}-g}) + (\frac{J_2}{b_2} + Mb_2)\frac{d^2}{dt^2}(\zeta\sqrt{1+(\frac{\ddot{\xi}}{\ddot{\zeta}-g})^2}). \quad (10)$$

It is easy to see that all variables of the system denoting by$(x_3, x_1, \xi, \zeta, T_1, T_2)$ can be expressed in function of $\xi$and ζ (the coordinates of the load) and of their derivatives until the order 4, this result is compatible with the principle of flatness.

4. Flatness and linearization

We are interested in this paragraph to appear a dynamic endogenous feedback.

In fact, by using expressions in subsection 3, we can put

$$\xi^{(4)} = v_1 \quad (11)$$

$$\zeta^{(4)} = v_2 \quad (12)$$

Then, the dynamic endogenous feedback can be calculated by identifying the derivatives of $\xi$and ζ until order 4 with their expressions in terms of inputsT_1 and T_2.

According to the equations in subsection 3 and by making the change of the control, we obtain:

$$w_1 = \frac{1}{H}(b_2J_1T_2 + \sin\varphi b_1J_2T_1 - \cos\varphi gJ_1J_2 + b_1^2b_2mT_2 - \cos\varphi gmJ_2b_1^2 - x_1\dot{\varphi}^2J_1J_2 - x_1\dot{\varphi}^2b_1^2mJ_2 \quad (13)$$

with,

$$H = J_1J_2 + J_1Mb_2^2 + b_1^2MJ_2 + b_1^2mJ_2 + b_1^2mMb_2^2 - \cos\phi^2Mb_1^2J_2 \quad (14)$$

Then, we will have:

$$\ddot{\xi} = w_1\sin\phi \quad (15)$$

$$\ddot{\zeta} = w_1\cos\phi + g \quad (16)$$

By deriving again $\xi$and ζ two times and using expressions in subsection (4),

we have:

$$\xi^{(4)} = \ddot{w}_1\sin\varphi + 2\dot{w}_1\dot{\varphi}\cos\varphi - w_1\dot{\varphi}^2\sin\varphi + w_1w_2\cos\varphi = v_1 \quad (17)$$

$$\zeta^{(4)} = \ddot{w}_1\cos\varphi - 2\dot{w}_1\dot{\varphi}\sin\varphi - w_1\dot{\varphi}^2\cos\varphi - w_1w_2\sin\varphi = v_2 \quad (18)$$

By reversing this linear system in relation to $w_1$and w_2, we will have:

$$\ddot{w}_1 = \sin\theta.v_1 + \cos\theta.v_2 + w_1\dot{\theta}^2 \quad (19)$$

$$w_2 = \frac{1}{w_1}(\cos\theta.v_1 - \sin\theta.v_2 - 2\dot{w}_1\dot{\theta}) \quad (20)$$

By this way, we construct a dynamic endogenous feedback by introducing a compensator whose state is given by$(w_1, \dot{w}_1, w_2)$

Then the problem that remains to solve consists to the generation of trajectory leaving from an initial position arriving to a final position.

5. Flatness and Generation of Trajectories

We suppose that we want to bring the load from the position of departure (ξ_i, ζ_i) at the instant t_i , without moving, to the final position (ξ_f, ζ_f) at the instantt_f, also without moving, and to make it pass by the point denoting by:$(\frac{\xi_f-\xi_i}{2}, 2\zeta_f - \zeta_i)$ which must be the maximum of this curve betweenξ_i and ξ_f.

This trajectory $\zeta(\xi)$ must verify the four conditions denoting by:

$$\zeta(\xi_i)=\zeta_i, \zeta(\xi_f)=\zeta_f, \zeta(\frac{\xi_f-\xi_i}{2})=2\zeta_f-\zeta_i,$$

$$\frac{d\zeta}{d\xi}(\frac{\xi_f+\xi_i}{2})=0$$

With the constraint $\frac{d^2\zeta}{d^2\xi}(\frac{\xi_f+\xi_i}{2})<0$ in order to have a local maximum in this point.

We can choose the polynomial of third degree in ξ denoting by:

$$\zeta(\xi)=\zeta_i+(\zeta_f-\zeta_i)(\frac{\xi-\xi_i}{\xi_f-\xi_i})(9-12(\frac{\xi-\xi_i}{\xi_f-\xi_i})+ 4(\frac{\xi-\xi_i}{\xi_f-\xi_i})^2). \quad (21)$$

which satisfied the last conditions.

It remains to construct the trajectory$\xi(t)$ which verifies:

$$\xi(t_i)=\xi_i, \xi^{(1)}(t_i)=0,\ldots,\xi^{(5)}(t_i)=0 \quad (22)$$

$$\xi(t_f)=\xi_f, \xi^{(1)}(t_f)=0,\ldots,\xi^{(5)}(t_f)=0 \quad (23)$$

Then, we will get the polynomial of degree 11denoting by:

$$\xi(t)=\xi_i+(\xi_f-\xi_i)\sigma^6(t)(462-1980\sigma(t)+ 3465\sigma^2(t)-3080\sigma^3(t)+1386\sigma^4(t)-252\sigma^5(t)). \quad (24)$$

with, $\sigma(t)=\frac{t\_t_i}{t_f-t_i}$

6. Flatness and Tracking of Trajectory

By using the expression of the dynamic endogenous feedback calculated in paragraph 4 and the expressions of (32)-(33), we will have the following curly system, with $w_1 \neq 0$

We have a curly system of order 8.

Thus, we can choose, if we measure all the state $(x_3, \dot{x}_3, x_1, \dot{x}_1, \emptyset, \dot{\emptyset})$, the following expressions corresponds of the new controls denoting by:

$$\vartheta_1=v_1^*-\sum_{j=0}^{3}k_{1j}(y_1^{(j)}-(y_1^*)^{(j)}) \quad (25)$$

$$\vartheta_2=v_2^*-\sum_{j=0}^{3}k_{2j}(y_2^{(j)}-(y_2^*)^{(j)}) \quad (26)$$

$v_1^*$and $v_2^*$are the inputs of references which corresponds of trajectories of referencesy_1^* andy_2^*.

The constantk_{1j} and k_{2j} are chosen in order to assure the stability of the systems denoting by:

$$e_1^{(4)}+k_{13}e_1^{(3)}+k_{12}e_1^{(2)}+k_{11}e_1^{(1)}+k_{10}e_1=0 \quad (27)$$

$$e_2^{(4)}+k_{23}e_2^{(3)}+k_{22}e_2^{(2)}+k_{21}e_2^{(1)}+k_{20}e_2=0 \quad (28)$$

and,

$$\xi_d=\xi(t)=\xi_i+(\xi_f-\xi_i).\sigma^6(t).(462-1980\sigma(t)+ 3465\sigma^2(t)-3080\sigma^3(t)+1386\sigma^4(t)-252\sigma^5(t)) \quad (29)$$

$$\zeta_d=\zeta(\xi)=\zeta_i+(\zeta_f-\zeta_i)(\frac{\xi-\xi_i}{\xi_f-\xi_i})(9-12(\frac{\xi-\xi_i}{\xi_f-\xi_i})+ 4(\frac{\xi-\xi_i}{\xi_f-\xi_i})^2) \quad (30)$$

Finally, we replaced $y_i^{(j)}$and$(y_i^*)^j$with their expressions in function of $(x_3, \dot{x}_3, x_1, \dot{x}_1, \emptyset, \dot{\emptyset}, w_1, \dot{w}_1)$ in order to assure the local exponential convergence of these last to their references.

7. Simulations

We present in this part the simulation of the controls and trajectories browsed by the load.

The conditions of simulation can be represented by:

$$\begin{cases} \xi = x_3 + x_1 \sin\phi \\ \zeta = x_1 \cos\phi \end{cases}$$

$x_{3i}=0$ corresponds to the position of the wagon att_i and $x_{3f}=10m$ corresponds to the position of the wagon at t_f . Concerning the length of the cable, we chose at t_i, $x_{1i}=-2m$ and at t_f, $x_{1f}=-1{,}5m$, we will have the initial position of the load $\begin{cases}\xi_i &= 0\\ \zeta_i &= -2m\end{cases}$ and the final position is defined by $\begin{cases}\xi_f &= 10\\ \zeta_f &= -1{,}5m\end{cases}$

Figures 2a and 2b show the crane position responses. It is clear that the tracking errors resulting for the two movements (horizontal axis and vertical axis) are acceptable. Figures 3a and 3b represent the corresponding control input based on flatness. Then the law control provides action control values that are suitable for the used actuators. This is yielded from the pole assignment, which imposes convenient dynamics for the closed loop system. Finally, it is obvious that the satisfactory output tracking performance has been achieved through the proposed control scheme. The controls are limited between a minimal intensity and a maximal one. Figure 4 shows the swing of the crane around the vertical and the last figure represent the motion of the crane on the phase.

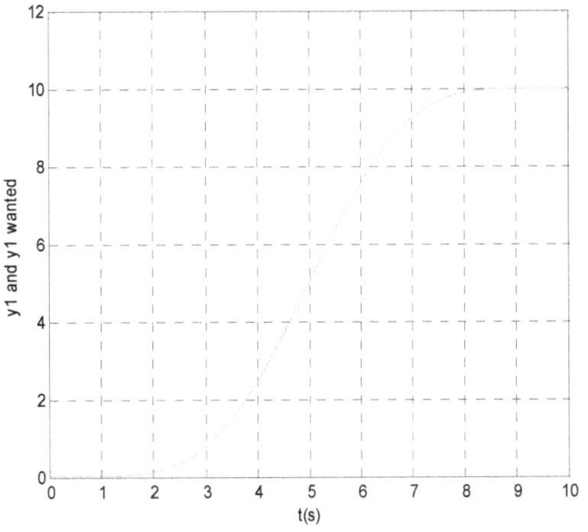

Fig. 2a. *The position responses(horizontal axis).*

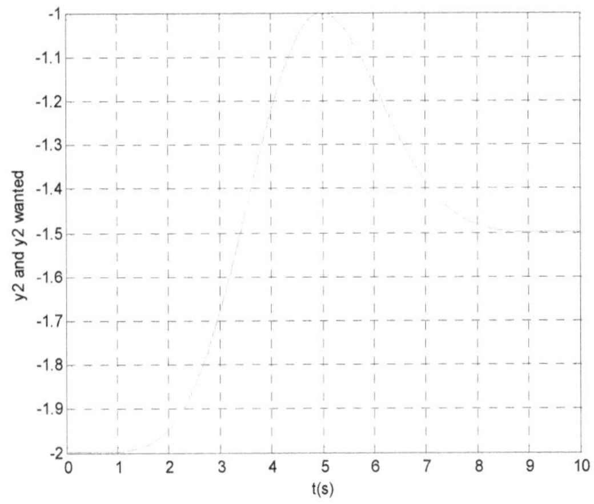

Fig. 2b. *The position responses(vertical axis).*

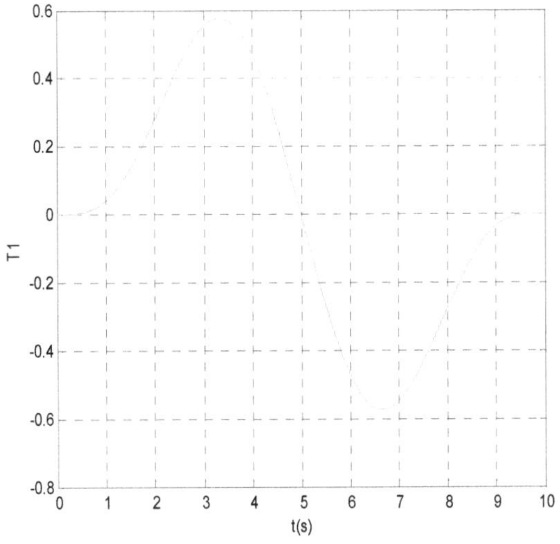

Fig. 3a. *The control variable(T1).*

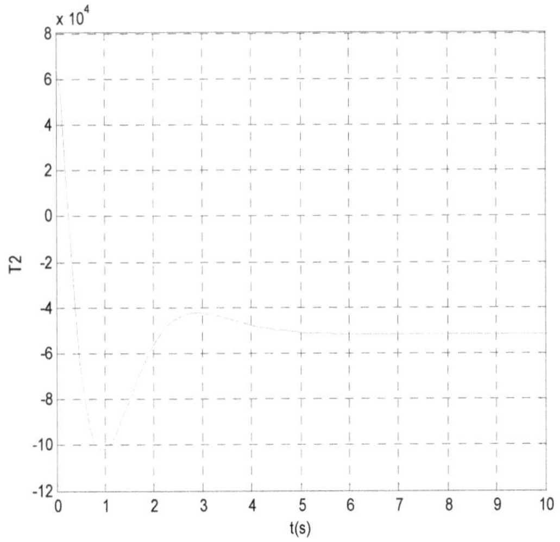

Fig. 3a. *The control variable(T2)*

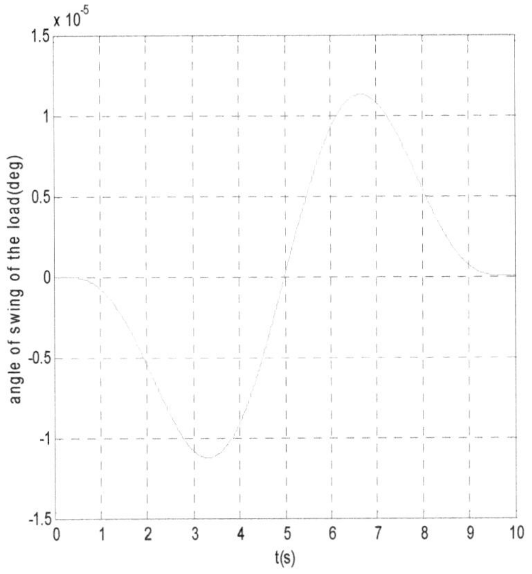

Fig 4. *The swing of the crane.*

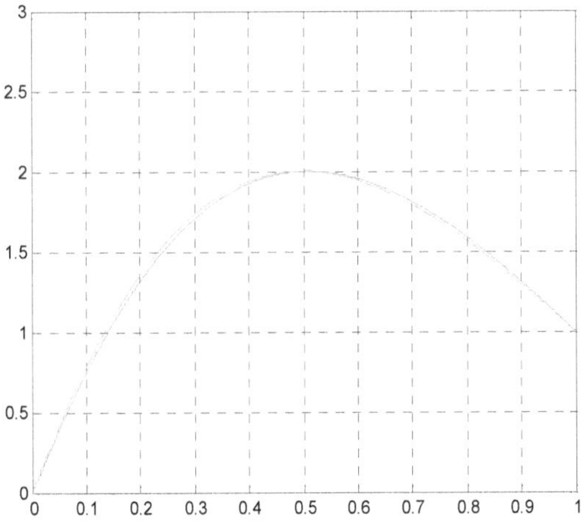

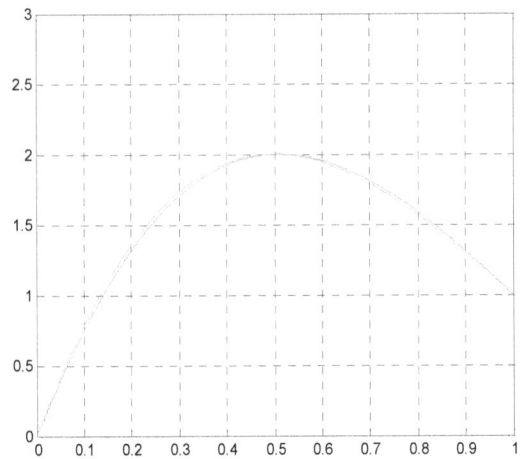

Fig 5. *The moving of the crane on the phase.*

8. Conclusion

In this paper we are proposed a flatness control of a crane. This control is used to generate the desired trajectory and to force the crane to follow it. Experimental results show the proposed algorithm efficiencies.

The real and the desired moving of the crane indicate that a similar procedure could be applied to controlling the luffing. The problem was also extended to take into account the case when working motion starts with non-zero swing of the payload or there are other disturbances. The swing-free stop control is frequently used in the case of overhead cranes, particularly those intended for transferring large payloads, or in the case of reloading works performed repeatedly. The proposed strategy of controlling the slewing motion used as a whole or only for stopping the swinging payload could increase work safety and improve work quality. Therefore, the objective of the control of the crane by flatness is to increase the productivity and the man's operational security on the one hand and to eliminate the undesirable swings of the load on the other hand.

References

[1] A. Chelouah, E. Delaleau, P. Martin and P.Rouchon, Differential flatness and control of induction motors, symposium on Control, Optimization and Supervision; Computational engineering in system applications, IMACS Multiconference, pp. 80-85, Lille, 9-12 July 1996.

[2] T. Burg, D. Dawson, C. Rahn, and W. Rhodes."Nonlinear control of an overhead crane via the saturating control approach of Teel". In Proceedings of the Internationl Conference on Robotics and Automation, pages 3155–3160, 1996.

[3] H. Butler, G. Honderd, et J. Van Amerongen." Model reference adaptive of a gantry crane scale model". IEEE Control system Magazine, pages 57-62, January 1991.

[4] J. Collado, R. Lozano, and I. Fantoni. "Control of a convey-crane based on passivity". In Proceedings of the American Control Conference, pages 1260–1264, 2000.

[5] M. Fliess, J. Lévine, P. Martin, P. Rouchon. "Flatness and defect of non-linear systems": introductory theory and examples, INT. J. Control,1995, Vol. 61, No. 6, 1327-1361.

[6] M. Fliess, J. Lévine, Ph. Martin and P. Rouchon, On differentially flat nonlinear systems, IFAC-Synopsium, NOLCOS'92 pp. 408-412, Borddeaux, 1992.

[7] D. Fragopoulos, M.P. Spathopoulos, and Y. Zheng. "A pendulation control system for offshore lifting operations". In Proceedings of the 14th IFAC Triennial World Congress, pages 465–470, Beijing, P.R. China, 1999.

[8] T. Gustafsson. "On the design and implementation of a rotary crane controller". European Journal of Control, 2(3):166-175, March 1996.

[9] K. S. Hong, J.H. Kim, et K.I Lee. "Control of a container crane: Fast trversing, and residual sway control from the erspective of controlling an underactuated system". In Proceedings of the American Control Conference, pages 1294-1298, Philadelphia, PA, June 1998. 1-305, 1995.ne: to the crane control system.

[10] J. Lévine, P. Rouchon, G. Yuan, C. Grebogi, B. Hunt, E. Ott, J. Yorke and E. Kostelich, On the control of US navy cranes, European Control Conference, ECC'97, Brussels, July 1997.

[11] Ph. Martin and P. Rouchon, Systèmes plats: planification et suivi de trajectoires, www.math.polytechnique.fr/xups/vol99.

[12] S. C. Martindale, D. M. Dawson, J. Zhu, et C. Rahn. "Approximate nonlinear control for a two degree of freedom overhead crane: Theory and experimentation". In Proceedings of the American Control Conference, pages 301-305, 1995.

[13] K.A.F Moustafa. "Reference trajectory tracking of overhead cranes". Journal of Dynamic Systems, Measurement, and Control, 123:139–141, March 2001.

[14] R. H. Overton. "Anti-sway control system for cantilever cranes". United States Patent, June 1996. Patent No.5, 526,946.

[15] F. Rotella, F. Carrillo and M. Ayadi, Digital flatness-based robust controller applied to a thermal process, IEEE international Conference on Control application, pp. 936-941, Mexico 2001.

[16] R. Rothfuss, J. Rudolph and M. Zeitz, Flatness based control of chemical reactor model, European Control Conference, pp. 637-642, Rome, September 1995.

[17] Y. Sakawa et Y. Shindo. "Optimal control of container cranes". Automatica, 18(3), 1981, 257-266.

[18] H. Souilem, H. Mekki, N. Derbel, "crane control by flatness", ninth International Multi-Conference on Systems, Signals & Devices, Chemnitz, Germany, March 2012.

[19] K. Yoshida and H. Kawabe. "A design of saturating control with guaranteed cost and its application to the crane control system". IEEE Transactions on Automatic Control, 37(1):121-127, 1992.

[20] J. Yu, F.L. Lewis, et T. Huang. " Nonlinear feedback control of a gantry crane". In Proceeding of the American Control Conference, pages 4310-4315, 1995.

2

A Comparative Evolutionary models for solving Sudoku

A. A. Ojugo.[1], D. Oyemade.[1], R. E. Yoro.[2], A. O. Eboka.[3], M. O. Yerokun[3], E. Ugboh[3]

[1]Department of Mathematics/Computer Sci, Federal University of Petroleum Resources Effurun, Delta State
[2]Department of Computer Science, Delta State Polytechnic Ogwashi-Uku, Delta State, Nigeria
[3]Department of Computer Sci. Education, Federal College of Education (Technical), Asaba, Delta State

Email address:
ojugo_arnold@yahoo.com(A. A. Ojugo), ojugoarnold@hotmail.com(A. A. Ojugo), davidoyemade@yahoo.com(D. Oyemade), rumerisky@yahoo.com(R. E. Yoro), an_drey2k@yahoo.com(A. O. Eboka), agapenexus@hotmail.co.uk(M. O. Yerokun), ugbohh@gmail.com(E. Ugboh)

Abstract: Evolutionary algorithms have become robust tool in data processing and modeling of dynamic, complex and non-linear processes due to their flexible mathematical structure to yield optimal results even with imprecise, ambiguity and noise at its input. The study investigates evolutionary algorithms for solving Sudoku task. Various hybrids are presented here as veritable algorithm for computing dynamic and discrete states in multipoint search in CSPs optimization with application areas to include image and video analysis, communication and network design/reconstruction, control, OS resource allocation and scheduling, multiprocessor load balancing, parallel processing, medicine, finance, security and military, fault diagnosis/recovery, cloud and clustering computing to mention a few. Solution space representation and fitness functions (as common to all algorithms) were discussed. For support and confidence model adopted ϖ1=0.2 and ϖ2=0.8 respectively yields better convergence rates – as other suggested value combinations led to either a slower or non-convergence. CGA found an optimal solution in 32 seconds after 188 iterations in 25runs; while GSAGA found its optimal solution in 18seconds after 402 iterations with a fitness progression achieved in 25runs and consequently, GASA found an optimal solution 2.112seconds after 391 iterations with fitness progression after 25runs respectively.

Keywords: Swarms, Agents, Elitist, Evolutionary Algorithms, Constraints, Fitness Function

1. Introduction

Soft Computing (SC) aims to harness the potentials of other disciplines via Artificial Intelligence. Thus, create a synergy dedicated to solve problems by exploiting numeric data and human knowledge simultaneously as mathematical models and symbolic reasoning, yielding a technique that is tolerant to imprecision, uncertainty, partial truth and noise in its data via optimization. Often termed evolutionary programming, SC performs quantitative data processing to ensure qualitative knowledge statements and experience using components such as genetic algorithm (GA), particle swarm optimization (PSO), artificial neural network (ANN) etc to mention a few (Abarghouei, Ghanizadeh and Shamsuddin, 2009).

SC has proven efficient in complex optimization. Ojugo (2012) notes 3-feats in their attempt to explore dynamic processes: (a) Continuous adaptation, (b) Flexibility and (c) Robustness. All evolutionary algorithms are derived from translating into mathematical models, principles of biological processing in the fastest time to yield implicit and predictive evolution of a model that stems from experience in its ability to recognize data feats and behaviours. And in turn, yield an optimal fitness of high quality and void of overfitting that will constantly affect any solution's quality (Coello, Pulido and Lechuga, 2002).

1.1. Sudoku Overview

Sudoku is a classical CSP task with variables whose permutation yields a unique solution to satisfy constraints. It is a logic-based combinatorial puzzle of 81-cells in 9X9 grid, each cell contains an integer 1 to 9, and further split into nine 3X3 sub-grids with these constraints in mind:

a. Each row and column of cell is only allowed to contain integers one through nine exactly once
b. Each 3X3 sub-grid is also allowed to contain integers one through nine exactly once.

A number of cells are predefined by the puzzle setter, so that each puzzle has a unique solution. Fig. 1 is a typical puzzle, whose solution is fig 2. Various algorithms have been used to solve Soduko (Santos-Garcia and Palomino, 2007; History of Sudoku, 2013). Some are easily solved via simple logic by mimicking how humans will solve it. Harder puzzles are solved via backtracking algorithms, whose demerit is that its efficiency depends on number of guesses required to solve puzzle (Mantere and Koljonen, 2007). Harder puzzles require longer time to solve, and its solution is via optimization.

5							9	8
		6	1		8			
2				5			1	6
	2		7			5		
			6			1		2
7				1			4	
6		8		4		9		1
	4	2			1		5	
3			9	6			8	

Fig. 1. *Typical Soduko Puzzle*

5	1	7	3	2	6	4	9	8
4	9	6	1	7	8	3	2	5
2	8	3	4	5	9	7	1	6
1	2	4	7	8	3	5	6	9
8	3	5	6	9	4	1	7	2
7	6	9	2	1	5	8	4	3
6	7	8	5	4	2	9	3	1
9	4	2	8	3	1	6	5	7
3	5	1	9	6	7	2	8	4

Fig 2. *Solution of Soduko Puzzle*

Studies exist that have employed stochastic optimization techniques. A major motivation of this study is that difficult puzzles can be solved efficiently as simple puzzles – due to the fact that the solution space is searched stochastically until a suitable solution is found. Thus, puzzle does not have to be logically solvable or easy for a solution to be reached efficiently (Lewis, 2007; Moraglio and Togelius, 2007). Stochastic methods can be used to find global optima for multipoint dependent tasks for which many local optima exists that requires systemic search, enabling a space (continuous or encoded discrete) whose solution via hill-climbing method often gets stuck at local minima, a function of their speed in time of finding the solution (global optima). Due to the nature of constraints in Soduko, it is likely to find a solution that satisfies some constraints (solution found is, local optima) but not all of them. Its stochastic nature allows solution space to be searched still (though local optimum is found) until global optima is found (Mantere and Koljonen, 2007).

This study explores the implementation of various stochastic evolutionary optimization methods. Each is implemented and tested on the puzzle in figure 1.

1.2. Solution Space Representation

Fig. 1 consists of 49-empty cells, corresponding to its solution space. Perez and Marwala (2011) notes this solution space can be represented as:

a. First method treats each one of 49 empty cells as separate variable, particle or agent so that each particle or agent requires its own swarm or population. Thus, the solution space consists of 49 separate population groups. The problem with this approach is that each agent can only be operated upon separately, which prevents the possibility of interaction between these individuals or particles. Thus, it is more computationally challenging and demanding.
b. Second, we treat as combination – 49 integers ranging between 1 and 9 (corresponding to the empty cells in fig. 1). As one individual or particle – so that the solution space instead of consisting of 49-different solution groups as in the first case, has only one population with each particle or individual having 49-dimensions or genes. This approach allows for greater interaction amongst the particles or individuals – since algorithm operations are carried out between all possible solutions. This approach is less more computationally demanding.
c. Third, represent an individual as a puzzle with all its cells filled while ensuring that one of the constraints mentioned above is always met. Thus, in initializing a population state, it is ensured that each 3X3 sub-grid in each of the puzzles contains the numbers 1-9 exactly once. Also, any operation carried out on an individual must ensure that this constraint is not violated. This, is less demanding (when compared to the first method) as individual is still represented as a complete puzzle (as opposed to one cell).

1.3. Fitness Function

A number of possibilities exist with regards to implementing a good fitness function. From arithmetic view, sum of each column-row-and-grid must equal 45 and its product, equals 362880. A possible fitness function to implement such must ensure that all constraints are met. Its demerit is non-repetition of same integer in same row, column or grid constraints is not guaranteed. A row with nine entries of 5 equals 45 – causing the algorithm to converge to local minimum and not meet all the constraints. Thus, different method is needed (Poli et al, 2006b). The fitness function implemented here involves determining whether an integer is repeated or not present in row, column or sub-grid. A fitness value is assigned to a possible solution, based on number of repeated or non-present integers. The more the repeated or non-present integers there are in a solution's rows and columns, the higher the fitness value assigned to that solution; while if third approach to solution space representation is considered, then only repetitions in rows and columns are considered; While if second approach is used, then repetitions in the sub-grid contributes to fitness value (Poli et al, 2006b).

1.4. Statement of Problem

Some evolutionary models use backtracking to offer systemic search (in discrete/continuous) spaces via hill-climbing method that often get them stuck at local minima (due to their speed). Thus, hybrids are designed to cub such defects.

1.5. Objective of Study

The study explores Soduko solved via optimization to find a solution space using the third approach in this study to avoid clumsy result presentation via: (a) Cultural Genetic Algorithm (CGA), (b) Genetic Algorithm Gravitational Search Algorithm, and (c) Genetic Algorithm Simulated Annealing respectively.

1.6. Significance of study

Application of this study will yield computational intelligence – veritable tool for dynamic multipoint search in CSPs, applied in areas such as image and video analysis, communication, control, antenna designs, VLSI, data route and compression, simulation, network design and reconstruction, multiprocessor load balancing, OS task scheduling and resource allocation, parallel processing, power generation, medical and pharmaceutical, finance and economics, security and military, engine design and automation, system fault diagnosis and recovery, forecasting and predictions, cloud and clustering computing etc.

1.7. Limitations of Study

Hybrid, though difficult to implement – are used to provide a means for better selection of search space, encoded via structured learning (to address the general problem of determining existing statistical dependencies amongst data variables) and yield better generation with crossover, mutation and temperature schedules etc.

2. Cultural Genetic Algorithm (CGA)

CGA is an evolutionary technique with individuals influenced both genetically and culturally (Reynolds, 1994), whose background is built on genetic algorithm (GA) as thus:

2.1. Genetic Algorithm

GA is a population optimization inspired by Darwinian evolution and genetics (survival of fittest and natural selection). It consists of a population (set of numeric data) chosen for natural selection that consists of potential solutions to a specific task with each potential solutions referred to as an individual (combination of genes). An optimal combination of genes can lie dormant in the population (from a combination of individuals). An individual with a genetic combination close to the optimal is described as being fit (Hassan and Crosswley, 2004).

A new pool is created by mating two individuals from the current pool. The fitness function is then applied to determine how close an individual is to the optimal solution. The selection function ensures that the genetic data from the fittest individuals is passed down to the next generation or pool – so that a fitter pool emerges. Eventually, the new population (as newer pools are created) will converge on the optimal solution or gets close to it as possible. GA operations are carried out in four steps namely:

a. Initialize – encodes chromosomes into a format suitable for natural selection and many encoding modalities exists (each with its own merits and demerits). An individual in population can be represented in binary (which requires more bits to do so). But, if decimal is used – it allows greater diversity in chromosome representation and greater variance of subsequent generations (Perez and Marwala, 2011). An issue with binary encoding is that populations are not naturally represented in binary due to length as it is computationally more expensive (Ojugo, Eboka, Okonta, Yoro and Aghware, 2012).

   Another allows individual to be encoded as floating point numbers or its combination and is far more efficient than binary encoding. Values encoded are similar with character and commands to represent an individual. *Encoding* scheme encodes data – so that each solution set consist candidates encoded as fixed length vector in one or more pools of different types. The *fitness* function sees a solution set from various candidates evaluated, to determine its goodness of fit. If a solution is reached, function is *good*; else, is *bad* and not selected for crossover. The fitness function is the only part with knowledge of the task at hand and the more solutions are found, the higher its fitness value (Heppner and Grenander, 1990). Ojugo et al (2012) notes the support and confidence fitness model is as thus:

If A then B,
Support = |A and B| / N
Confidence = |A and B| / |A|
Fitness = w1 * support + w2 * confidence

b. Selection – First, a fitness function is used to determine how close an individual is to an optimal solution. After which, individuals are selected for mating. Two selection methods are: (a) *Roulette* method first sums the fitness of all individuals. Then, selects random number between 0 and the summed result. The fitness's are summed again until the random number is reached or just exceeded, from which last individual to be summed is selected, and (b) *tournament* selects a random number of the individuals in pool and the fittest individual is selected. The larger the number of individuals selected, the better the chances of selecting a fittest individual. It continues until one is chosen, from last two or three solutions remaining, to become selected parents to create the new offspring. Selection ensures that the fittest individuals are selected and more likely chosen for mating but also allows for less fit individuals from the pool and the fittest to be selected. A selection function which only mates the fittest is termed *elitist* and often leads to the algorithm converging at a local optima. Here, the tournament algorithm is adopted (it is easier and more efficient to code) as it works on parallel architectures, allowing selection pressure to be easily adjusted (Ojugo et al, 2012) as thus:
Algorithm: Tournament Selection {}
1. Input: Population of chromosome
2. Output: Selected Chromosome for crossover
3. Randomly select 3-chromosomes from pool
4. Pick best 2-solution based on fitness value
5. Return the selected two solution
6. Apply Crossover | Select best solution as parent

c. Crossover – involves the reproductive process in which two individuals exchange their genetic materials to yield a new, fitter individual while ensuring that genes of fit individuals are mixed in an attempt to create a fitter new generation. There are various types of crossover depending on encoding type, two of which are stated as: (a) simple crossover on binary encoded pool, involves choosing multi- or particular-point and all genes are from one parent, and (b) arithmetic crossover in which the new pool is created by adding percentages of one individual to another (Kilic and Kaya, 2001; Ojugo et al, 2012).

d. Mutation – A child's chromosome, gene sequence is slightly altered by either (changing its genes or its sequence) – to ensure the pool converge to a global minimum (instead of local optimum). Algorithm stops once an optima is found. Though computationally expensive, GA can also stop when a number of new pools are created or once no better solution is found. A gene may or may not change depending on mutation rate. Mutation improves diversity needed in reproduction (Ojugo et al, 2012).
Algorithm for Mutation
1. Input: A chromosome rule
2. Output: Mutated solution, a fns of mutation rate
3. Set mutation threshold (between 0 and 1)
4. For each network attribute in chromosome
5. Generate a random number between 0 and 1
6. If random number > mutation threshold then
7. Generate Random value for N-Queen
8. Set solution attribute value with
9. Generated attribute value
10. End if
11. End For Each

2.2. Cultural GA (CGA)

Cultural GA is one of the many variants of GA with a belief space categorized as: (a) Normative (where there is a particular range of values to which an individual is bound), (b) Domain (data about the domain of the task is available), (c) Temporal (data about events in the space is available) and (d) Spatial (topographical data of space is available).

In addition to a belief space, an influence function is needed for CGA (Reynolds, 1994) to form interface between the pool and belief space, to help alter individuals in the pool to conform to its belief space. CGA is chosen, as the model must yield individuals that cannot violate its belief space and reduces number of possible individuals GA needs to generate until an optimum is found. Thus, it is best for Sudoku than other variants (Mantere and Koljonen, 2007). Two CGA methods used in Sudoku:

a. First – implements the second solution space in section 1.2 (each individual consist 49- genes, each gene corresponding to a non-fixed cell in puzzle). Population of 55-individuals is randomly initialized and each contain genes to conform to belief space defined as: (1) Normative (individuals contain genes ranging from 1-to-9), (2) Domain (individual contain genes, as integers) and (3) Spatial (individual contains genes that do not result in repetition of a fixed cell value within same row, column and grid as defined in fig 1). Third belief has topographic knowledge of the space (i.e. fixed cell values). An influence function ensures a belief space is adhered to, and only random numbers between 1 and 9 are used to initialize puzzle. It also implements a rounding function to ensure that the values are all integer and checks that the random numbers generated are not repetitions of one of the fixed numbers in the same row, column and grid.
Once problem is initialized, fitness function determines the fitness of each individual in the pool. From which a sub-pool of 30 individuals are

selected for reproduction via tournament, to determine which individuals will mate.

In reproduction, both crossover (simple single point) and mutation is carried out – in which a number between 1 and 49 is randomly generated from a Gaussian distribution, corresponding to the point of crossover. All genes before this point come from one parent; while the other parent contributes the rest. A new individual, whose genetic makeup is a combination of both parents is thus, reproduced. The new individual also undergoes mutation from which three random genes are selected for mutation and are allocated new random values that still conforms to the belief space. The new individuals replace ones in the pool, with low fitness values (creating a new pool). This continues until an individual with a fitness value of zero (0) is found – to imply that the solution to the puzzle has been reached (Cantu-Paz and Goldberg, 2000).

b. Second – uses third solution space (each individual is a complete puzzle: each 3X3 grid in each puzzle contain numbers 1 to 9 exactly once so that each 3X3 grid is a gene). Pool of 100 such individual randomly initialized and their fitness computed. The best individual and fitness in the pool at each generation is tracked. The fitness function as described above is used where only repetitions in the rows and columns, contributes to an increase in the fitness value. Thus, no repetitions in the grid as this will also help ensure each contains genes that conform to its belief space which are: (1) Normative (each 3X3 grid contains entries 1 and 9), (2) Domain (each 3X3 grid contain integer entries), (3) Spatial (each 3X3 grid must have integers 1 to 9 just once), and (4) temporal (with mutation, it cannot alter fixed cell values).

This process only implements mutation on each individual separately so that in a 3X3 grid (randomly selected), two unfixed cells in grid are randomly selected and switched. During reproduction, the mutation is applied on each individual population. Number of mutation applied on an individual depends on how far CGA has progressed (how fit is the fittest individual in the population). Thus, number of mutations implemented equals the fitness of the fittest individual divided by 2. If fittest individual equals 31, number of mutation equals 16.

Thus, at initialization – it is ensured that the first three (3) beliefs are met; while mutation ensures the fourth belief is met. In addition is an influence function in which best fitness helps influence how many mutations takes place. Thus, knowledge of solution (how close puzzle is from being solved) has direct impact on how the algorithm is implemented; and thus, the algorithm terminates when the best individual has a fitness of 0 – to imply that the solution has been found (Reynolds, 1994).

3. GA-Gravitational Search (GAGSA)

GAGSA is a powerful optimization method, which explores GA's parallel ability to search a space via multiple individual and GSA's speed and flexibility in finding a better optimal point even when a local minimum is found. Both are essential to find solution, to a Sudoku task (Perez and Marwala, 2011)

3.1. Gravitational Search Algorithm

GSA is based on laws of gravity and motion of isolated masses, with each mass representing a solution space and states that particle attracts each other and gravitational force between them, is directly proportional to their masses product and inversely proportional to their distance. Thus, an agent's performance depends on its mass (agents with heavier masses attracts those of smaller masses). GSA uses exploration to navigate its space and guarantee the choice of values by these agents are not violated; and uses exploitation to find optima in shortest time – with agents of heavier masses, moving slowly in order to attract those of lesser mass (Ojugo, 2012). Agents are randomly initialized. At time *t*, a gravitational force of mass *j* acts on mass *i* based on R_{ij} Euclidean distance between any two masses as thus:

$$F_{ij}^{d} = G(t)\frac{M_i(t) * M_j(t)}{R_{ij}(t) + \varepsilon}\left(X_j^d(t) - X_i^d(t)\right) \quad (1)$$

G (gravitation constant) decreases in time to control its accuracy with ε is small constant. Total force is:

$$F_i^d = \sum_{j \in kbest, j \neq 1} rand_j * F_{ij} \quad (2)$$

rand – randomizes agents' initial states at intervals [0,1]. The acceleration of agent i, at time t in dimension d is directly proportional to force acting on that agent, and inversely proportional to its mass, given by:

$$A_i^d(t) = \frac{F_i^d(t)}{M_{ij}(t)} \quad (3)$$

Next agent's velocity is a function of its current velocity and its current acceleration computed as:

$$V_i^d(t+1) = rand_i * V_i^d(t) + A_i^d(t) \quad (4)$$

$$X_i^d(t+1) = X_i^d * V_i^d(t+1) \quad (5)$$

$V_i^d(t)$ is agent velocity in *d*th dimension at time t, and rand is between [0,1]. Masses are calculated via fitness function, as agents of heavier masses keeps attracting those of lesser mass. Masses are updated:

$$M_i(t) = \frac{Fit_i(t) - worst(t)}{best(t) - worst(t)} \quad (6)$$

$fit_i(t)$ is fitness value of agent i at time t. Best(t) and worst(t) indicates strongest and weakest agent according to fitness. For minimization task via reverse engineering, best(t) and worst(t) are defined as:

$$best(t) = \min_{j\in\{1,2..N\}} Fit_i(t) \quad (7)$$

$$worst(t) = \max_{j\in\{1,2..N\}} Fit_i(t) \quad (8)$$

At start, agents are located as solution points in the search space such that with each cycle, the positions and velocities of agents are updated via Eq. (4) and (5), and masses M is updated via Eq. (6). The iteration is stopped when an optimal is found. Thus, we seek agents of lower masses (reverse engineering).

3.2. GAGSA as in Sudoku

The initial use of GA will help achieve a low fitness – so that once a better individual is not found by GA after a number of generations, the best individual is chosen for a series of random walks via its structured learning till an optimal solution is found. Factors defined for GAGSA includes (with GA), how many number of runs is there, how is population representation, its size and reproduction function – must be addressed.

As used in Sudoku, a population of 10 puzzles are initialized to represent the third solution scheme of section 1.2 is met with each 3X3 grid containing integer 1 to 9 exactly once. The reproduction function for the GA, only mutation is implemented to randomly select a grid and randomly swap two unfixed cells in the grid. Number of mutations corresponds to best fitness; and, best fitness and individual, is tracked until a fitness of 2 is found (experimentally, it is found that GA found a fitness of 2 very quickly). Perez and Marwala (2011) notes GA found individuals with low energy of 2, which enters a GSA cycle fairly late. Thus, if GA yields an individual with fitness close to optimal – gravitational force and force acting on each particle is computed with Eq. (1) and (2), to accept individuals with masses lower or equal to current (state's) mass. This runs, until a state with the mass of 0 is reached (solution is found).

4. GA-Simulated Annealing (GASA)

A background of simulated annealing as thus:

4.1. Simulated Annealing

SA as inspired by annealing process used to strengthen glass and crystals – such that a glass is heated until it liquefies and then, allowed to cool slowly so that the molecules settles into lower energy states. Thus, it rather tracks and alters the state of an individual, continuously evaluating its energy using an energy function. Its optimal point is found by running series of Markov chain under different thermodynamic states: neighbouring state is determined by randomly changing an individual's current state by implementing the neighbourhood function. If state with lower energy is found, individual moves to it. Else, if neighbourhood state has a higher energy, then the individual will move to that state only, if an acceptance probability condition is met. If not met, the individual remains at the current state (Perez and Marwala, 2011).

The acceptance probability is difference in energies between current and neighbouring states, and temperatures. Temperature is initially set high, so individual is more inclined towards higher energy state – allowing the individual to explore a greater portion of the space and preventing it from being trapped in local optimum. As algorithm progresses – temperature reduces with cooling so that individuals converge towards lowest energy states and thus, an optimum point (Perez and Marwala, 2011). The algorithm is as thus:

Simulated Annealing Algorithm:

1. Initialize an individual state and energy
2. Initialize temperature
3. Loop until temperature is at minimum
4. Loop until maximum number of iterations reached
5. Find neighbourhood state via neighbourhood function
6. If neighbourhood state has lower energy than current
7. Then change current state to neighbouring state
8. Else if the acceptance probability is fulfilled
9. Then move to the neighbouring state
10. Else retain the current state
11. Keep track of state with lowest energy
12. End inner loop
13. End outer loop

4.2. Hybrid GA-SA

GASA is a powerful model that combines GA's parallel search to explore the space via multiple individuals and SA's flexibility to find a better optimal point, even when a local minimum is found. Both are essential in finding solution to a Sudoku task (Perez and Marwala, 2011).

Initial use of GA helps achieve a low fitness – so that once a better individual is not found after a number of runs, the best individual is chosen for a series of random walks until an optimal solution is found. These factors must be defined for GASA: (a) On GA: number of runs, population representation, size and reproduction function, and (b) On SA (with GA complete), SA is run on the fittest individual until a solution is found and what is the neighbourhood size and function.

As applied to Sudoku, a population of 10 puzzles are initialized to represent the third solution scheme of section 1.2 is met with each 3X3 grid containing integer 1 to 9 exactly once. The reproduction function for the GA, only mutation is implemented to randomly select a grid and randomly swap two unfixed cells in the grid (for which, if GA produces an individual with a fitness close to the optimal – then temperature schedule is omitted and only a single Markov chain is run). The number of mutations will correspond to the best fitness, with the best fitness and individual tracked until a fitness of 2 is found (experimentally, it is found that GA found a fitness of 2 very quickly). Perez and Marwala (2011) notes since GA found individuals with low energy, they enter into the SA cycle fairly late so that no temperature schedule is needed. Instead, a simple moderated Markov chain is used, which accepts the states with energies that are lower or equal to the current state's energy. This runs until the state with the energy of 0 is reached (solution is found). The SA and GA shares the same fitness function; while SA neighbourhood function is same as mutation function used in GA.

5. Result Discussion

After testing all three (3) models on figure 1 puzzle, the results are presented as follows:

5.1. CGA Result

CGA took 32 seconds to find the solution after 188 iterations or generations (at best). CGA was run 25 times (to eradicate non-biasness) and it was able to find an optimal solution every time – and the time taken varied significantly between 32 seconds and 8 minutes, as CGA convergence time depends on how close the initial population is to the solution and on the random mutation applied to the individuals in the pool and is supported by Perez and Marwala (2011).

5.2. GAGSA Result

GSAGA solves puzzle (at best) 18seconds after 402 iterations with a fitness progression achieved across GSA and GA as well. The GA cycle achieved a fitness of 2 in 90 iterations and GSA implemented a gravitational pull and mass update of 332 iterations before finding a solution. GSAGA was run 25 times and solved the puzzle each time on a range between 12seconds and 11minutes – due to its stochastic nature so that convergence time depends on initialization and gravitational pull cum mass updates.

5.3. GASA Result

GASA solves puzzle at 2.112seconds after 391 iterations with fitness progression across GA and SA. GA achieved a fitness of 2 in 90 iterations and SA used Markov chain of 301 iterations to find a solution. With 25 runs, GASA solved puzzle every time on a range between 4seconds and 3minutes – due to its stochastic nature as convergence time depends on initialization as well as the random swaps and is supported by Perez and Marwala (2011).

6. Conclusion / Recommendation

The Sudoku is solved efficiently via stochastic method: three of which are used in this work. Solution space representation and fitness functions (as common to all algorithms) were discussed, and support/confidence model adopts $\varpi 1=0.2$ and $\varpi 2=0.8$ to give better convergence. Other values, led to a slower convergence or non-convergence.

References

[1] Abarghouei, A., Ghanizadeh, A and Shamsuddin, S., (2009): Advances in soft computing methods in edge detection, J. Advance Soft Comp. Applications, ISSN: 2074-8523, 1(2).

[2] Cantu-Paz, E and Goldberg, D.E., (2000): Efficient parallel genetic algorithms: theory and practices, Computer methods in applied mechanics and engineering, 186(2-4), pp 221-238.

[3] Coello, C. A., Pulido, G. T and Lechuga, M.S., (2002): Handling multiple objectives with particle swarm optimization, Evo. Comp., Vol. 8, pp 256–279.

[4] Hassan, R and Crosswley, W., (2004): Variable population-based sampling for probabilistic design optimization and with a genetic algorithm, AIAA-2004-0452), 42nd Aerospace Science meeting, Reno, NV.

[5] Hassan, R., Cohanin, B., De Wec and Venter, G., (2006): Comparism of PSO and GA, American Institute of Aeronautic and Astronautics (AIAA-2006), 44th Aerospace Science meeting, Washington–DC.

[6] Heppner, H and Grenander, U (1990): A stochastic non-linear model for coordinated bird flocks, In Krasner, S (Ed.), The ubiquity of chaos, (pp. 233–238). Washington: AAAS.

[7] History of Sudoku, Conceptis Editoria, [online]: www.conceptispuzzle.com/articles/sudoku, last accessed 17-01-2013.

[8] Kilic, A. and Kaya, M.A (2001): A new local search algorithm based on genetic algorithms for the n-queens problem, Proc. Genetic and Evo. Comp. conf. (GECCO-2001), 158 – 161

[9] Lewis, R., (2007): Metaheuristics can solve Sudoku, J. Heuristics Archive, 13(8), pp 387 – 401

[10] Mantere, T and Koljonen, J., (2007): Solving and rating Sudoku puzzles via genetic algorithm, Proc. Congress on Evol. Comp.,1382-1389.

[11] Moraglio, A and Togelius, J., (2007): Geometric particles swarm optimization for Sudoku puzzle, http://julian.togelius.com/Moraglio2007Geomet-ric.pdf, last accessed 16-January-2013.

[12] Ojugo, A., Eboka, A., Yoro, E., Okonta, E and Aghware, F.O., (2012): Genetic algorithm rule-based intrusion detection system, J. Emerging Trends in Comp. Info. Syst., ISSN: 2079-8407, 3(8), pp 1182-1194

[13] Ojugo, A.A., (2012): Gravitational search neural network algorithm for rainfall runoff modeling, Unpublished PhD thesis, Abakiliki: Ebonyi State University, Nigeria.

[14] Perez, M and Marwala, T., (2011): Stochastic optimization approaches for solving Sudoku, Proc. IEEE Congress on Evo. Comp., pp 256–279, Vancouver: Piscataway.

[15] Poli, R., Wright A., McPhee, N and Langdon, W., (2006b): Emergent behaviour, population based search and low-pass filtering, Proc. on Comp. Intelligence and Evo.Comp., pp395-402, Vancouver: Piscataway

[16] Reynolds, R., (1994): An introduction to cultural algorithms, Proc. of 3rd Annual Conf. on Evo. Programming, River Edge: New Jersey, World Scientific, pp 131-139.

[17] Santos-Garcia, G and Palomino, M., (2007): Solving the Sudoku puzzle with rewriting rules, Notes on Theo. Computer Sci., 17(4), pp79-93

Ant colony optimization with genetic operations

Matej Ciba, Ivan Sekaj

Institute of Control and Industrial Informatics, Bratislava, Slovakia

Email address:
bigmato@centrum.sk(M. Ciba), ivan.sekaj@stuba.sk(I. Sekaj)

Abstract: This paper attempts to overcome stagnation problem of Ant Colony Optimization (ACO) algorithms. Stagnation is undesirable state which occurs at a later phases of the search process. Excessive pheromone values attract more ants and make further exploration hardly possible. This problem has been addressed by Genetic operations (GO) incorporated into ACO framework. Crossover and mutation operations have been adapted for use with ant generated strings which still have to provide feasible solutions. Genetic operations decrease selection pressure and increase probability of finding the global optimum. Extensive simulation tests were made in order to determine influence of genetic operation on algorithm performance.

Keywords: Ant Colony Optimization, Genetic Operations, Crossover, Mutation, Minimal Path Search

1. Introduction

Wide range of problems like Routing problem, Assignment problem, Scheduling problem and others can be transformed into graph representation. Exact algorithms for instance Dijkstra or Bellman-Ford appear to be slow and inefficient on large scale graphs. In this case some heuristic information which guide search process is useful. One of the well-known graph search algorithm that utilizes a heuristic is A* search [1] or ACO algorithm.

Ant colony optimization represents an efficient tool for optimization and design of graph oriented problems. It is a multi-agent meta-heuristic approach and was first purposed by M. Dorigo et al. [2] as Ant system (AS) algorithm.

During the search process each ant sets off from ant colony (start position) and moves to search food (destination). The aim is to find the shortest path. As ants are passing the terrain (graph) they mark used routes (arcs of the graph) by chemical substance called pheromone. On their way back they use the same way from which abundant loops has been removed, but the amount of pheromone (1) $\Delta\tau^k_{ij}(t)$ they produced is inversely proportional to the tour length $L^k(t)$.

$$\Delta\tau^k_{ij}(t) = \begin{cases} Q / L^k(t) & if\,(i,j) \in T^k(t) \\ 0 & if\,(i,j) \notin T^k(t) \end{cases} \tag{1}$$

$T^k(t)$ is the tour generated by ant k, Q is a constant and tuple (i, j) denotes beginning and termination node of an arc.

All pheromone tracks (2) are preserved by arcs of the graph

$$\tau_{ij}(t+1) = (1-\rho)\tau_{ij}(t) + \sum_{k=1}^{m} \Delta\tau^k_{ij}(t) \tag{2}$$

where $\rho \in (0,1)$ is the pheromone persistence ($1 - \rho$ is evaporation rate) and m is the number of ants. Evaporation rate is a user adjusted parameter and affects pheromone durability; i.e. how long the acquired information will be available. Too high values causes random search, too low values get algorithm stock in local optimum.

An ant in each node has to make a decision which arc to take. At the beginning when no pheromone values are available heuristic values η_{ij} takes dominance. Later the ant uses probability selection rule to choose the next arc according to

$$p^k_{ij}(t) = \frac{p_{ij}(t)}{\sum_{j \in N^k_i} p_{ij}(t)} \tag{3}$$

where $p^k_{ij}(t)$ is probability the ant k chooses the arc (i, j) from the neighborhood N^k_i of node i except the node visited previously. The more pheromone is located on particular arc, the more attractive it appears. The probability $p_{ij}(t)$ of choosing the particular arc (i, j) depends on pheromone $\tau_{ij}(t)$ and the heuristic η_{ij} values which are associated with the arc (4).

$$p_{ij}^{k}(t) = \frac{\tau_{ij}^{\alpha}(t) + \eta_{ij}^{\beta}}{\sum_{j \in N_i^k} \tau_{ij}^{\alpha}(t) + \eta_{ij}^{\beta}} \tag{4}$$

Symbols α and β are weight parameters and represents balance between ant's gathered knowledge and the user preferred area. Heuristic values η_{ij} affect probability only at the beginning when pheromone values are low.

2. Modification of ACO Algorithm

Disadvantages of ACO algorithms are (i) many user parameters and (ii) the selection pressure. The first point is a property of the algorithm, while the second point has had many papers devoted to it. Let's mention ant colony system (ACS) with *pseudo-random proportional* rule [3] in which random uniformly distributed variable $q \in (0,1)$ is compared with a tunable parameter $q_0 \in <0,1>$. If $q > q_0$ then

$$p_{ij}^{k}(t) = \begin{cases} 1 & \text{if} \quad j = \arg\max p_{ij} \\ 0 & \quad \textit{otherwise} \end{cases} \tag{5}$$

else the probability selection rule (3) is applied; random selection applied to AS_{rank} [4] where *random selection rate r* is the probability of random selection and represents a user parameter which adjusts the balance between exploration and exploitation; prevention of quick convergence (i) and stagnation avoidance (ii) mechanisms applied to AS [5].

The mechanism for prevention of quick convergence (i) is based on *pseudo-random proportional* rule [3], but the tunable parameter q_0 is dependent on algorithm iteration (6)

$$q_0 = \frac{\log_e(NC)}{\log_e(N_{\max})} \tag{6}$$

where NC is the current iteration and N_{max} is the termination iteration.

The stagnation avoidance mechanism (ii) is based on the comparison of a randomly generated quantity $q \in (0,1)$ with probability $p^k_{ij}(t)$ of selected arc. If $q \geq p^k_{ij}(t)$, then choose the next node randomly. This occurs in later stages of the search process, where pheromone values tend to be high, and thus the chance of further exploration is low.

3. Genetic Operations Applied on ACO

Genetic algorithms (GA) were proposed by Holland (1975). The original GA is known as simple genetic algorithm (SGA). GA belongs to adaptive stochastic optimization class and is typically used for combinatorial problems. The four main components of GA are representation (i), mutation (ii), crossover (iii) and selection (vi) mechanism. Each component is adapted in order to provide feasible solution for ACO algorithm.

In ACO algorithms representation (i) of genotype space is sequence of nodes:

$$n^L; \quad n \in N \tag{7}$$

where gene n is graph node and L is path length. The population size is given by the number of genomes, i.e. the number of ants m which generate the set of paths within one generation.

Mutation (ii) mimics random gene changes. In each genome each gene is changed with the equal probability. The simplest form is one point mutation on Fig. 1. In ACO adaptation the first and the last node is excluded from mutation. For feasibility reason the replacement node n_r (new gene) is such a node from the node n_i neighborhood N_i, to which an arc from n_i predecessor n_p to n_i successor n_s exists (Fig. 2). If more such nodes occur, random selection is applied. If no such node exists, another gene is randomly picked up from the list.

In ACO algorithm crossover position is represented by a common node of parental strings except the first and the last node (Fig. 3). If more of such nodes exist, random selection is applied. Crossover operation makes sense only if both child strings differ from their parents.

In GA many selection (vi) mechanisms are available, like roulette-wheel selection, tournament selection, stochastic universal sampling or reward-based selection [7]. Since optimization process is primary done by ants cooperative behavior, the selection process has purely random concept and genetic operations serve just for selection pressure decrease.

On both figures survivor strategy where parents are replaced by their children is shown. No string is allowed to take the same genetic operation more than once.

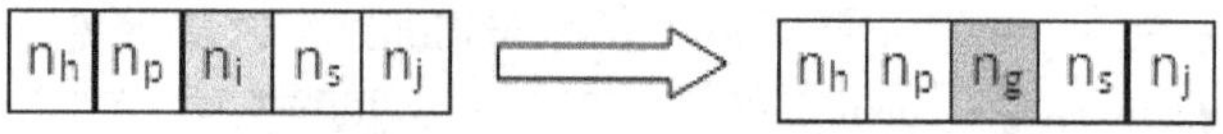

Figure 1. One point mutation.

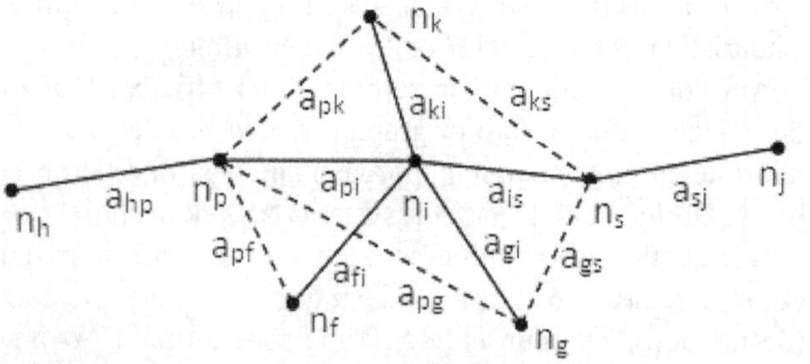

Figure 2. Candidates which can replace node n_i are n_g and n_k only.

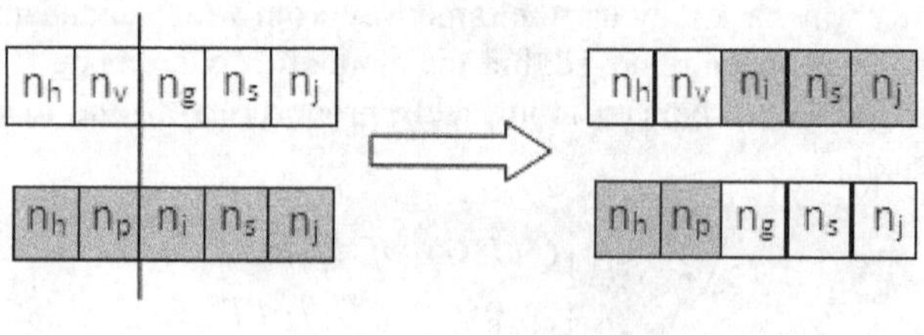

Figure 3. One point crossover.

4. ACO Algorithm with Genetic Operations

The above described genetic operations have been applied to one of the best performing ACO algorithms of Kumar, Tiwari and Shankar (ACO_{KTS}) [5]. At the end of each cycle *t*, when all the ants finish their tours $T^k(t)$, genetic operations are applied on the $T^k(t)$ strings which represents the list of nodes. Prior the genetic operations all loops are removed from the tours.

At first mutation is applied. It is applied on random selected tour $T^k(t)$ in random selected node. If mutation is not feasible, another node is chosen. If more candidates by which the selected node can be replaced occur, the new node is random chosen from the candidates. If mutation fails on all nodes of the tour, another tour is chosen.

After all mutation operations are performed, crossover operations are applied. Parent strings are random selected. If crossover operation is not feasible, another second string is selected. If no tour has common node with the first selected tour, another first tour is selected and the random selection process is repeated.

Since genetic operations may produce strings with loops, in ACO framework prior and immediately after each genetic operation a loop removal procedure is performed. After all genetic operations are executed fitness evaluation and pheromone update are scheduled.

Genetic operations do not have to be necessarily feasible. Feasibility of genetic operations depends on the graph and generated tours. For this purpose ACO_{GO} algorithm has embedded user feedback which represents a ratio between accomplished and required genetic operations. Two of such rates are provided; one for each genetic operation.

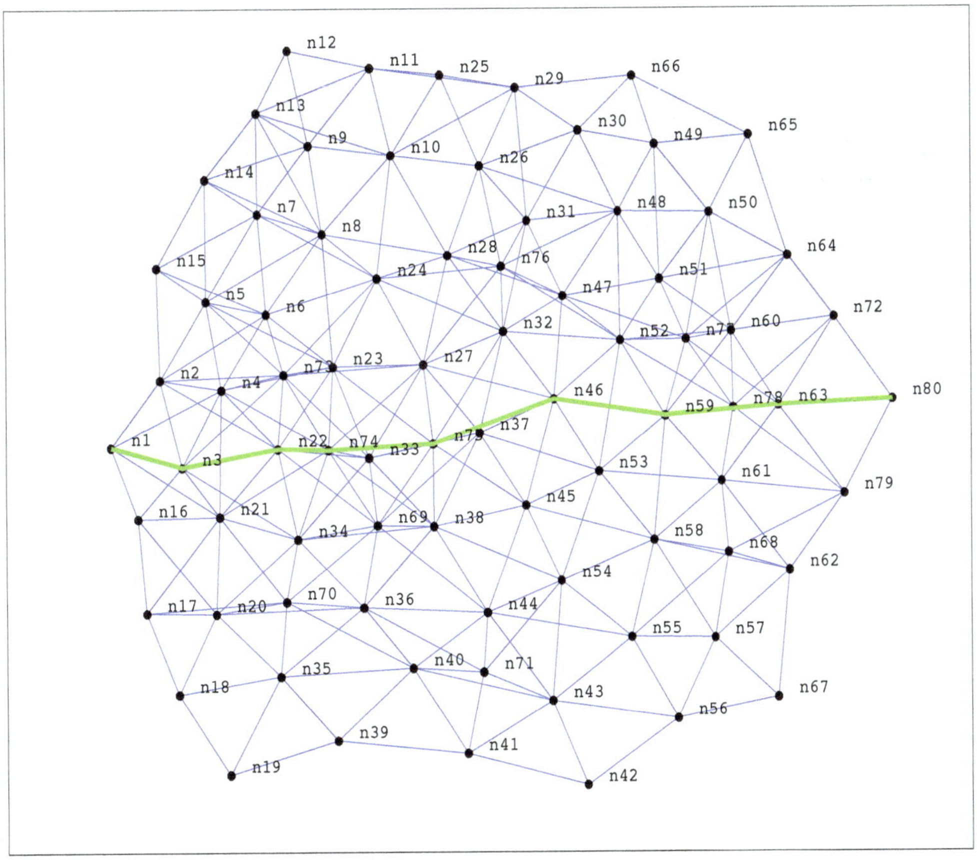

Figure 4. *The 80 node graph with dashed minimal path*

5. Case Study

The above described ACO_{GO} algorithm has been tested on a random generated graph. Common ACO parameter values were set in accordance with [8] and are listed in the Table 1.

The value for the number of cycles represents three macro cycles of ACO_{MC} [9] for the same graph and parameters.

Test graph is a symmetrical multi-graph with 80 nodes and 300 arcs (Fig. 4). Node coordinates *x,y* fall in range $<0,1>$ and arc's values c_{ij} represent the arc lengths. The task is to find the shortest path between start node $n_s = 1$ and end node $n_e = 80$.

Variable parameters were set to determine the influence of the genetic operations quantity on algorithm performance and effect of distribution of mutation operations between paths. For each setting 500 trials were performed. For test reconciliation probability *n [%]* of finding the global optimum ($T^* = [1\ 3\ 22\ 74\ 75\ 46\ 59\ 63\ 80]$) was evaluated.

Table 1. Common ACO parameters settings

Parameter name	Value
Initial pheromone value $\tau_{ij}(0)$	0.1
Weight of pheromone information α	0.5
Heuristic values η_{ij}	0.1
Weight of heuristic information β	0.1
Pheromone persistence ρ	0.05
Number of ants m	10
Number of cycles	200

6. Results

Simulation results were divided into three groups according to number of crossover pairs and are listed in the Table 2.

Table 2. Simulation results for various GO settings

Mutation paths	Mutations per paths	Probability of finding the global optimum [%]		
		No crossover operation	One crossover pair	Two crossover pairs
0	0	5.6	4	6.2
	1	6.4	6	6
	2	6.8	8.8	3.6
1	3	6	6	7
	4	7.6	5.6	7
	5	6.2	7.2	7.4
	1	6.4	8.6	6.2
	2	6.4	7.4	7.2
2	3	7.8	7	7.2
	4	8.8	6.2	8
	5	7.6	10	9.6
	1	7.4	5.6	6.4
	2	8.2	5.8	8.4
3	3	9.2	7.4	7.8
	4	9	9	9.4
	5	8.2	7.4	9.6
	1	6.4	8.2	7.6
	2	8.8	8.8	9
4	3	10.4	13	8.4
	4	10	9.6	10.2
	5	9.4	8.2	11.6
	1	8.8	8	6.6
	2	9.6	8.6	9
5	3	11.2	8	9.2
	4	11.2	10.2	12
	5	11.2	9	10

6.1. Mutation effect

The reference value of n *[%]* was received without any genetic operation and is 5.6 (Table 2, row 1). The results received with GO are better almost in any case. It can be seen that the higher number of mutation operations, the better the performance is (Tables 2).

For better results representation three graphs are provided. Their color map was set to show green - blue when the results are worse than reference value and yellow to red otherwise (Fig. 5 – 7).

The outcome with different mutation distribution is asymmetric. Results received without crossover operation have higher values along with the Mutation paths axis (Fig. 5). However, results received with two crossover pairs have higher values along with the Mutations per path axis (Fig. 7). I.e. with no crossover pair certain amount of mutation operations should be spread out among more paths, but with 2 crossover pairs concentration of mutation operation on less paths tends to perform better. This behavior may be caused by the execution order of the GO: crossover is applied after mutation, thus crossover may re-distribute mutation substrings between more paths.

The results for one crossover pair show different behavior. It does not have the highest value on edges of the surface where the highest amount mutation operation is. The highest value 13% was received with four mutation paths with three mutation operation per path.

Genetic operations where nearly always feasible; ratio accomplished / required mutation operations is 100% and for the crossover operation over 99%.

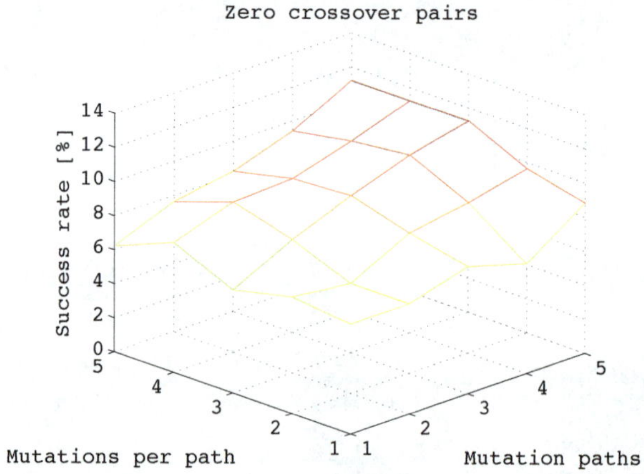

Figure 5. Zero crossover pairs results

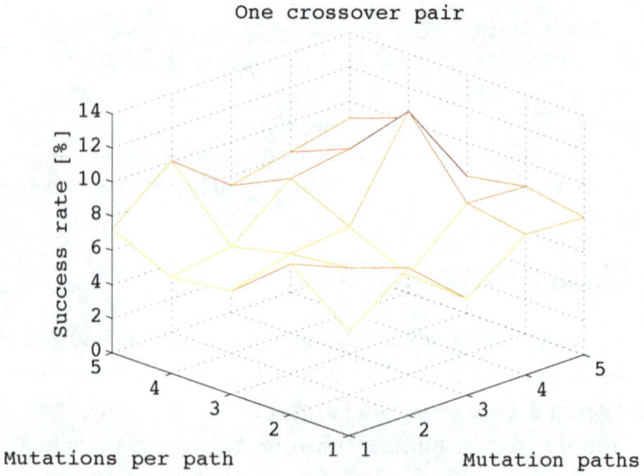

Figure 6. One crossover pair results

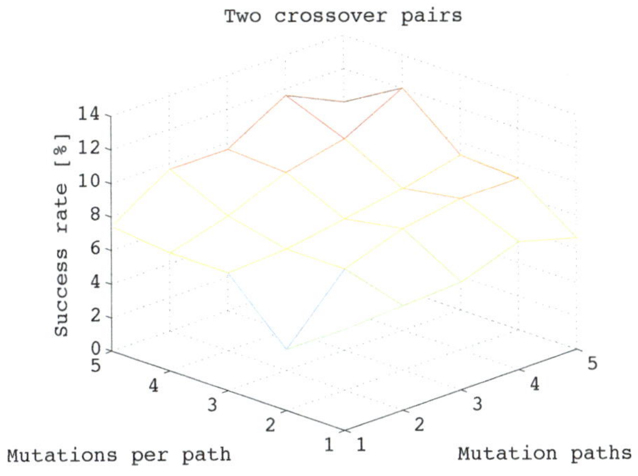

Figure 7. *Two crossover pairs results*

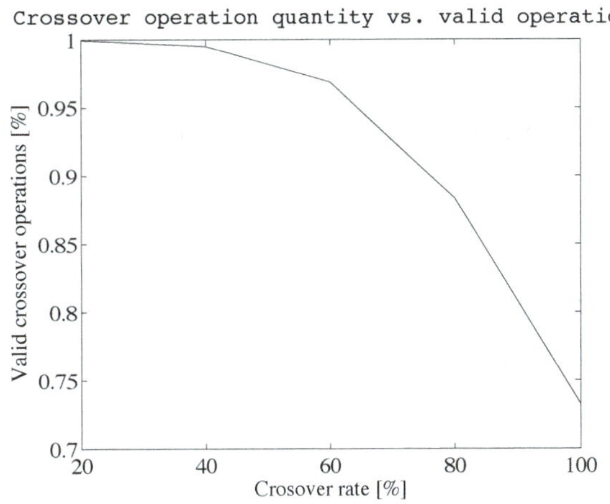

Figure 9: *Crossover operation quantity vs. valid operations*

6.2. Crossover effect

In order to determine the effect of crossover operation crossover rate was let to grow up to 100% (Table 3). To prevent interference, no mutation operation was allowed.

The results vary (Fig. 8); the highest output was gained for 60% of crossover rate. Beyond 60% threshold ants foraging behavior is suppressed by crossover overload. As the crossover rate increases, ratio accomplished / required operation decreases (Fig. 9). This is caused by the search space dimension. It is too large for ten ants to meet.

GO does not affect the length of the search process. The mean value of the cycle when the best value was found is 109.081 with standard deviation 2.617.

Table 3. *Crossover operation results*

Crossover rate [%]	Probability of finding global optimum [%]	Valid crossover operations [%]
0	*5.6*	*N/A*
20	*4*	*0.99953*
40	*6.2*	*0.99498*
60	*7.4*	*0.96859*
80	*5*	*0.88374*
100	*5.8*	*0.73246*

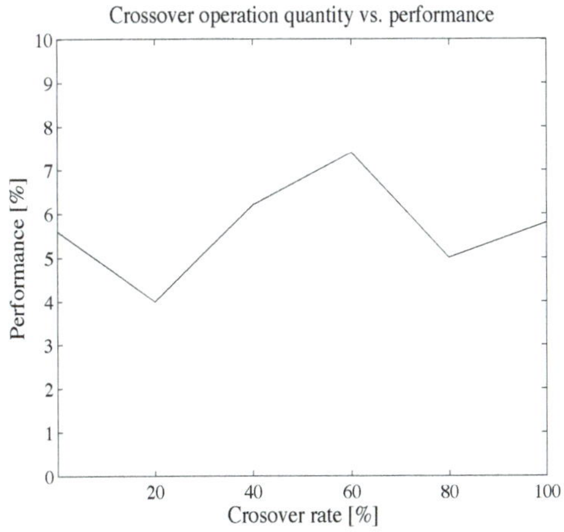

Figure 8: *Crossover operation quantity vs. performance*

7. Conclusion

It has been proved that genetic operations increase ACO algorithm performance. Even small number of any genetic operation causes positive effect.

Limit of crossover is 60% of crossover rate. The higher the crossover rate, the lower the accomplished / required ratio. Mutation operation causes better results than crossover operation. This can be explained by the nature of the mutation operation which creates new paths whilst crossover operation can only combine already existing solutions.

The higher amount of mutation operations the higher the performance gain is. No limit for amount of mutation operation was found during the simulation. Without crossover operation distributed mutation operation has better performance, but with two crossover pairs concentrated mutation operation on less paths tends to perform better. The impact of the GO execution order on the mutation operation distribution needs to be verified.

The results are promising; GO improves ACO algorithm performance more than twice. Further research and more experiments are needed to determine the distribution and optimal amount of mutation operation with respect to the number of ants and length of the paths.

Acknowledgments

Thanks to Science Publishing reviewers for valuable feedback and provided comments which increased the paper quality.

References

[1] P. E. Hart, N. J. Nilsson and B. Raphael, A Formal Basis for the Heuristic Determination of Minimum Cost Paths. IEEE Transactions on Systems Science and Cybernetics SSC4 4(2), 1968, 100–107

[2] M. Dorigo, G. Caro and L. Gambardella, Ant algorithms for discrete optimization, Artificial Life, 5(2), 1999, 137-172

[3] L. Gambardella and M. Dorigo, Solving symmetric and

asymmetric TSPs by ant colonies, In Proceedings of the IEEE Conference on Evolutionary Computation, ICEC96, IEEE Press, 1996, 622–627

[4] Y. Nakamichi and T. Arita, Diversity control in ant colony optimization, In Abbas HA (ed) Proceedings of the Inaugural Workshop on Artificial Life (AL'01), Adelaide, Australia, Dec 11, 2001, 70-78

[5] R. Kumar M. K. Tiwari and R. Shankar, Scheduling of flexible manufacturing systems: An ant colony optimization approach, proc. Instn. Mech. Engrs Vol. 217 Part B: J. Engineering Manufacture, 2003, 1443–1453

[6] J. H. Holland, Adaptation in Natural and Artificial Systems: An Introductory Analysis with Applications to Biology, Control, and Artificial Intelligence, University of Michigan Press, 1975

[7] I Sekaj, Evolucne vypocty a ich vyuzitie v praxi, IRIS Press, 2005

[8] M. Becker and H. Szczerbicka, Parameters influencing the performance of ant algorithms applied to optimization of buffer size in manufacturing, IEMS Vol. 4, No. 2, December 2005, 184–191

[9] M. Ciba, ACO algorithm with macro cycles, Proceedings on 14th Conference of Doctorial Students on Elitech'12, Slovak Technical University of Bratislava, May 2012

4

An analysis of the reciprocal collision avoidance of cooperative robots

A. Fratu[1], M. Dambrine[2, 3, 4], L. Vermeiren[2, 3, 4], A. Dequidt[2, 3, 4]

[1]Transilvania University of Brasov, 500036 Brasov, Romania
[2]Univ. Lille Nord de France, F-59000 Lille, France
[3]UVHC, LAMIH, F-59313 Valenciennes, France
[4]CNRS, UMR 8530, F-59313 Valenciennes, France

Email address:
fratu@unitbv.ro(A. Fratu), michel.dambrine@univ-valenciennes.fr(M. Dambrine), laurent.vermeiren@univ-valenciennes.fr(L. Vermeiren), antoine.dequidt@univ-valenciennes.fr(A. Dequidt)

Abstract: This paper presents a formal approach that addresses the reciprocal robots collision avoidance, where two robots need to avoid collisions with each other, while moving in a common workspace. Based on our formulation, each physical robot acts fully independently, communicating with the corresponding virtual prototype and imitating its behavior. Each physical robot reproduces the pathway of its virtual prototype. With a view to collision avoidance, it is necessary to detect a possible collision. This action includes the potentially intersecting regions test of the corresponding virtual prototypes. The estimation of the collision-free actions on the virtual robots and the collaborative work of the physical robots which imitate their virtual prototypes are the original ideas. Based on potentially intersecting regions of the virtual robots, we identified a collision-free motion corridor for two cooperative robots. Using the definition of velocity obstacles, we derived sufficient conditions for the collision-free motion of the two virtual robots. We tested the present approach on several complex simulation scenarios involving two virtual robots and estimating collision-free actions for each of them during the cooperative tasks. The focus of this paper is the identification of the collision-free actions for two virtual robots and their behavioral imitation by the physical robots.

Keywords: Virtual Robots, Collision Detection, Reciprocal Collision Avoidance, Behavioral Imitation

1. Introduction

Cooperative robots systems are designed to achieve tasks by collaboration. Collaborative robots which we can see deployed nowadays in research or industries are permanently endangered to collide. Therefore, installations with multiple robots in the real world, such as collaborative work and maintenance of the good state of the production line, require collision avoidance methods, which take into account the mutual constraints of the robots.

A key requirement for their efficient operation is good coordination and reciprocal collision avoidance.

Collision avoidance is a fundamental problem for cooperative robots. Collision avoidance is a highly advanced robot control option that automatically detects collisions and quickly causes the robot to stop and back up to release the pressure. Not only does it reduces the force of the collision, but also prevents the robot and its tooling from being pressed against an object after a collision.

The contact of the robot with an obstacle must be detected and it will cause the robot to stop quickly and thereafter back off to reduce forces between the robot and the environment. The problem of the contact with obstacles imposes the null velocity in the moment of the impact.

The problem of the contact detection is better analyzed on the virtual prototypes in the virtual environment, where the virtual objects can be intersected.

The problem of the contact detection in the virtual environment using virtual robots is important for the reason that this built-in function is proven superior to mechanical collision detection devices. It detects collisions in all directions, protecting not only the end-effectors, but also the work pieces and the robot itself.

The problem of the collision avoidance can generally be defined in the context of an autonomous robot operating in an environment with obstacles, and /or other moving entities, where the robot employs a continuous acting cycle.

For each cycle, an action for the robot must be computed based on local observations of the environment, so that the robot stays free of collisions with the obstacles and the other moving entities, while making progress towards a goal.

The problem of local collision-avoidance differs from motion planning, where the global environment of the robot is considered to be known and a complete pathway towards a goal configuration is planned at once. Therefore, the collision detection simply determines if two geometric objects are intersecting or not.

The intersecting or mutually penetrating of two objects is possible in the virtual world, where the intersecting of virtual objects is possible, and where there is no risk of destruction.

The ability of predicting the behavior of cooperative manipulators is important for several reasons: for example in design, designers want to know whether the manipulator will be able to perform a certain typical task in a given time frame; in creating feedback control schemes, where stability is a major problem, the control engineer cannot risk a valuable piece of equipment by exposing it to untested control strategies. Therefore, a facile strategy for collision avoidance, capable of predicting the behavior of a robotic manipulator or of the system at whole becomes imperative.

In the real world, like collision detection, where robots need to interact with their surrounding, it is important that the computer can simulate the interactions of the cooperative participants with the passive or active changing environment, using virtual prototyping.

In this paper, we address the more necessary but less studied problem of two-robot reciprocal collision avoidance, where collisions need to be avoided for multiple tasks.

This problem has important applications in many areas in robotics, such as multi-robot cooperation and coordination. It is also a key component in modeling and behavioral simulation of robots, for computer graphics and Virtual Reality.

In this paper, we propose a new method that simultaneously determines actions for two virtual robots that each may have different objectives or will cooperate for a common objective. The actions are computed for each virtual robot and are transferred to the corresponding physical robot, with a central coordination for the collaborative tasks. Yet, we prove that our method guarantees the collision-free motion for each of the robots.

We assume that each virtual robot end-effectors can move in any direction, such that the control input of each robot is given by a three-dimensional velocity vector. Also, we assume that the algorithm is able to deduce the exact shape, position and velocity of obstacles and of the virtual robots, in the virtual environment.

The present simulation method is based on the velocity approach, which provides a sufficient condition for each robot to be collision-free for at least a fixed amount of time into the future. That implies that each robot takes into account the observed velocity of other robots in order to avoid collisions with them. Also, each robot selects its own velocity from its velocity space in which certain regions are marked as "forbidden" because of the presence of another robot.

The formulation "reciprocal collision avoidance", supposes for each robot that there are a lot of velocities, within the velocity-space, which may to be selected in order to guarantee the collision avoidance.

In this paper we develop and formal analysis of a new collision avoidance strategy for a group of two cooperative robots.

The remainder of this paper is organized as follows: in Section 2, we review prior work in collision avoidance. Section 3 describes collision detection of the virtual objects.

The model we used for reciprocal collision avoidance is described in Section 4. Collision detection through the animation of the virtual robots is detailed in Section 5. A possible strategy of the programming pathway of physical robots, based on a patent is proposed in Section 6. Finally, in Section 7, we present an experimental configuration for physical robot programming by imitation of the virtual prototype and we discuss the limitations and possible extensions of our model to answer complex situations, before concluding.

2. State-of-the-Art and Possible Extensions

The problem of collision avoidance has been extensively studied. Many approaches assume the observed obstacles to be static (i.e. non-moving), and compute an immediate action for the robot that would avert collisions with the obstacle, in many cases taking into account the robot's kinematics and dynamics.

If the obstacles are also moving, such approaches typically repeatedly "plan again" based on new readings of the positions of the obstacles. This may work well if the obstacles move more slowly than the robot, but among fast obstacles, the velocity of the obstacles needs to be specified. The problem of the obstacles moving at high speeds is generally referred to as "asteroid avoidance", and approaches typically extrapolate the observed velocities in order to estimate the future positions of obstacles [10].

However, such approaches are insufficient for multi-robot location, where the robot encounters other robots that also make decisions based on their surroundings: considering them as moving obstacles neglects the fact that they react to the robot in the same way as the robot reacts to them, and inherently causes adverse actions in the motion of the robots [11].

Velocity Obstacles (VO) [1, 3, 19] have been a successful velocity-based approach to avoid collisions with moving obstacles; they provide a sufficient and necessary condition for a robot to select velocity that avoids collisions with an obstacle moving at a known velocity.

Besides the Velocity Obstacle approach, many other methods have been proposed for collision-avoidance, navigation, and planning among moving obstacles [12, 13, 18, 19]. There are also Recursive Velocity Obstacles [10] and Common Velocity Obstacle methods [15, 18].

However, most of the existing work does not take into account that the obstacles' motion may be affected by the presence of the robot. Such approaches are generally not able to plan safe paths among obstacles moving at high speeds [11].

There is also an extensive amount of literature on multi agent navigation, in which each agent navigates individually among the other agents, which are considered as obstacles, e.g. [3, 4, 9 - 16, 18, 19]. Most of these techniques have focused on multitude simulation. Also, in these cases, the other agents are assumed to be either passively moving obstacles or static obstacles. A number of approaches follow the Velocity Obstacle concept to avoid the collision between agents.

In multi-agent planning, the composite configuration space of the agents is considered, and a path is centrally planned in this space [17, 20]. These works focus on different aspects of the problem (e.g. finding optimal coordination) and are frequently not suited for on-line real-time application.

The problem of collision avoidance becomes harder when the obstacles are also intelligent decision-making entities that try to avoid collisions as well. The reactive nature of the other entities must be specifically taken into account in order to guarantee that collisions are avoided.

However, the robot may not be able to communicate with other entities and may not know their intents. It is the case of the Reciprocal Velocity Obstacles (RVO) problem [10, 11], in which robots are partially given the responsibility of avoiding the collisions. This formulation only guarantees collision-avoidance under specific conditions, and does not provide a sufficient condition for collision-avoidance in general.

To overcome this limitation, there exists the Optimal Reciprocal collision Avoidance (ORCA) [15], which provides a sufficient condition for multiple robots to avoid collisions with one another, and thus can guarantee the collision-free operating. However, decoupled multi-agent operation is not only computationally impractical; it also requires central coordination among robots.

Many approaches that in fact guarantee collision avoidance have so far been limited to robots with specific and simple dynamics.

In this paper, we propose a new strategy, which is supposed to detect the collision using the virtual prototypes and to transfer the trajectory of virtual robots to physical robots, in the real environment, assuming that each physical robot can perfectly imitate the movement /behavior of its virtual prototype.

3. Collision Detection

Collision detection frequently arises in various applications including virtual prototyping, dynamic simulation, interaction and motion planning. Collision detection has been exhaustively researched for more than four decades. Most of the commonly used algorithms are based on spatial partitioning or Bounding Volume Hierarchies (BVHs).

Typically, for a simulated environment consisting of multiple moving virtual objects, collision problems consist of two phases: the "broad phase", where collision is performed to reduce the number of pair intelligent tests, and the "narrow phase", where the pairs of objects in proximity are checked for collision.

In this section, we present a study based on the collision detection algorithm for computing all the contacts between multiple moving virtual objects in a large virtual environment. It uses the visibility reducing algorithm described in [2]. The overall algorithm is general and applicable to all environments.

Algorithms for narrow phase can be further subdivided into efficient algorithms for convex objects and general purpose algorithms based on spatial partitioning and BVHs, for polygonal models [2]. However, these algorithms often involve pre-computation and are mainly designed for rigid models.

The performance of collision detection depends on the input model complexity and the problem output, which is the number of colliding or overlapping primitives. However, existing algorithms may not achieve interactive performance on large models consisting of thousands of triangles, due to their high complexity and output of the problem. Moreover, the memory requirements of these algorithms are typically very high.

Based on our strategy, we compute a Potentially Colliding Set (PCS) of virtual objects that are either overlapping or are in close proximity [2].

If an object O_i does not belong to the PCS, it implies that O_i does not collide with any object in the PCS. Based on this property, we can reduce the number of virtual object pairs that need to be checked for exact collision.

This is similar to the concept of computing the potentially visible set (PVS) of primitives from a viewpoint for spatial relation. We perform visibility computations between the objects in the image space to check whether they are potentially colliding or not.

Given a set S of virtual objects, we test the relative visibility of an object O with respect to set S, using an image space visibility query. The query checks whether any part of O is spatially intersected by S, rasterizing all the objects belonging to set S.

The object O is considered fully-visible if all the fragments generated by the rasterizing of O have a depth value less than the corresponding pixels in the frame buffer. We do not consider the self masking of a virtual object O in checking its visibility status.

If an object does not intersect in either of the two passes, then it does not belong to the PCS. Each pass requires the object representation for an object to be rendered twice. We can either render all the triangles used to represent an object or a bounding box of the object. Initially, the PCS consists of all the objects in the scene.

We perform these two passes to reduce objects from the PCS. We will check if an object potentially intersects with a set of objects or not.

It should be noted that our method based on reducing algorithm [7] is quite different from algorithms that reduce PCS using 2D overlap tests.

Many applications need to compute the exact overlapping features (e.g. triangles) for collision response. Instead of testing each object pair in the PCS for exact overlap, we again use the visibility formulation to identify the potentially intersecting virtual regions among the objects in the PCS.

Specifically, we use a fast global reducing algorithm to localize these regions of interest. We perform object level reducing to compute the PCS of objects.

Initially, all the objects belong to the PCS. Firstly, we perform the reduction along each coordinate axis by using the axis aligned bounding boxes as the object's representation for collision detection. The reduction is performed till the PCS does not change between successive iterations.

We also use the object's triangulated representation for further reducing the PCS.

The size of the resulting set is expected to be small and we use all-pair bounding box overlap tests to compute the potentially intersecting pairs.

If the size of this set is large, then we use the sweep-and-reduce technique [5] to reduce this set instead of performing all-pair tests.

We decompose each object into sub-objects. We have extended this approach at sub-object level and computed the potentially intersecting areas.

4. Reciprocal Collision Avoidance

For two robot end-effectors A and B, the velocity obstacle, $VO^{\tau}_{A|B}$ (read: the velocity obstacle for A induced by B for time window τ) is the set of all relative velocities of A with respect to B, which will result in a collision between A and B at some moment before time τ [9]. It is formally defined as follows.

Let be an A robot end-effector with radius r_A, positioned at $\boldsymbol{p}_A$ on a horizontal disc. For a configuration of two robot end-effectors A and B, the horizontal disc of radius $(r_A + r_B)$ is centered at $(\boldsymbol{p}_B - \boldsymbol{p}_A)$ in the Cartesian space.

Let $S(\boldsymbol{p}, r)$ denote an open horizontal disc of radius r centered at vector position $\boldsymbol{p}$ and defined by (1).

$$S(\boldsymbol{p},r)=\{\boldsymbol{s}\in S,\|\boldsymbol{s}-\boldsymbol{p}\|\leq r\} \quad (1)$$

Then the velocity obstacle is defined as:

$$VO^{\tau}_{A|B}=\{v\ |\exists t\in[0,\tau],vt\in S(\boldsymbol{p}_B-\boldsymbol{p}_A,r_A+r_B)\} \quad (2)$$

Let $\boldsymbol{v}_{tA}$ and $\boldsymbol{v}_{tB}$ be the current operational velocities of the robots' end-effectors A and B, respectively. The definition of the velocity obstacle implies that if $(\boldsymbol{v}_A - \boldsymbol{v}_B) \in VO^{\tau}_{A|B}$, or equivalently if $(\boldsymbol{v}_B - \boldsymbol{v}_A) \in VO^{\tau}_{B|A}$, then A and B will collide at some moment before time τ if they continue moving at their current velocity.

On the other hand, if $(\boldsymbol{v}_A - \boldsymbol{v}_B) \notin VO^{\tau}_{A|B}$, the two robot end-effectors A and B are guaranteed to be collision-free for at least τ time. Robot end-effector A will collide with robot end-effector B within τ time if its velocity $\boldsymbol{v}_A$ is inside $VO^{\tau}_{A|B}$ and it will be collision-free for at least τ time if its velocity is outside the velocity obstacle.

For an articulated robot arm, the robot end-effectors velocity vector is calculated as, $v = \dot{X}_t$ where $\boldsymbol{X}_t$ is the Cartesian position vector of the robot end-effector. The vector $\boldsymbol{X}_t$ can be described as a function of robot joints variables vector, $\boldsymbol{q}$:

$$X_t = \mathrm{f}(\boldsymbol{q}) \quad (3)$$

Equation (3) can be obtained easily with the help of the Denawit - Hartenberg operators.

The robot end-effectors velocity vector **v** can be obtained as:

$$v = \dot{X}_t = \boldsymbol{J}_t(\boldsymbol{q})\dot{\boldsymbol{q}} \quad (4)$$

where $\boldsymbol{J}$ is a Jacobean matrix, and $\dot{\boldsymbol{q}}$ is the robot joints velocities vector.

Given a trajectory that each moving robot end-effector will travel, we can determine the exact collision time. Please refer to [18] for more details. If the path that each robot end-effector travels is not known in advance, then we can calculate a lower bound on the collision time. This lower bound on the collision time is calculated adaptively to speed up the performance of the dynamic collision detection.

5. Collision Detection through Animation of two Virtual Robots

Our proposed scheme for dynamic simulation or animation, using the distance computation algorithm, is an iterative process which continuously inserts and deletes the object pairs from a stack according to their approximate time of collision, as the objects move in a dynamic environment.

Our simulation method purely exploits the spatial arrangement of the two end-effectors without any other information.

The end-effector pair which has a small separation is likely to have an impact within the next few time instances, and those virtual pairs which are far apart from each other cannot possibly come to interfere with each other until a certain moment.

For spatial tests to reduce the number of virtual pairs judicious comparisons, we assume the environment is quite free and the end-effectors move in such a way that the geometric coherence can be preserved, i.e. the assumption that the motion is essentially continuous in the time domain.

Planning scene formulation for cooperative tasks is presented in the figures below.

In the cooperative task studies, the simulation is used to find whether it is possible to avoid the collision between a particular part of the robot arm and diverse objects in the work space and so to find one possible free path.

In this section we will describe how to render a planning scenario in the form of constraints for the constraint-based planning framework.

Our visibility using the PCS computation algorithm is based on a hardware visibility query which determines if a primitive is fully-visible or not.

By assuming that the geometry representing the robots and obstacles is given, the motion is prescribed for obstacles over time.

Our system then defines constraints that will restrict the motion of the robots to meet the design specifications, and also guides the robots to complete the planning tasks so that the collision will be avoided.

Our virtual system was implemented on a programming platform, using the Delphi object-oriented programming language.

In this section we have tested our system of two robots for collision detection in the following scenes for the virtual prototyping applications.

5.1. Scene 1: Individual Task for each Robot

The individual task for each robot is studied in scene 1.

Each of the two robot arms is composed of rigid components that are held together by constraints. For all of the components of the robot arm, the planner must compute paths to ensure the joint constraints do not collide with the obstacles as the conveyors and lead the end-effectors along the prescribed path.

In this scene (an example seen in Fig. 1), two articulated robot arms, with six degrees of freedom, are used to transfer (manipulate) cubic objects from one conveyor belt to another conveyor belt.

Each robot arm follows a path over the conveyor body while avoiding obstacles.

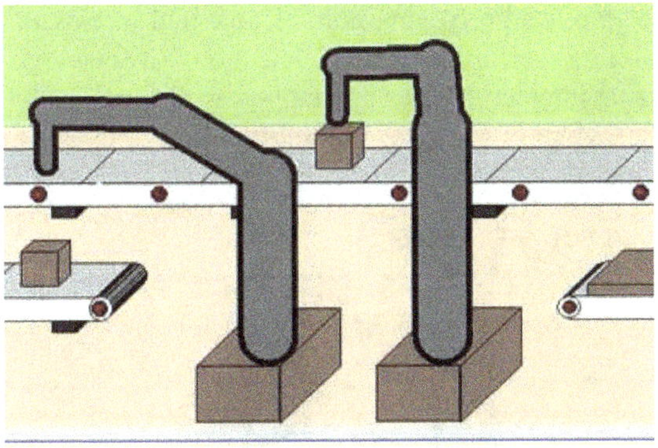

***Figure1**. Scene for individual tasks*

5.2. Scene 2: Cooperative Mission

The cooperative mission for two robots is analyzed below. In our example, a second scene shown in Fig. 2, the end-effectors of the left robot and of the right robot respectively, avoid each other in a firm behavior during a cooperative task. In this scene, two articulated robot arms, with 6 degrees of freedom, are used to manipulate together a rigid object for an assembly operation. The robot arms avoid the moving belt to get in touch with a plate object passing it on the assembly conveyer belt. The goal to manipulate together the same object in a cooperative mission requires both robots to maneuver around each other without colliding.

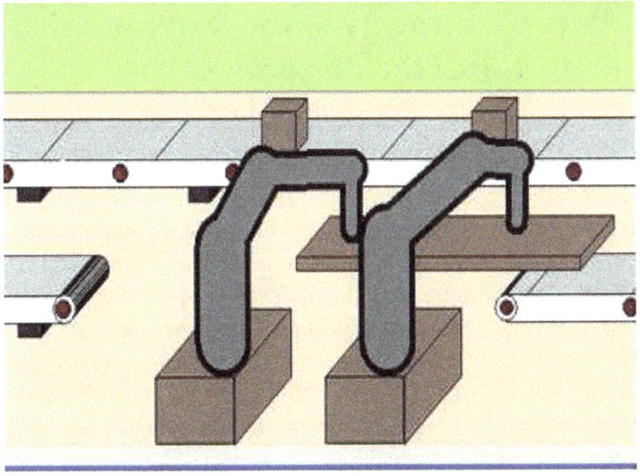

***Figure 2**.Scene for cooperative mission*

5.3. Scene 3: Assembly Line Planning

In this example, shown in Fig. 3, two robot arms must assemble a plate object with two cubical objects, on a conveyer belt. Both robots must be moved simultaneously around each other to position the plate object on top of the two cubical objects and to avoid collision. The assembly line contains a support structure that is moving over the conveyer belt in the same direction as the assembled parts.

The moving structure may become an obstruction that causes the robots to reactively modify their path, to avoid collision. In spite of the rapid progress in the performance of the graphics processing units, it may not be possible to visualize or perform collision detection between massive models at interactive rates on graphics hardware.

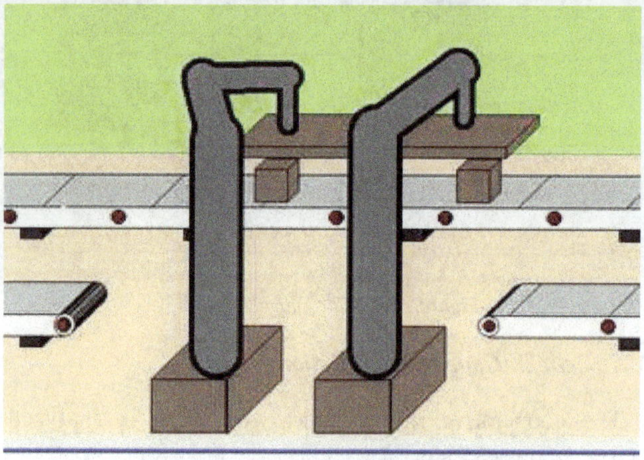

Figure 3*. Scene for assembly line planning*

Since the usual models of the cooperative robots are rather complex and may have thousands of components, the algorithm as described in [10, 11], becomes essential to generate a realism of motion.

The visual results indicate that the approximated implicit integration, mixed with the post-step inverse dynamics process, achieves simulation of rigid objects very well.

6. Possible Strategy of the Programming Pathway of the Physical Robots

We propose an original idea that allows us to transfer the joint trajectory of each virtual robot to the corresponding joint of the real (physical) robot. With other worlds we prepare the free-trajectories for the pair of the virtual cooperative robots; these trajectories can be transferred to their corresponding pair of the physical cooperative robots.

In the real world, the programming pathway of the physical robots can be realized using the pathway of the virtual robot prototypes [9]. Therefore, the virtual robot behavior must be specifically taken into account in order to guarantee that collisions are avoided between corresponding physical robots. Each real robot may be able to communicate with her virtual corresponding entity and may imitate their intents. We call this problem *reciprocal collision avoidance using the virtual robots prototypes*, and it is the focus of this paper.

The strategy that we realized in this paper, for the virtual robots, is formally defined as follows. Let there be a set of two virtual partners-robots sharing a virtual environment. Each robot has a current position and a current velocity. These parameters are part of the robot's external state, i.e. we assume that they can be estimate in the virtual environment.

Furthermore, each virtual robot has a maximum speed and a preferred velocity, which is the velocity the robot would assume had no other robot/object been in its way. We consider these parameters part of the internal state of the robot, and can therefore not be observed by the other robot.

The task is for each virtual robot to independently (and simultaneously) select a new velocity for itself such that both virtual robots are guaranteed to be collision-free for at least a fixed amount of time, when they would continue to move at their new velocity.

As a secondary objective, the virtual robots should select their new velocity as close as possible to their preferred velocity. The virtual robots are not allowed to communicate with each other, and can only use observations of the partner-robot's current position and velocity.

In our collaborative work systems, the virtual robot prototypes are used mainly as an intermediate result for calculating the "nearest neighbors" and the potentially intersecting areas in a possible collision. We compute a PCS of virtual objects that are either partly covered or in close proximity. If an object does not belong to the PCS, it implies that this object does not collide with any object. We initially compute the PCS of virtual objects based on the above algorithm (section 3 and 4). As an alternative of testing each object pair in the PCS for partial cover, we use the visibility formulation to identify the potentially intersecting virtual areas among the objects.

Based on this property, we developed a programming platform for a pair real (physical) robots based on a cooperative virtual robot's pair, which needs to be checked for exact collision detection. This platform needs to compute the exact overlapping area of the virtual cooperative robots, as collision response.

7. Experimental Configuration for Programming by Imitation of an Anthropomorphic Robot

A physical anthropomorphic robot is controlled to follow the arbitrary path reference with a predefined velocity profile over time. The physical robot follows an imaginary robot path which is ideally generated by the virtual robot. The physical robot must follow the virtual robot's trajectory.

On the basis of robot kinematics equations, a robot control system is presented in Fig. 4.

Fig. 4 also displays (left image) the user interface for a virtual anthropomorphic robot arm, which has been created by the motion simulation system.

The proposed architecture provides libraries and tools focused on 3D simulation of the dynamics systems and on a control and planning interface that provides primitives for motion planning by imitation. This architecture is composed of several modules and has separate nodes launched as executables which communicate via message passing. Via the numerical interface NI, one transfers the joint angles data from a motion capture system to the kinematic model for an anthropomorphic robot.

To generate the desired motion sequence for the real robot, we capture the motions from a virtual robot model and map these to the joint settings of the physical robot.

Initially, a set of virtual postures is created to the virtual robot arm, VRA and the pictures' positions are recorded for each posture, during motion [9].

These recorded pictures' positions provide a set of Cartesian points in the 3D capture volume for each posture.

To obtain the final robot posture, the virtual pictures' positions are assigned as positional constraints on the physical robot. To derive the joint angles, standard inverse kinematics (IK) routines are used.

The IK routine then directly generates the desired joint angles on the robot for each posture.

We assume the use the virtual robot prototypes and the motion capture systems to obtain the reference motion data, which typically consist of a set of trajectories in the Cartesian space.

The data is obtained using a motion capture channel taking into account the joint motion range. The symbolic spatial relations specifying the virtual environment can be used for the automatic planning of the possible virtual path as reference for the real robotic arm, RRA, which may guide the motion process during execution.

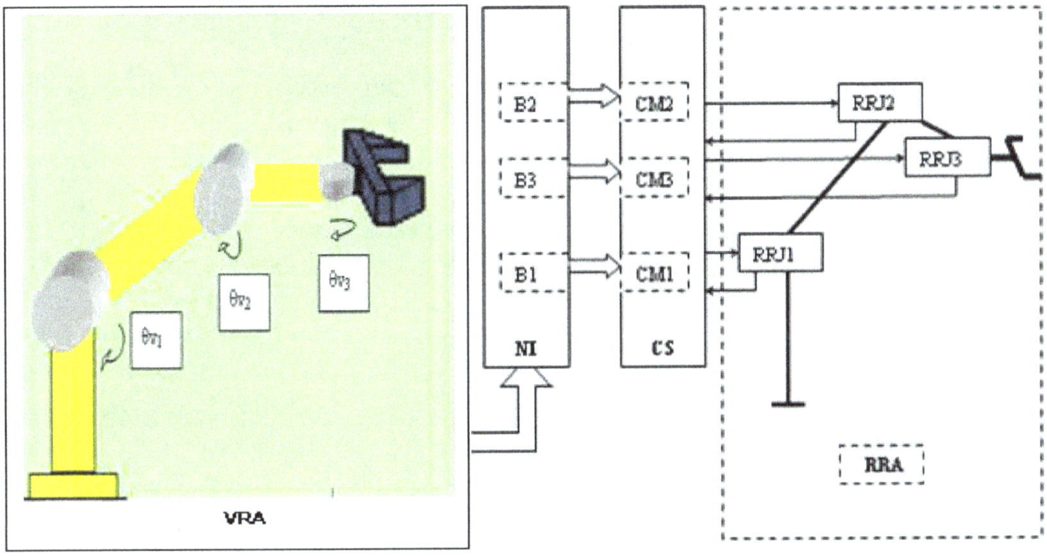

***Figure 4**. Programming platform with the corresponding virtual robot and the real robot face-to-face*

The easiest way to generate the spatial relations explicitly is the interactively programming of the behavior of the virtual prototype in its virtual environment in order to specify suitable position coordinates θ_{v1}, θ_{v2}, θ_{v3}. These position coordinates are used by the physical robot as reference position coordinates.

This kind of specification provides an easy to use interactive graphical tool to define any kind of robot path; the user has to deal only with a limited and manageable amount of spatial information in a very comfortable manner.

An automatic robot programming system has to recognize the correct robot task type and should map it to a sequence of robot operations [6]. The desired pathways are automatically transferred and parameterized in the NI, using the path planner. The physical robot receives the position coordinates of the virtual robot through NI.

Fig. 4 shows a simple path-following system which keeps a constant communication between the virtual robot's path and the control system, CS. The control system is designed to force the real robot to follow the reference path.

The main program simply defines "start" and "goal" positions. After moving the virtual robot to the 'start' position in the joint "interpolation" mode, the real robot is moved in the "following" mode, while a "monitor function" has been activated.

The "monitor function" is reading the reference path values, which are used in closed loops to compute the physical joint torques.

In similar ways, any functional dependencies of some path properties (speed, distance etc.) can be specified in a textual programming manner.

The trajectories are sent to visualization, so that users can see the results of the animation.

The robot's control system is connected via the Transmission Control Protocol (TCP) to a PC, equipped with the interface card; the PC is running the simulation and control process. The robot control system receives and executes each 16 ms, an elementary move operation.

The communication protocol between the virtual robot and the physical robot uses the CAN bus. This application can be coded with just a few lines of the DELPHI code, presented below.

```
unit ControlCAN;
interface
uses
 WinTypes;
const
 DLL_NAME = 'ControlCAN.dll';
type
//1.ZLGCAN
VCI_BOARD_INFO = Record
         hw_Version : WORD;
         fw_Version : WORD;
         dr_Version : WORD;
         in_Version : WORD;
         irq_Num : WORD;
         can_Num : BYTE;
         str_Serial_Num : array[0..19] of CHAR;
         str_hw_Type : array[0..39] of CHAR;
         Reserved : array[0..3] of WORD;
END;
PVCI_BOARD_INFO=^VCI_BOARD_INFO;
//2.
VCI_CAN_OBJ = Record
         ID : UINT;
         TimeStamp : UINT;
         TimeFlag : BYTE;
         SendType : BYTE;
         RemoteFlag : BYTE;
         ExternFlag : BYTE;
         DataLen : BYTE;
         Data : array[0..7] of BYTE;
         Reserved : array[0..2] of BYTE;
END;
PVCI_CAN_OBJ = ^VCI_CAN_OBJ;
//3.
VCI_CAN_STATUS = Record
         ErrInterrupt : UCHAR;
         regMode : UCHAR;
         regStatus : UCHAR;
         regALCapture : UCHAR;
         regECCapture : UCHAR;
         regEWLimit : UCHAR;
         regRECounter : UCHAR;
         regTECounter : UCHAR;
         Reserved : DWORD;
END;
PVCI_CAN_STATUS = ^VCI_CAN_STATUS;
//4.
VCI_ERR_INFO = Record
         ErrCode : UINT;
         Passive_ErrData : array[0..2] of BYTE;
         ArLost_ErrData : BYTE;
END;
PVCI_ERR_INFO = ^VCI_ERR_INFO;
//5.
VCI_INIT_CONFIG = Record
         AccCode : DWORD;
         AccMask : DWORD;
         Reserved : DWORD;
         Filter : UCHAR;
         Timing0 : UCHAR;
         Timing1 : UCHAR;
         Mode : UCHAR;
END;
PVCI_INIT_CONFIG = ^VCI_INIT_CONFIG;
//6.
function VCI_OpenDevice ( DeviceType : DWORD;
 DeviceInd : DWORD;
 Reserved : DWORD) : DWORD;
 stdcall;
 external DLL_NAME;
function VCI_CloseDevice ( DeviceType : DWORD;
 DeviceInd : DWORD) : DWORD;
 stdcall;
 external DLL_NAME;
function VCI_InitCAN ( DeviceType : DWORD;
```

The applicable robot tasks are designed and the desired pathways are programmed off-line and stored in the buffers B1, B2, B3.

The following errors are delivered by the comparative modules CM1, CM2, CM3. The controllers interpret following errors and generate corresponding variables, which are transmitted to the actuators.

Process changes from disturbances result in new sensor signals, identifying the state of the process, to be transmitted again to the controller. A control loop, including sensors, control algorithms and actuators, is arranged for each joint in such a way as to try to regulate the position variables at reference positions value to obtain the desired closed loop performances.

While motion execution is in progress, the real robot joints RRJ1, RRJ2, RRJ3 are activated into the real environment. Each time, a skill primitive is executed by the CS; it changes the robot joints state. As no time limit for the motion is specified, the real robot imitates the behavior of the virtual robot.

In our laboratory we are currently developing Cartesian control architecture able to interpret the physical robot commands in the above given form. The basis of our implementation is a flexible and modular system for robot programming by imitation.

In our experimental configuration, in order to prove the correctness of the robot programming by imitation, we have chosen an anthropomorphic robot arm, with 3 DOF equipped with electrical actuators, mounted on the real robot's joints.

The designed control algorithm proved stable and robust to the errors when following the reference path, to input and output noises and to other disturbances.

8. Conclusion and Future Work

This paper gives an application of the collision detection algorithm described in [2, 10, 11], for virtual manipulation planning with virtual robots. We have applied this algorithm to perform collision detection in a virtual environment.

This algorithm has also been utilized for dynamic simulation and its practicality has been demonstrated for different applications.

The distance computation method, described in [11] has been used in the dynamics simulator written in the Delphi language.

Our vision of this dynamic simulator is the ability to simulate small mechanical parts of a robot arm. It reduces the frequency of the checks significantly, so as to help speed up the calculations.

We revealed the potential of the Reciprocal Velocity Obstacle approach by applying it to scenarios in which two virtual robots accomplish their tasks, independently or in cooperation, in a complex environment.

We would like to extend the current method, allowing it to handle various types of time-varying data sets used in animation process.

Furthermore, we would like to apply our collision detection framework to several applications including the motion planning of physical robots while passing near each other.

In our formulation, the real robots must have exactly the same dynamics model as virtual robots in order to be able to imitate the behavior of the latter.

Virtual robots could be handled using an abstraction of the dynamics model of their real homologue. Real robots will imitate the virtual robot's behavior and will move according to it.

For virtual robots, we have implemented this algorithm in 2D and 3D. In 3D, we used an immersive environment, to be able to virtually manipulate the animated objects. We ordered the robots to grip the object and move it around a scene containing different obstacles situated on a ground plane.

It is important to note here that this virtual model is not dynamic, but rather a succession of static postures, which greatly limits its applications. We experimented with our approach on several complex simulation scenarios, containing virtual interoperations. As each robot is independent, we can fully parallelize the simulation of the actions for each robot to realize the animation.

Actually, the method and installation described in [9] are currently under testing to be eventually integrated into a real collision environment, developed at Transilvania University of Brasov.

In addition to discussing the original contributions, this paper presents a set of directions to be considered for future work.

The authors intend to extend experiments to investigate these ideas and examine the possibility of how to implement them.

The authors expect fully automated robot programming by imitation based on this method, using robust enough system to be applied in industrial applications, will not become true before the end of this decade.

Acknowledgements

The authors wish to thank the entire team at LAMIH Laboratory of University of Valenciennes-France, for cooperation and engagement in the research activity in the field of robotics.

This cooperation gave us the opportunities to work in a multidisciplinary team and provided us important experiences to extend our research in future projects.

References

[1] C. Fulgenzi, A. Spalanzani, C. Laugier, "Dynamic obstacle avoidance in uncertain environment combining PVOs and occupancy grid". In Proc. IEEE Int. Conf. on Robotics and Automation, pp.1610–1616, 2007.

[2] N. K. Govindaraju, S. Redon, M. C. Lin and D. Manocha, "CULLIDE: Interactive Collision Detection Between Complex Models in Large Environments using Graphics Hardware", M. Doggett, W. Heidrich, W. Mark, A. Schilling (Editors), Graphics Hardware, 2003.

[3] D. Hennes, D. Claes, W. Meeussen, K. Tuyls, "Multi-robot collision avoidance with localization uncertainty", In: Proceedings of the 11th International Conference on Autonomous Agents and Multiagent Systems, Conitzer, Winikoff, Padgham, and van der Hoek (eds.), June, 4– 8, 2012, Valencia, Spain.

[4] Y. Abe, M.Yoshiki, "Collision avoidance method for multiple autonomous mobile agents by implicit cooperation". IEEE/ RSJ, Int. Conf. Intelligent Robots and Systems, pp.1207-1212, 2001.

[5] K. O. Arras, J. Persson, N. Tomatis, R. Siegwart, " Real-Time Obstacle Avoidance for Polygonal Robots With a Reduced Dynamic Window" in Proc. IEEE Int. Conf. on Robotics and Automation, Washington DC, May 2002, pp.3050-3055.

[6] N. Galoppo, "Animation, Simulation, and Control of Soft Characters using Layered Representations and Simplified Physics-based Methods" Dissertation submitted to the

faculty of the University of North Carolina Chapel Hill, 2008.

[7] S.E. Yoon, "Interactive Visualization and Collision Detection using Dynamic Simplification and Cache-Coherent Layouts". Dissertation submitted to the faculty of the University of North Carolina, Chapel Hill, 2006.

[8] A. Fratu, L. Vermeiren, A., Dequidt, "Using the Redundant Inverse Kinematics System for Collision Avoidance. The 3rd International Symposium on Electrical and Electronics Engineering - ISEEE- 2010, 16-18 sept. Galati, Romania, Proceedings ISBN 978-1- 4244-8407-2, pp. 88-93.

[9] A. Fratu, "Method and installation for joints trajectory planning of a physical robot arm" (proposal - patent) unpublished.

[10] J. van den Berg, S. Guy, M. Lin, D. Manocha, " Reciprocal n-body collision avoidance". In: Proc. Int. Symposium on Robotics Research, 2009

[11] J. van den Berg, S. J. Guy, M. Lin, and D. Manocha, C. Pradalier, R. Siegwart, and G. Hirzinger, "Reciprocal n-body collision avoidance", Robotics Research, The 14th International Symposium ISRR, Springer Tracts in Advanced Robotics, vol. 70, Springer-Verlag, May 2011, pp. 3-19.

[12] J. Snape, J. van den Berg, S. J. Guy, and D. Manocha, "Independent navigation of multiple mobile robots with hybrid reciprocal velocity obstacles", IEEE/RSJ Int. Conf. Intelligent Robots and Systems, St. Louis, Mo., 2009.

[13] J. Snape, S. J. Guy, J. van den Berg, S. Curtis, S. Patil, M. Lin, and D. Manocha, "Independent navigation of multiple robots and virtual agents". In Proc. of the 9th Int. Conf. on Autonomous Agents and Multi agents Systems (AAMAS 2010), Toronto, Canada, May 2010.

[14] S. J. Guy, J. Chhugani, C. Kim, N. Satish, M. Lin, D. Manocha, and P. Dubey, " Clear Path: Highly Parallel Collision Avoidance for Multi-Agent Simulation" In: Proc. ACM SIGGRAPH/Eurographics Symposium on Computer Animation (SCA), Aug. 2009.

[15] J. Snape, J.van den Berg, S.J. Guy, D. Manocha, "S-ORCA: Guaranteeing Smooth and Collision-Free Multi-Robot Navigation Under Differential-Drive Constraints". In: Proc. IEEE Int. Conf. Robotics and Automation, 2010.

[16] D. M. Stipanovic, P. F. Hokayem, M. W. Spong, D. D. Siljak, "Cooperative Avoidance Control for Multiagent Systems". In: ASME Journal of Dynamic Systems Measurement and Control, Vol.129, pp. 699–707, 2007.

[17] R. Diankov and J. Kuffner, "Openrave: A planning architecture for autonomous robotics". Technical report, CMU-RI-TR-08-34, The Robotics Institute, Carnegie Mellon University, 2008.

[18] M. Turpin, N. Michael, V. Kumar," Trajectory planning and assignment in multi robot systems". Proc. Workshop on Algorithmic Foundations of Robotics, 2012.

[19] J. van den Berg, D. Wilkie, S. Guy, M. Niethammer, D. Manocha, "LQG-Obstacles: Feedback control with collision avoidance for mobile robots with motion and sensing uncertainty." IEEE Int. Conf. on Robotics and Automation, River Centre, Saint Paul, Minnesota, USA, May 14-18, 2012, pp. 346- 353.

[20] Y. Li and K. Gupta, "Motion planning of multiple agents in virtual environments on parallel architectures," in Proc. IEEE Int. Conf. on Robotics and Automation, 2007, pp. 1009–1014.

5

The intelligent forecasting model of time series

Sonja Pravilović[1, 2], Annalisa Appice[2]

[1]Montenegro Business School, "Mediterranean" University, Podgorica, Montenegro
[2]Dipartimento di Informatica, Università degli Studi di Bari Aldo Moro, Bari, Italy

Email address:
sonja.pravilovic@uniba.it (S. Pravilovic), annalisa.appice@uniba.it (A. Appice)

Abstract: Automatic forecasts of univariate time series are largely demanded in business and science. In this paper, we investigate the forecasting task for geo-referenced time series. We take into account the temporal and spatial dimension of time series to get accurate forecasting of future data. We describe two algorithms for forecasting which ARIMA models. The first is designed for seasonal data and based on the decomposition of the time series in seasons (temporal lags). The ARIMA model is jointly optimized on the temporal lags. The second is designed for geo-referenced data and based on the evaluation of a time series in a neighborhood (spatial lags). The ARIMA model is jointly optimized on the spatial lags. Experiments with several time series data investigate the effectiveness of these temporal- and spatial- aware ARIMA models with respect to traditional one.

Keywords: Time Series Analysis, Arima, Auto. Arima, Lag. Arima

1. Introduction

In recent years, globalization has significantly accelerated the communication and exchange of experience, but has also increased the amount of data collected as a result of monitoring the spread of economic, social, environmental, atmospheric phenomena. In such circumstances, it is necessary a useful tool for analyzing data that represents the behavior of these phenomena and drawing useful knowledge from these data to predict their future behavior.

Accurate forecasts for the phenomenon behavior can make anticipation of the actions (for example, the prediction of wind speed in a region allows us to define the better strategy to maximize profit in the energy market).[16]

In the last two decades, several models of (complex) time series dynamics have been investigated in statistical analysis.[3] Forecasting algorithms must determine an appropriate time series model, estimate the parameters and compute the forecasts. They must be robust to unusual time series patterns, and applicable to large numbers of series without user intervention. The most popular automatic forecasting algorithms are based on the ARIMA model, which optimizes the parameters of the model for a single univariate time series.[6] In this way, multiple univariate time series, such that, geo-referenced time series, which are generated by several sensors of a network, are modeled separately. This naive approach neglects the spatial component of time series, which refer points placed at specific spatial locations. When analyzing geo-referenced data, we frequently the phenomenon of spatial autocorrelation. [18]

Spatial autocorrelation is the correlation among the values of a single variable (i.e., object property) strictly attributable to the relatively close position of objects on a two-dimensional surface, introducing a deviation from the independent observations assumption of classical statistics. Intuitively, it is a property of random variables taking values, at pairs of locations a certain distance apart, that are more similar (positive autocorrelation) or less similar (negative autocorrelation) than expected for pairs of observations at randomly selected locations (Moran 1950).[20] Positive autocorrelation is common in spatial phenomena (Goodchild 1986).[18] Spatial positive autocorrelation occurs when the values of a given property are highly uniform among spatial objects in close proximity, i.e., in the same neighborhood. In geography, spatial autocorrelation is justified by Tobler's first law (Tobler 1970)[17], according to which "*Everything is related to everything else, but near things are more related than distant things*".[17] This means that by picturing the spatial variation of some observed variables in a map, we may observe regions where the distribution of values is smoothly continuous, with some boundaries possibly marked by sharp discontinuities. This suggest that a forecasting model is smoothly continuous in neighborhood and inappropriate treatment of data with spatial dependencies, where spatial a autocorrelation is ignored, can obfuscate important insights in the models.

In this paper we formulate an inference procedure that allows us to and to obtain a robust and widely applicable automatic forecasting algorithm which optimizes the traditional ARIMA model by "jointly" estimating forecasting parameters for several time series lags.[12] These lags can be consecutive seasons of a single time series or multiple time series with spatial dependence. This algorithm has been implemented in the forecast package for R. For each (temporal- or spatial-aware) lag we apply all models that are appropriate, optimizing the parameters of the model in each case and selecting the best model according to the AIC. The point forecasts can be produced by using the best model (with optimized parameters) for as many steps ahead as required.

The paper is organized as follows. In the next Section we revise basic concepts and background of this work. In Section 3, we describe the algorithm *lag.arima* in its temporal and spatial formulation. In Section 4, we report results of an empirical evaluation on real-world times. Finally, some conclusions are drawn in Section 5.

2. Background

The Box-Jenkins approach is one of the most popular and powerful forecasting technique. It is an ARIMA model [2], which is a generalisation of an ARMA model [12].

The autoregressive–moving-average (ARMA) model describes a (weakly) stationary stochastic process in terms of two polynomials, one for the auto-regression and the second for the moving average.[21]

$$X_t = c + \varepsilon_t + \sum_{i=1}^{p} \varphi_i X_{t-i} + \sum_{i=1}^{q} \theta_i \varepsilon_{t-i} \tag{1}$$

The auto-regressive model of order p. refers to the moving average model of order p.

$$X_t = c + \sum_{i=1}^{p} \varphi_i X_{t-i} + \varepsilon_t \tag{2}$$

where φ_i are parameters, c is a constant, and the random variable ε_t is white noise.

MA(q) refers to the moving average model of order q.

$$X_t = \mu + \varepsilon_t + \sum_{i=1}^{q} \theta_i \varepsilon_{t-i} \tag{3}$$

□$_i$ are parameters of the model, μ is the expectation X_t (often assumed to equal 0), and the t, ε_t, ε_{t-1},... are white noise error terms.

Gershenfeld and Weigand [5] indicate that the selection of ARMA order (p, q) models is not simple. ARMA models in general can, after choosing p and q, be fitted by least squares regression to find the values of the parameters which minimize the error term. It is generally considered good practice to find the smallest values of p and q which provide an acceptable fit to the data [3] recommend using AICc for finding p and q.

Time series were mainly studied under a deterministic aspects, until in 1927 Yule [15] introduced the notion of stochasticity. According to him, every approach to time series can be regarded as the realization of a stochastic process. This idea of stochastic process launched a different number of time series methods, varying in parameter estimation, identification, forecasting and checking method.

Box and Jenkins in their publication Time Series Analysis: Forecasting and Control [3] integrated the existing knowledge and made a breakthrough in the area creating a coherent and versatile approach identifying the three stage iterative cycle for time series: identification, estimation and checking diagnostic. The autoregressive integrated moving average (ARIMA) models by the evolution of computers made the use more popular and applicable in many scientific fields.

Auto.arima function [9] conducts a search over possible model within the order constraints provided and returns the best ARIMA model according to either AIC, AICc or BIC value. Non-stepwise selection can be slow, especially for seasonal data. Stepwise algorithm outlined in Hyndman and Khandakar [9] except that the default method for selecting seasonal differences is now the OCSB test rather than the Canova-Hansen test. *Auto.arima* function that is within the library (forecast) in the R programming language has the task to define the parameters p, d and q having a minimum AIC criterion. [9].

The combination of forecasts is a widely investigated issue in the statistical field. Many researchers have recognized the value of combining forecasts produced by various techniques as a means of reducing forecast error [4]. Armstrong's meta analysis [14] marked that combining of different model can be more useful for short range forecasting, where random errors also can be more significant. As because these errors are off setting, a combined prediction methods should reduce them. Many combining forecasts techniques have been introduced over the past years, varying in level of success and complexity.

Timmerman argue that simple combinations that ignore the correlation between the predicted errors often dominate over refined combination schemes aimed at assessing theoretically optimal combination weights. The use of a simple average has proven to do as well as more sophisticated approaches. Nevertheless, there are situations where one method is more accurate than another. If such cases can be identified in advance, simple averages would be inefficient [14].

Applying Box-Jenkins methodology optimization parameter there are available various packages to find the right parameters for the ARIMA model. In this paper, we used R functions included in the standard stats package, which includes arima function documented in 'ARIMA Modelling of Time Series' [7,8]. The ARIMA(p,d,q) function also includes seasonal factors, an intercept term, and exogenous variables called 'external regressors'. The 'forecast' package with the auto arima() function in R automatically select the best ARIMA model for a given time series.

The forecast package for the R system for statistical computing (R Development Core Team 2008) is part of the forecasting bundle [7,8,9,10,11] is available from the Comprehensive R Archive Network at http://CRAN.R-project.org/package=forecasting. Version 2.15 of the package was used for this paper. The forecast package contains functions for univariate forecasting and implements automatic forecasting using exponential smoothing, ARIMA models, the Theta method (Assimakopoulos and Nikolopoulos 2000), cubic splines [10,11], as well as other common forecasting methods.

But in classical statistics, correlation analysis has as a basic assumption independence between the variables. This assumption does not exist in spatial statistics in which the observed values depends with each other. In fact, it is assumed that the observations or measured values in near places are more connected and are very similar. In other words, what the station which measured variables are further, it would measured values more different to each other.

Automatic prediction algorithms must determine the appropriate time-series model to estimate the parameters and calculate the forecast, but the most popular prediction algorithms and techniques only treat space or time dimension, a rarely time and space simultaneously.

The idea is to use these methods of predicting univariate and multivariate time series, which however do not take into account the spatial dependence to get more information needed for better prediction.

This article presents a new methodology that provides short-term forecasts of natural phenomena first form and then dividing time series in specific lags and exploring the unique and best predicting parameters model valid for each lag (for all sensors in closely area) of the time series. This is because, our intention is experimentation and applying algorithms that use the methodology and techniques of time series predictions explaining the fact that the statistical properties of the recorded data (prediction in time series behaviour) can be applicable not only to the dimension of time, but also space. In doing so, the goal of this empirical study is to analyze accuracy and efficiency of our proposed algorithm, dividing the time series in lags and calculating the mean absolute prediction errors and the mean of residuals for all lags, instead of calling *auto.arima* which gives only one group of best prediction parameters for all-time series. By comparing the result of prediction using those parameters and the result of *auto.arima* for all-time series we demonstrate that our model is more accurate and provides a detailed analysis of future behaviour of time series. Our research aims to take advantage of the methodology of univariate time series forecasting techniques such as arima, *auto.arima*, and others in predicting measurable values of sensors that are distributed in spatial dimension by grouping in zones, and predict the measurable values of the sensors founded in their neighbourhood.

3. Algorithm

It is possible to take advantage of the methodology of time series forecasting techniques such as *arima*, *auto.arima*, and others in predicting measurable values (for example sensors) that are distributed in spatial dimension to predict measurable values in the neighbourhood.

The main task of this paper was to show that dividing time series in the time intervals of specified length (lags) and then finding the best and unique model parameters valid for all lags needed to make predictions of future values provides a big advantage. By comparing the result of prediction using those parameters and the result of *auto.arima* for all-time series we demonstrate that our model is more accurate and provides a detailed analysis of future values of time series.

The first idea to solve problem of forecasting of such formed time series with application of *auto.arima* function was dividing time series in the lags, and then for each lag call *auto.arima* function that finds the best parameters. But, dividing the time series at lags and then using the function *auto.arima* for each lags occur two kinds of problems.

1. For each lag, *auto.arima* finds the best parameters *p, d* and *q*, regardless of whether that part of the interval of lag is stationary or not. Acceptance of the parameters that give the optimal solution in one lag may prove to be a complete failure in the other (in case that the part of the series in the next lag is not stationary or has completely different flow). Thus, the obtained model and parameters *p, d* and *q*, that are commonly used for each lag gives the results, but the solution is not general, since it can simultaneously apply more equal number of times or obtained solution is not applicable for each lag of time series. Some lags of time series can also be non-stationary and the parameters *p, d* and *q*, for all lags in general do not give a meaningful result).
2. Inefficient multi calling function *auto.arima* and different number of parameters for each lag.

Thus, our algorithm has the task to applied modified *auto.arima* function that defines the length of the time interval (after that period the time series repeats or has a very similar behavior). This is because the time series consists of data obtained from the sensors that are located closely and measured very similar values of the phenomenon in the same period of time.

Our algorithm examine the unique parameters that are valid for each lag of whole time series and looking for those parameters for which average AIC is minimal for all lag, since they allow to obtain unique and best predicting parameter of future behavior of the time series. The idea is that the algorithm which uses the methodology of time series forecasting techniques, while explaining the fact that the statistical properties of the recorded data can be applicable not only to the dimension of time, but also space.

Thus, as the input data set there are more short time series of measured data on spatially distributed sensors, which are located closely and is formed only one time series for the same period of time.

Our algorithm apply modified *auto.arima* model for the entire time series, which is divided into lags (values measured on the sensor in near neighborhoods) and find a unique and best model that gives the minimum average value of AIC's which passed all lags of formed time series. In the cases that are possible more than one and same model (parameters p, d and q) to apply to all lags (time intervals) will be selected that one which for the entire time series, for each lag, chooses the one that give the minimal average AIC.

In order to conduct a more precise forecasting of longer time series as at specific time intervals show repeats flow, in this paper, we propose a new prediction algorithm as follows:

First, we display the entire time series and observed length of time interval after which the time series behaves about the same or similar way.

Second, we call modified function *auto.arima* within the R software package, which we called *lag.arima* whose task is to identify the ARIMA model with the same parameters p, d, q, P, D, Q, which are valid for each lag of time series. In the event that there are multiple models that are valid for each lag, *lag.arima* will selected as the optimal model and its parameters one whose average AIC is minimal for all lag.

As we analyzed at the time series in the following figure 1, we can see the characteristic movement repeated similar values 12 times. Since the timestamp series contains 4032 values related to the half-hourly demands for electricity, when divided by 12, we can get the length of a period or 336 values.

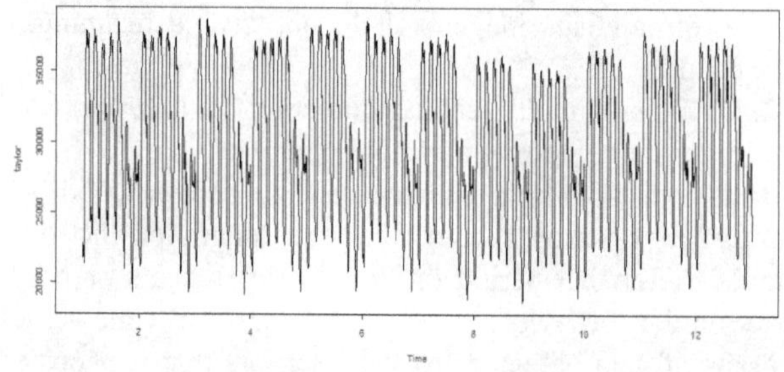

Figure 1. *The time series 'Taylor'*

Having established the length of a lag, we called the modified functions *auto.arima*, which we called *lag.arima*.

The time series is formed in such way to take elements of the sensor that is located in the center, and below all other values of sensors located closely (immediate neighborhood) to the central sensor for a defined period of time (example 24 hours), then form input data set, i.e. the time series W as follows:

Sensor ID	Wind[t1]	Wind[t2]	...	Wind[24]	...
Sensor 1	w1[2]	w1[2]	...	w1[24]	...
Sensor 2	w2[1]	w2[2]	...	w2[24]	...
...	...	...	...	...	...
Sensor k	wk[1]	wk[2]	...	wk[24]	...

The first period (lag) of time series can be represented by figure 2.

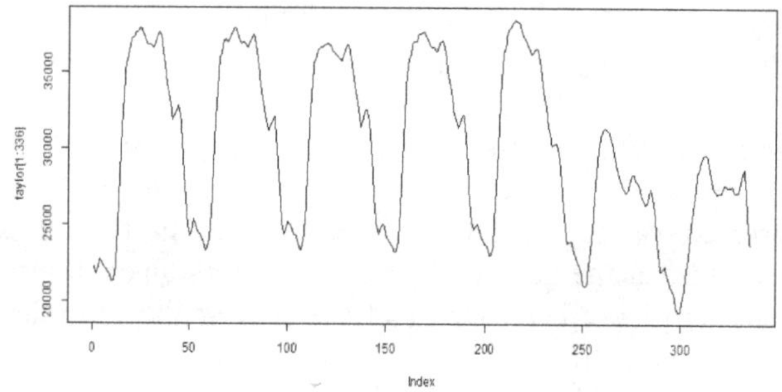

Figure 2. *The first lag of time series*

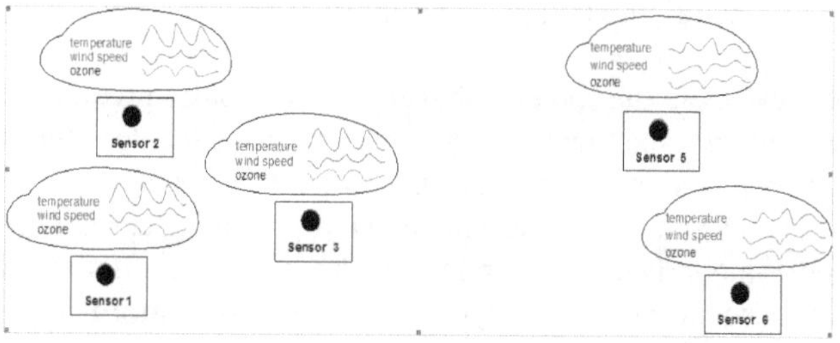

Figure 3. *The spatial distribution of the sensor with the time series of measured parameters*

Time series: w1[1], w1[2], ...,w1[24], w2[1], w2[2], ...,w1[24], and lag=24.

As has already been explained at the beginning of this paper, *auto.arima* function of R software package is based on the minimum value of AIC calculating optimal ARIMA model parameters *p,d,q,P,D,Q* for the entire time series.

Our algorithm sets all possible parameter values (as it does *auto.arima* for the whole time series choosing as optimal one with minimum of AIC). Only those parameters of model that are valid for each period are preserved in the matrix results which besides the parameters p, d, q, P, D, Q, contains also the average value of AIC for each period of the time series. So, while the output values of *auto.arima* function are ar1, ar2, ..., ma1, ma2, ... intercept, the minimum AIC, BIC, estimated sigma^2, log likelihood, the output values of our algorithm are the various models parameters *p, d, q, P, D, Q*, constant and average value of

AIC for all lags and valid for whole time series.

Based on these parameters, and in particular based on the average value of AIC for each model conclusion are the following:

- if there are some period of time series stationary or not;
- which ARIMA model can be applied to the whole, and not on any lag (or time interval) period of the time series;
- evaluate the accuracy of predictions of future values of the time series.

Though our approaches we gave adequate successful forecasts separately to demonstrate accurateness in predictions of future value of time series for a short time period. An application of the ARIMA model would result for predicted values are higher than the actual data and can be explained by the fact that the ARIMA modelling in forecasting is mainly based on the recent historical data.

4. Experimental Results

The *auto.arima()* function in R uses a variation of the Hyndman and Khandakar algorithm which combines unit root tests, minimization of the AICc and MLE to obtain an ARIMA model. The algorithm follows steps as follows:

- The number of differences d is determined using repeated KPSS tests.
- The values of p and q are then chosen by minimizing the AICc after differencing the data d times. Rather than considering every possible combination of p and q, the algorithm uses a stepwise search to traverse the model space.

Auto.arima tries all models for time series (finding one with the smallest AICc) and then selecting the best from the following four. ARIMA(2,d,2), ARIMA(0,d,0), ARIMA(1,d,0), ARIMA(0,d,1). If d=0 then the constant c is included; if d≥0 then the constant c is set to zero.

The best model considered becomes the new model until no lower AICc can be found.

Modeling procedure consists of fitting an ARIMA model to a set of time series data. A useful general approach can be provides following next procedure.

- Plot the data and identify any unusual observations;
- Transform the data using a Box-Cox transformation to stabilize the variance and if necessary;
- If the data are non-stationary take the first differences of the data until are stationary;
- Examine the ACF/PACF to understand if is an AR(p) or MA(q) model appropriate;
- Try chosen model using the AICc to search a better model.
- Check the residuals from chosen model by plotting the ACF of the residuals and making various tests of the residuals. If they do not look like white noise, try another modified model.
- Once the residuals look like white noise then is advisable, calculate forecasts.

In a phase of choosing the order of Box-Jenkins ARMA processes it is necessary automatic procedure using statistically based set of rules. In this sense, the proposed numerous criteria for model selection the most commonly used is AIC criterion.

The Akaike information criterion (AIC) is a measure of the relative goodness of fit of a statistical model. The AIC is grounded in the concept of information entropy, in effect offering a relative measure of the information lost when a given model is used to describe reality. AIC values provide a means for model selection in analysis of time series.

In the general case, the AIC is *AICek-2kn(L)* where k is the number of parameters in the statistical model, and L is the maximized value of the likelihood function for the estimated forecasting model of time series.

The values p and q in our algorithm are chosen to minimize mean AIC(p,q) for all lags. For evaluation of unknown parameters in time series models commonly are used method of maximum likelihood. On the renewal of the sample $(x_1,\ldots,x_n)$ is formed maximum likelihood function.

In this article, we primarily used the ARIMA modelling approach to automatic forecasting and we describe the implementation of this methods in the forecast package, along with other features of the package with some our transformation in the *auto.arima* model.

The forecast package in R language contains the function Arima() which is largely a wrapper to the *arima()* function in the package stats. The Arima() function makes easier including a drift term when d +D = 1. Setting *include.mean=TRUE* in the *arima()* function from the stats package will only work when d + D = 0. The facility provides fitting function in ARIMA model to a new data set and the ARIMA models gives a possibilities to one-step forecasting available via fitted() function.

Therefore, this paper introduces a methodology for combining the two methods in this special case of forecasting, taking into account the above mentioned facts. The methodology relies on the use of the two predictions of the most suitable ARIMA model measured by the Mean Absolute Error (MAE) and mean of residuals (MR) incorporated into model as actual data in order to construct a new one.

MAE and MR are frequently used measure of the differences between predicted values by a model or an estimator of the actually observed values. The individual differences called residuals present the calculations performed over the data sample used for estimation. There are called prediction errors when computed out of sample. The MAE and MR as the errors in predictions are good measure of accuracy.

Our hypothesis is that the short term forecast of many similar or equal lags produced by the ARIMA models are better than the one produced by *auto.arima* model at this stage of the process, which is validated in the present work. As this forecast is anticipated to be over-optimistic, the predicted values are taken into consideration in order to construct a new model, which will be more effective than the second one. After the calculation of the new model's parameters, the first two values are recalculated. These forecasts incorporate the effects of the ARIMA and *auto.arima* models which anticipated to be more precise than each approach separately. This procedure uses the simple averages approach, which would not be sufficient in cases of only ARIMA or *auto.arima* models and produce notice able to improve short-term predictions.

The use of weights as a method of combinations is also avoided, as it involves personal judgement regarding their value or evaluation of correlations between forecast errors that can change from period to period.

4.1. Experimental Data

For the experiments, we took three groups of data. The first is a time series of half-hourly electricity demand in England and Wales from Monday 5 June 2000 to Sunday 27 August 2000. Discussed in Taylor [13], and provided by James W. Taylor.

The second one is atmospheric concentrations of CO_2 expressed in parts per million (ppm) reported in the preliminary 1997 SIO manometric mole derived from in situ air samples collected at Mauna Loa Observatory, Hawaii.ftp://cdiac.esd.ornl.gov/pub/maunaloa-co2/maunaloa.co2. It is a time series of 468 observations registered monthly from 1959 to 1997. There are missing values for February, March and April of 1964 and they have been obtained by linear interpolation between the values for January and May of 1964. The monthly values have been adjusted to the 15th of each month. Missing values are denoted by -99.99. The 'annual' average is the arithmetic mean of the twelve monthly values. In years with one or two missing monthly values, annual values were calculated by substituting a fit value (4-harmonics with gain factor and spline) for that month and then averaging the twelve monthly values. [21]

The third group of data relates to the Australian monthly gas production in the period 1956-1995.

4.2. Experimental Methodology

The accuracy of each approach of research and applications of prediction model is measured by deviation error from the real value. Our approach is therefore also evaluated on base of two measures of accuracy, mean absolute error (MAE) and mean of residuals (MR) to verify the superiority of our approach versus only *arima* or *auto.arima* model for the entire time series. This article presents a new methodology that provides short-term forecasts of behavior of natural phenomena that divides the time series at specific time intervals (lags), and explores the best model parameters values valid for each lag of the time series.

The goal of forecasting of time series analysis starts from the available data from the past necessary to formulate and evaluate the time-series model and then used it to predict future values of the series. In doing that are used series of statistical tests and criteria that verifies the validity of the evaluated model.

In this paper, in the analysis and prediction are applied the class of autoregressive moving average model ARIMA (p,d,q). In this class of models is the assumption that the current value (element) of series depends: (1) on the value of previous members of the series, (2) the current value of the random process and (3) the previous value of the random process. In time series with observed effect of the trend, cyclical and seasonal components, the application of these models includes prior removal of that influence. To eliminate the influence of systematic components from the time series is used the operator of differentiation d.

For the identified model, the next step was a recursive procedure of estimation of model parameters or fitting the model. The basic approaches of fitting models are method of nonlinear least squares errors and the method of maximum likelihood.

Once adopted the appropriate model requires estimations of parameters and methodology requires verification of the residuals. If the residuals are random, the model is appropriate. Several tests like the Box-Pierce statistics can be proposed to determine the randomness of the residuals. Otherwise, it is necessary to return to the stage of identifying the model and try to find a better one.

As at any lag the time series can be not stationary, it is necessary differencing of all time series. Stationary time series lag has a constant mean value and variance. Non-stationary at any lag is corrected by corresponding differencing of all values of time series. In this case, the first order of difference and first order of seasonal differences are sufficient to achieve a stationarity of all time series.

The goal is to find such forecasting parameters p, d, q that give the lowest mean values of AIC applicable to each further lag of time series, and on the basis of whose results it is possible to forecast behavior in the coming intervals that show similar or the same behavior, not examining the whole time series always from the beginning, thus giving more accurate predictions.

Table 1-8. *MAE for training - testing data set of two models*

co2 (1:1) Training/ Testing set	**Training set x<-co2[1:228] Lag=12 x 19**	**Testing set x<-co2[229:456] Lag=12 x 19**
Forecast **arima(2,1,0)**	Mae	Mae
Abs error of prediction	0.4373796	0.5992418
Mean of residuals	0.02405235	0.03207524

Auto.arima: Training set x<-co2[1:228] - without division in lags - Training set		Abs error of prediction
Mean of residuals	-0.005477198	0.1242049
Arima (2,1,1): Testing set co2[229:456] - without division in lags		Abs error of prediction
Mean of residuals	0.1708991	0.5816793

co2 (1:1) Training/Testing set	**x<-co2[1:216] Lag=24 x 9**	**co2[217:432] Lag=24 x 9**
Forecast **arima(1,1,1)**	Mae	Mae
Abs error of prediction	0.3452741	0.2722938
Mean of residuals	0.04391378	0.04335611

Auto.arima: Training set x<-co2[1:216] - without division in lags - Training set		Abs error of prediction
Mean of residuals	-0.005578176	0.1938634
Arima (2,1,1): Testing set co2[217:432] - without division in lags		Abs error of prediction
Mean of residuals	0.169955	0.248319

Taylor (1:1) Training/Testin g set	**x<-taylor[1:1680] Lag=5 x 336**	**x<-taylor[1681:3360]Lag= 5 x 336**
Forecast **arima(2,0,2)**	Mae	Mae
Abs error of prediction	188.4699	188.4699
Mean of residuals	3.590077	3.129715

Auto.arima: Training set x<-taylor[1:1680] without division in lags- Training set		**Abs error of prediction**
Mean of residuals	0.7800593	459.88
Arima (4,0,4): Testing set taylor[1681:3360] - without division in lags		Abs error of prediction
Mean of residuals	0.5106114	103.1996

gas (1:1) Training/Testing set	**x<-gas[1:216] Lag=24 x 9**	**x<-gas[217:432] Lag=24 x 9**
Forecast **arima(1,1,0)**	Mae	Mae
Abs error of prediction	185.4429	741.2236
Mean of residuals	35.64954	117.0666

Auto.arima (2,1,2): Training set x<-gas[1:216] - without division in lags - Training set		Abs error of prediction
Mean of residuals	32.97419	1061.73
Arima (2,1,2): Testing set gas[217:432] - without division in lags		Abs error of prediction
Mean of residuals	283.3677	1788.014

In each of the three group of experiments we divided the data into training set / testing set 1:1. After that, our algorithm find the arima model parameters (p, d, q, P, D, Q, constant, mean (aic)) for the entire time series of data which are identical for each lag, and which at the same time give the minimum mean value of AIC. With these output parameters of our algorithm the absolute predictive and the mean value of the residual is calculated. It was the second part of our algorithm. The results are shown in the table.

In order to be able to compare the results of our experiment, the third part of our algorithm calls *auto.arima* function and through it we get the predictive and mean value of the residual for all sets of data (training set) without dividing in lags.

As can be seen from the table 1. (for a group of data co2 with a lag=12 in the training set) residual values vary from 0.02405235 with our algorithm to -0.005477198 with *auto.arima,* and the predictive value from 0.4373796 with our algorithm to 0.1242049 with *auto.arima.*

In the taylor data group (Training set with a lag=336) the mean residual value changes with our algorithm the absolute error of prediction changed from 188.4699 to 459.88 with the *auto.arima* function, which is significantly larger deviation with the *auto.arima* function.

The biggest differences are apparent in the fourth group of data (gas) where lag=24 and where in the training set the mean residual value with our algorithm changed from 35.64954 to 32.97419 (little smaller with *auto.arima* function), while the absolute prediction error increased from 741.2236 to 1061.73 with the *auto.arima* function.

Another important observation is that the forecasting accuracy of the ARIMA model diminishes gradually at this stage of the growth process, from period to period. Even though the forecasting power of the methodology seems limited, it should be taken into consideration that the forecasting improvement is for one day or one-month horizon. This single day or month's improved forecast could make the difference in the sense of competition, as this knowledge is a useful guideline for the upcoming day or month's strategy programming.

5. Conclusions

This paper presented a new methodology that delivers short-term forecasts of the natural phonemes measured for a period of time in the spatial distributed sensors that measure different parameters. After obtaining enough actual data to construct a time-series, a diffusion model and an ARIMA model are applied over the sample and the first forecasts of the ARIMA model are used to perform an improved short-term prediction using a conventional aggregate model. Since the two categories of modelling are of completely different concepts and implementations, the choice of combining their forecasts in the most suitable manner has been made. Therefore, it can be concluded that our algorithm that divides the time series into lags and calculates the best parameter values of the ARIMA model (*p, d, q, P, D, Q, constant, mean(aic)*), valid at the same time for each lag, gives more accurate and significantly better solutions in forecasting procedures than the *auto.arima* function that calculates the best model parameters by treating the whole time series (table 1-8) for several data set.

This methodology can be probably applied over all cases for obtaining future forecasts. Its main limitations consist of the prerequisite for having enough historical data points in order to create a time-series and that the diffusion process should be at the time point when the take off stage of the diffusion process is initiated. The study was limited to a forecasting horizon of one day or one month ahead. Future research in this topic includes the application of the methodology in other cases of space model forecasting, as well as the further investigation of its use in other stages of the process and for other forecast elements.

The accuracy of each approach's forecast is measured in terms of MAE in our research and applications, as noted earlier in this paper. This approach have been also compared based on the other two measures of accuracy, the Mean of residuals Error (MRE) and the Mean Absolute Error (MAE) in order to confirm the superiority of our approach opposite of the *auto.arima* model and this observation of the other measures resulted in the same conclusions as well.

Our intention in future research in this topic will include the application of the methodology in other cases of space model forecasting, as well as further investigation of its use in other stages of the process and for other forecast elements. As a future work, we plan to investigate a combination of the multivariate time series forecasting technique and polynomial regression with a combination of linear interpolation techniques like inverse distance weighting or Kriging that include not only temporally but also spatially distributed data.

References

[1] J.S. Armstrong, "Combining forecasts: the end of the beginning or the beginning of the end?", Int. Journal Forecast. 5, 1989, pp.585–588,

[2] G.E.P. Box, G.M. Jenkins, Time series analysis: Forecasting and control. Holden Day, San Francisco, 1970.

[3] P.J. Brockwell, R.A. Davis, Time Series: Theory and Methods, 2nd ed. Springer, 2009.

[4] D.W. Bunn, "Combining forecasts", European Journal of the Operational Reseach. 33, 1988, pp. 223-229.

[5] N.A. Gershenfeld, A.S. Weigand, "The Future of Time Series", Learning and Understanding. Time Series Prediction. Forecasting the Future and Understanding the Past. In: Eds. A.S.Wigand and Gersehenfeld, N.A. SFl Studies in the Sciences Complexity, vol. IV, Addison Wesley, 1993, pp. 1-70.

[6] S.L. Ho, "The use of ARIMA models for reliability forecasting and analyses", Computers and industrial engineering. 35 (1-2), 1998. pp. 213-216.

[7] R.J. Hyndman, Data from the M-Competitions. R package version 1.11, http://CRAN.R-project.org/package= forecasting , 2008.

[8] R.J. Hyndman, M. Akram, B.C., Archibald, "The Admissible Parameter Space for Exponential Smoothing Models", Annals of the Institute of Statistical Mathematics, 60 (2), 2008, pp. 407-426.

[9] R.J. Hyndman, Y. Khandakar, "Automatic time series forecasting", The forecast package for R. Journal of Statistical Software, 26(3), 2008.

[10] R.J. Hyndman, "Data Sets from\Forecasting: Methods and Applications By Makridakis", Wheelwright & Hyndman 1998, R package version 1.11.http://CRAN.R-project.org/ package=forecasting, 2008.

[11] R.J. Hyndman, "Forecasting Functions for Time Series", R package version 1.11, http://CRAN.R-project.org/package =forecasting , 2008.

[12] P. Newbold, "ARIMA model building and the time-series analysis approach to forecasting", Journal Fore cast. 2, 1983, pp. 23–35.

[13] J.W. Taylor, "Short-term electricity demand forecasting using double seasonal exponential smoothing", Journal of the Operational Research Society. 54, 2003, pp. 799-805.

[14] A. Timmermann, Chapter 4: forecast combinations. Handbook. Econ. Forecast. 1, 2006.

[15] G.U. Yule, "On the method of investigating periodicities in disturbed series, with special reference to Wolfer's sunspot numbers", Philos. Trans. Roy. Soc. London Ser. A 226, 1927, pp. 267–298.

[16] O. Ohashi, L. Torgo, Wind speed forecasting using spatio-temporal indicators. In L. D. Raedt, C. Bessiere, D.

Dubois, P. Doherty, P. Frasconi, F. Heintz, P. J. F. Lucas, editors, 20th European Conference on Artificial Intelligence. Including Prestigious Applications of Artificial Intelligence (PAIS-2012), SystemDemonstrations Track, volume 242 of Frontiers in Artificial Intelligence and Applications, pp. 975–980. IOS Press, 2012.

[17] W. Tobler., A computer movie simulating urban growth in the Detroit region". *Economic Geography*, 46(2), 2012, pp. 234-240.

[18] M. F. Goodchild, *Spatial autocorrelation*. Norwich, England: GeoBooks. 1986.

[19] Y. C Lee, L. Tong, Forecasting time series using a methodology based on autoregressive integrated moving average and genetic programming. Knowledge- Based Systems, (24), 2011, pp. 66–72.

[20] P.A.P. Moran, "Notes on Continuous Stochastic Phenomena," Biometrika, 37, 1950, pp. 17–23.

[21] C.D. Keeling, T. P. Whorf, "Scripps Institution of Oceanography (SIO)", University of California, La Jolla, California USA 92093-0220.

6

Dynamic obstacle avoidance in multi-robot motion planning using prediction principle in real environment

Suparna Roy[1], Dhrubojyoti Banerjee[1], Chiranjib Guha Majumder[1], Amit Konar[1], R. Janarthanan[2]

[1]ETCE Department, Jadavpur University, Kolkata-700032
[2]Jaya College of Engineering, Chennai, Tamil Nadu

Email address:
suparna.ry03@gmail.com (S. Roy), dhrubo_jyoti_banerjee@yahoo.co.in (D. Banerjee), cguhamajumder@yahoo.com (C. G. Majumder), konaramit@yahoo.co.in (A. Konar), srmjana_73@yahoo.com (R. Janarthanan)

Abstract: This paper provides a new approach to the multi-robot path planning problem predicting the position of a dynamic obstacle which undergoes linear motion in the given workspace changing its direction at regular intervals of time. The prediction is done in order to avoid collision of the robots with the dynamic obstacle. First the work is done in simulation environment then the entire work has been implemented on Khepera II mobile robot. The performance of the above mentioned approach has been found to be satisfactory compared to the classical non-predictive approaches of dynamic obstacle avoidance.

Keywords: Linear Prediction, Particle Swarm Optimization, Multi-Robot Motion Planning

1. Introduction

Mobility is an important aspect of modern robots. Several approaches to mobility management of a mobile robot have been studied over the last four decades. Some of these popular methods for path planning include Voronoi Diagram, A* Heuristic Algorithm, Neural Networks, Fuzzy algorithms. Since the beginning of this decade, researchers are taking keen interest to study the scope of optimization algorithms in mobility management of robots. The justification of using optimization techniques arises particularly when the motion planning of a number of robots are considered together in the same workspace. This paper attempts to overcome one fundamental problem in multi-agent robotics.

One interesting problem in multi-agent robotics is Multi-robot Motion planning[1][2], where the robots have to determine their trajectory of motion from predefined starting point to fixed goal point without hitting any obstacles in the environment. Most of the multi-robot motion planning algorithms developed so far only considered static obstacles. The multitude of the Multi-robot Motion Planning problem grows significantly, when one or more dynamic obstacles are present in the scenario. This paper addresses one such problem of Multi-robot Motion Planning taking into consideration the predictive motion of the dynamic obstacle.

The prediction logic employed here assumes a linear motion of the dynamic obstacle within a short span of time. When the speed of the dynamic obstacle is relatively slower than that of the robots, we may presume that the linear motion of the obstacle is maintained within two successive sampling instances of the obstacle by the robots. This consideration provides a new opportunity to formulate the multi-robot motion planning as an optimization problem.

The formulation of the problem is concerned with designing an objective function considering two important issues. The first issue refers to determining the minimum distance between each robot and its respective goal without hitting any static obstacle during the trajectory planning by the robot team. The second issue deals with maximization of the distance between a dynamic obstacle and its nearest robot. These two objectives are put together to construct a single objective optimization function, which here has been optimized by the well known Particle Swarm Optimization (PSO) algorithm.

After predicting the location of the dynamic obstacle, the robot takes the decision of the next position accordingly, the advantage being improved efficiency. The distributed approach to solve the path planning problem has been undertaken in this paper. Here we consider n-iterative algorithms for n robots, and the ith algorithm determines the next position for the ith robot, satisfying the necessary constraints.

The rest of the paper is organized as follows. Section II considers a formulation of the problem presuming static and dynamic obstacles separately, and then combines them together to construct a general objective function for the problem. The prediction approach has been incorporated here. The principle of PSO is outlined in section III. Section IV provides the algorithm to solve both the static and dynamic obstacle avoidance problem by using PSO. Computer simulation over simulated platforms is given in section V. Online experimental details with snapshots as well as the overview of the Khepera Robot used in the experiment are discussed in section VI. Performance analysis with respect to two standard metrics is undertaken in section VII. Inferences drawn from the paper are given in section VIII with a tinge of references at the end.

2. Formulation of the Problem

Here we evaluate the next position of the robots from their current position in a given robot's world map with a set of static obstacles and one dynamic obstacle. A set of principles listed below is first developed to formalize the path-planning problem by a uniform treatment.

A. Pre-assumptions

Current position of each robot is known with respect to a given reference in the Cartesian-coordinate system. The robots have a fixed set of actions for motions. A robot can select one action at a given time. Obstacles are detected by their colour which is known to the robots. The robots can come to know of the next position of the dynamic obstacle by using a prediction approach.

B. Principles

i. A robot always attempts to align itself towards the goal position by calculating a optimal path [3] calling PSO.

ii. In each step the robot tries to predict the location of the dynamic obstacle moving in a linear path. The introduction of the linearity in the path for a small amount of time of the dynamic obstacle is done so that the prediction-based approach is satisfied.

iii. On detecting a static obstacle near it, the robot deviates from its current position to a next obstacle free position following a minimal path obtained by PSO algorithm.

iv. After predicting a dynamic obstacle, the robot has to move from its current position to a next obstacle free position following a minimal path obtained by the PSO algorithm.

Let (x_i, y_i) be the current position of the ith robot at time t, $(x_{i'}, y_{i'})$ be the next position of the same robot at time (t+1). V_i be the current velocity of the ith robot.

Let (x_{ig}, y_{ig}) be the goal position of the ith robot. It is evident from Fig.1 that

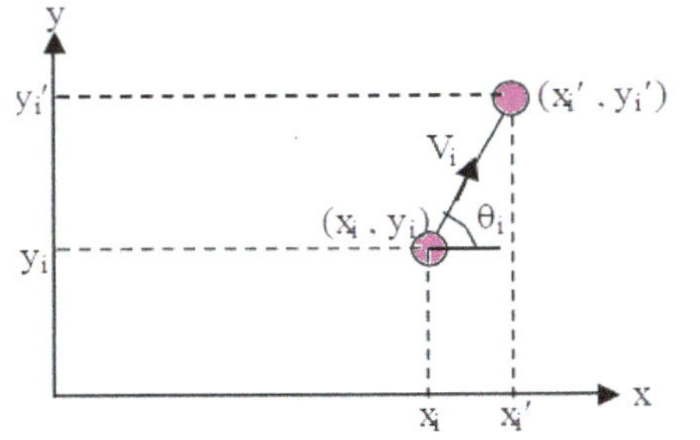

Figure 1. *Current and next position of the ith robot.*

$$x_i' = x_i + v_i \cos\theta_i \Delta t \quad (1)$$

$$y_i' = y_i + v_i \sin\theta_i \Delta t \quad (2)$$

Where Δt =1,

C. Distributed Planning

The constraint for the ith robot can be formulated as follows:

Let F be an objective function for the ith robot that determines the length of the trajectory. For n number of robots, then

$$F = \sum_{i=1}^{n} \{v_i + \sqrt{\{(x_i + v_i \cos\theta_i - x_{ig})^2 + (y_i + v_i \sin\theta_i - y_{ig})^2\}}\} \quad (3)$$

The robot is predicting its next position avoiding the dynamic obstacle. The objective function incorporating the prediction principle is calculated as

$$F' = F + \sum_{i=1}^{n} \sqrt{(x_p - x_{ig})^2 + (y_p - y_{ig})^2} \quad (4)$$

Where (x_p, y_p) is predicted next position of the dynamic obstacle. Now $x_p = x_{i'} + u\Delta t'$ and $y_p = y_{i'} + v\Delta t'$.

Where (x_i, y_i) is the current position of the dynamic obstacle.

$\Delta t'$ is the sampling time where $\Delta t = arr[i] / arr[j]$, in arr[i] the total time needed for the movement of the dynamic obstacle is kept and the number of steps undertaken by the dynamic obstacle is stored in arr[j]. Here u stands for the velocity of the dynamic obstacle. The prediction made here is linear because extrapolation of the path of the dynamic obstacle is made considering the direction of motion constant over a given interval of time. The distance between robots at any point of time should not be less than a predefined threshold to avoid collision; this logic is used as a primary constraint to this problem.

In fig.3 D_i represents the current position of the dynamic obstacle [8] whereas D_p denotes the predicted position of the same. R denotes the robot centroid whereas G denotes the goal centroid. The distance between the dynamic obstacle's predicted position and robot's current position is to be maximized to avoid any collision. Thus it becomes an objective function maximization problem. Let d_{ij} be the distance between i^{th} and j^{th} robots' current positions, $d_{i'j'}$ be the distance between i^{th} and j^{th} robots' next positions, then the constraint that the robot will not hit its kin is given by $d_{i'j'} - 2r \geq \in$, where r denotes the radius of the robots and $\in$ (>0) denotes a small threshold.

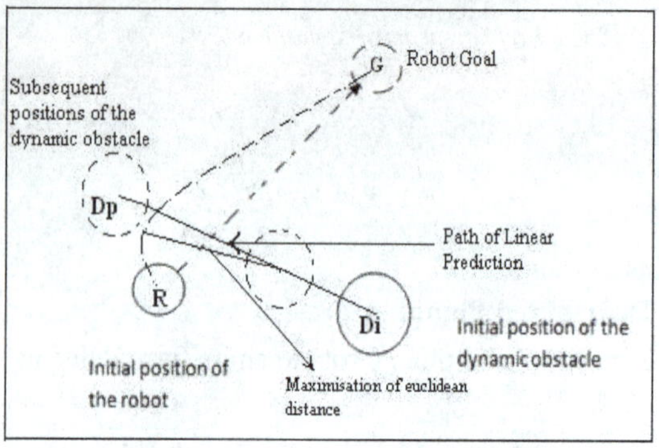

Figure 3. Linear prediction of the path of the dynamic obstacle and the maximization of the Euclidean distance between robot current position and predicted next position of dynamic obstacle.

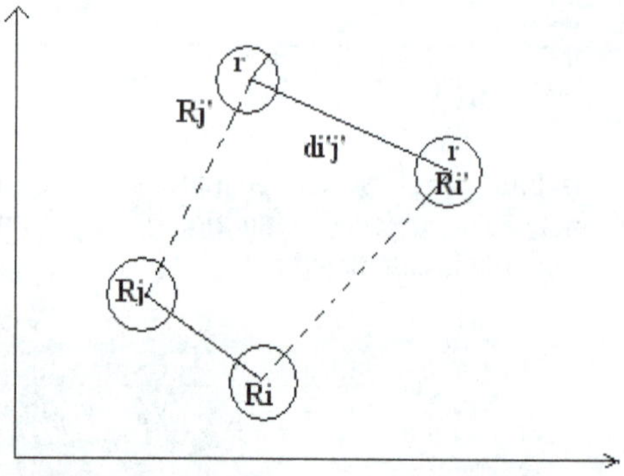

Figure 4. Diagram showing the constraint such that no two robots can collide with each other

The multi-robot path-planning as an optimization problem includes an objective function, concerning minimization of the Euclidean distance between the current positions of the robots with their respective goal positions, constrained by obstacles and other robots on the path. Thus, the constrained optimization problem in the present context for the ith robot is given by,

$$F_i = \sum_{i=1}^{n} \{v_i + \sqrt{((x_i + v_i \cos\theta_i - x_{ig})^2 + (y_i + v_i \sin\theta_i - y_{ig})^2)}\}$$
$$+ f_{dp} \sum_{i'j'=1}^{n(n-1)/2} (\min_{\forall_i}(0,(d_{i'j'} - (2r+\varepsilon)))^2 + f_{st} / d_{i-obs} \qquad (5)$$
$$+ f_{dy} / (n \times (R+r)) + 1 / \sum_{i-1}^{n} \sqrt{(x_p - x_{ig})^2 + (y_p - y_{ig})^2}$$

where f_{dp} (>0) and f_{st} (>0) denote scale factors to the second and third terms in the right hand side of expression (5) d_{i-obs} represents the distance of the obstacle from the i-th robot, R is the radius of the dynamic obstacle.

1. Static obstacle

Consider the robot Ri is initially located at (x_i, y_i) .It needs to select point $(x_{i'}, y_{i'})$, i.e. next position of the robot, such that the line joining $\{(x_i, y_i), (x_{i'}, y_{i'})\}$ and $\{(x_{i'}, y_{i'})\ (x_g, y_g)\}$ do not touch the obstacle, as shown in Figure. 2. This is realized with PSO algorithm, such that it will always select a minimal path [9] to reach the respective destinations. To take case of static obstacles in the environment, we add one penalty function to the constrained objective function (5).Thus the present constraint optimization problem is transformed to –

$$F = \sum_{i=1}^{n} v_i + \sqrt{((x_i + v_i \cos\theta_i - x_{ig})^2 + (y_i + v_i \sin\theta_i - y_{ig})^2)}$$
$$+ f_{dp} \sum_{i'j'=1}^{n(n-1)/2} (\min(0,(d_{i'j'} - (2r + \in))))^2 + f_{st} \qquad (6)$$

Where f_{st} = positive constant when a static obstacle is present on the planned local trajectory = 0, otherwise.

2. Dynamic obstacle

The main objective of the prediction principle is to anticipate the next position of the dynamic obstacle [4]. Thus the robot can take action accordingly and call PSO to move back to its optimal path.

Consider the robot Ri is initially located at (x_i, y_i) .It needs to select point $(x_{i'}, y_{i'})$, i.e. next position of the robot, such that the line joining $\{(x_i, y_i), (x_{i'}, y_{i'})\}$ and $\{(x_{i'}, y_{i'}),\ (x_g, y_g)\}$ do not touch the obstacle, as shown in fig. 2.

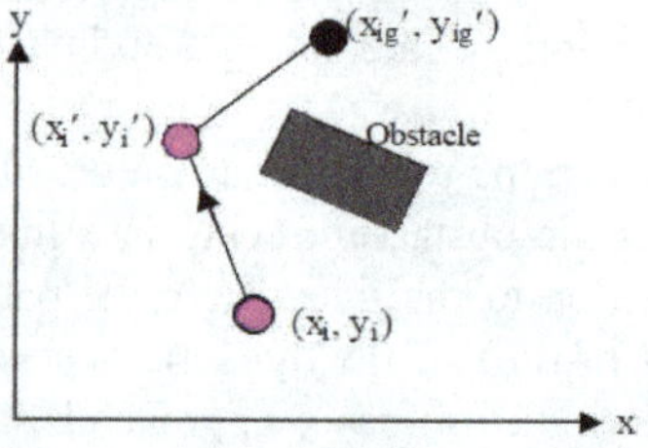

Figure 2. Selection of (x_i', y_i') *from* (x_i, y_i) *to avoid collision with an obstacle.*

To take case of dynamic obstacles in the environment, we add one maximization function. Thus the present constraint optimization problem is transformed to

$$F = \sum_{i=1}^{n} v_i + \sqrt{((x_i + v_i \cos\theta_i - x_{ig})^2 + (y_i + v_i \sin\theta_i - y_{ig})^2)}$$
$$+ 1 / \sum_{i-1}^{n} \sqrt{(x_p - x_{ig})^2 + (y_p - y_{ig})^2} + f_{dy} \qquad (7)$$

f_{dy} = a positive constant when a dynamic obstacle is predicted =0, otherwise.

3. The Particle Swarm Optimization

The PSO scheme has the following algorithmic parameters:

1) V_{max} or maximum velocity which restricts $\vec{V}_i(t)$ within the interval $[-V_{max}, V_{max}]$.

2) An inertial weight factor ω.

3) Two uniformly distributed random numbers φ1 and φ2 which respectively determine the influence of $\vec{p}(t)$ and $\vec{g}(t)$ on the velocity update formula.

4) Two constant multiplier terms C_1 and C_2 known as self confidence and swarm confidence respectively [5].

Initially the settings for $\vec{p}(t)$ and $\vec{g}(t)$ are $\vec{p}(0) = \vec{g}(0) = \vec{x}(0)$ for all particles. Once the particles are initialized, the iterative optimization process begins where the positions and velocities of all the particles are altered by the following recursive equations. The equations are presented for the d-th dimension of the position and velocity of the i-th particle.

$$V_{id}(t+1) = \omega V_{id}(t) + C_1\varphi_1(P_{id}(t) - X_{id}(t)) + C_2\varphi_2(g_d(t) - X_{id}(t)) \quad (8)$$
$$X_{id}(t+1) = X_{id}(t) + V_{id}(t+1)$$

The first term in the velocity updating formula represents the inertial velocity of the particle. The second term involving $\vec{P}(t)$ represents the personal experience of each particle and is referred to as 'cognitive part'. The last term of the same relation is interpreted as the 'social term' which represents how an individual particle is influenced by the other members of its society. Typically, this process is iterated until some acceptable solution has been found by the algorithm. Once the iterations are terminated, most of the particles are expected to converge to a small radius surrounding the global optima of the search space.

4. Solving the Constraint Optimization

Problem Using Pso

Pseudo Code:

Input: Initial position (x_i, y_i), goal position (x_{ig}, y_{ig}) and velocity v_i for n robots where 1<=i<=n and a threshold value €.

Output: Trajectory of motion P_i for each robot R_i from (x_i, y_i) to (x_{ig}, y_{ig})

Begin

Set for all robot i

$x_{icurr} \leftarrow x_i$; $y_{icurr} \leftarrow y_i$ //current position in both x & y coordinate of ith robot//

For robot i=1 to N

Repeat

Check obstacle ( ) //at each (x_{icurr}, y_{icurr}) the robot rotates 360° with radius n*(R+r) [n >1] to search for obstacle//

IF static obstacle

Move away and call PSO // to find the next obstacle free optimal position //

Predict dynamic obstacle's next position

Compute sampling time // delta t that is time taken to undergo one step//

Update xp // next position of obstacle(dynamic) //

$$x_p = x_i' + u\Delta t \quad y_p = y_i' + u\Delta t .$$

Maximize the distance between robot(current->pos) and dynamic obstacle(predicted next->pos). // to avoid collision//

Call PSO

Move to -> (x_{inext}, y_{inext}) ;

// moves to next obstacle free position. //

$x_{icurr} \leftarrow x_{inext}$, $y_{icurr} \leftarrow y_{inext}$; //update the position//

Until $\|curr\_i - G_i\| \le \varepsilon$ // $curr\_i$ = (x_{curr_i}, y_{curr_i}), Gi = (x_{ig}, y_{ig})//

End for;

End.

Procedure PSO

$(x_{icurr}, y_{icurr}, pos - vector)$

Begin

initialize 10 particles with random position and velocity;

For k < Maxiter do

Begin

1. update V_i and X_i i by (8);

2. determine local best position and global best position of

the particles;

End for;

Update:

$$x_{curr-i} \leftarrow x_{curr-i} + v_i \cos\theta_r$$

$$y_{curr-i} \leftarrow y_{curr-i} + v_i \sin\theta_r$$

Return;

End.

5. Experiments Using Computer Simulation

In this section, we provide the results of computer simulations of the proposed scheme of multi-robot motion planning avoiding both static and dynamic obstacles.

The multi-robot path-planning is realized in Turbo C environment on a Pentium processor. The number of robots (n) was varied from two to fourteen and the performance of the system was evaluated. The static obstacles are represented by Dark gray colour while the dynamic obstacles are represented by Orange colour in the world-map. The following figures show the screenshot of the program. The configuration of the world map with robots, dynamic obstacles and five static obstacles at different instant of time during the execution of the code is depicted in the following figures. The number of robots and the velocity of robots are changed to get different conditions and make sure in no cases the robots collide with any obstacle. The trajectories of the robot path avoiding both static and dynamic obstacles are clearly visible. The intermediate positions as well as the final position of the world map are shown where the robots are seen to reach their predefined goal safely.

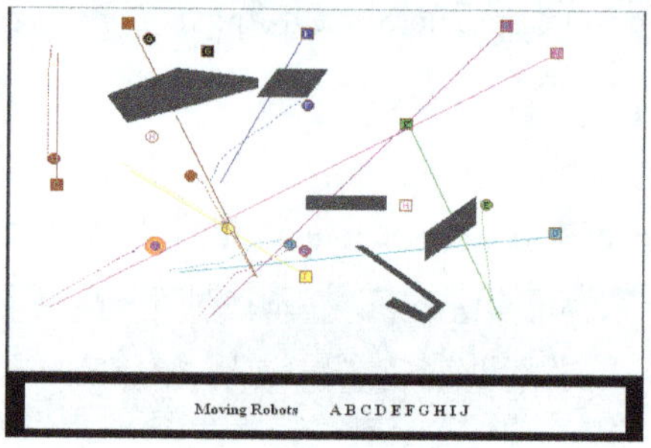

Figure 5. *Screenshot showing the Collision of the robot J with the orange coloured Dynamic obstacle in the world-map in classical non-prediction approach.*

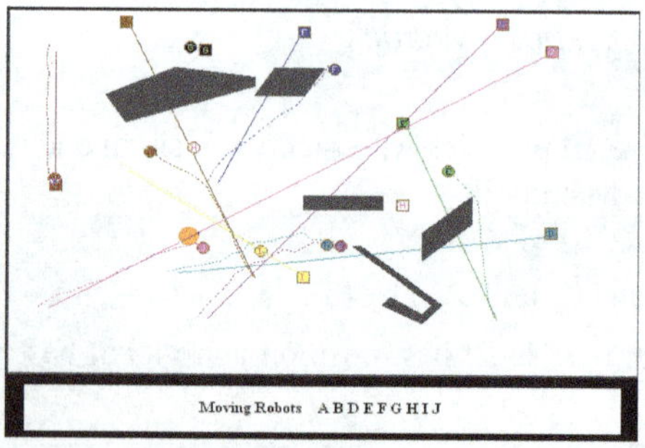

Figure 6. *Screenshot showing the path of Dynamic obstacle avoidance by the robot J in the world-map when 10 robots together are considered and prediction principle is used.*

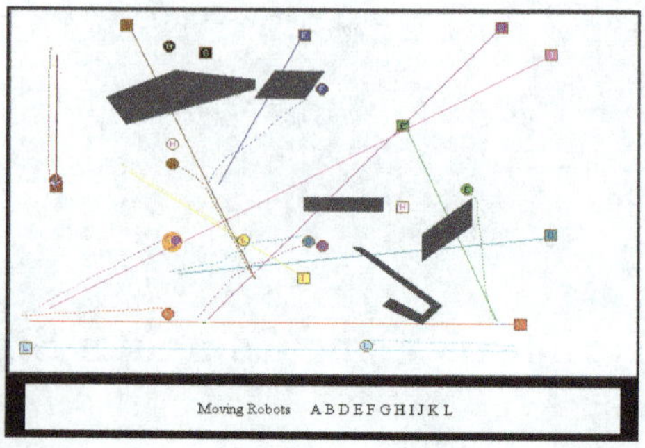

Figure 7. *Screenshot showing the Collision of the robot J with the orange coloured Dynamic obstacle in the world-map when 12 intelligent robots are considered at a time, in classical non-prediction based approach.*

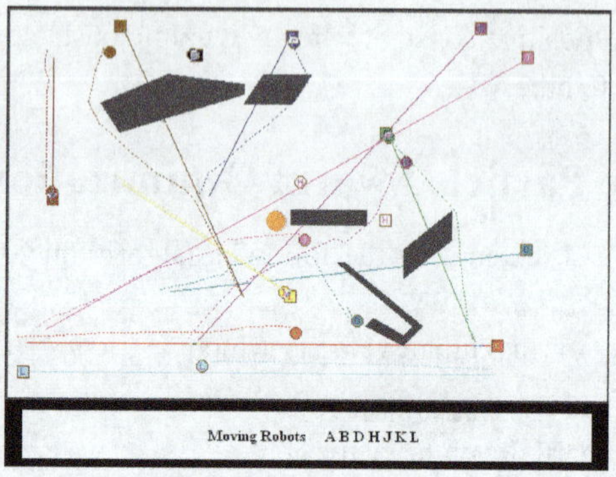

Figure 8. *Screenshot showing the path of Dynamic obstacle avoidance by the robot J and trajectories of eleven other robots in the world-map when 12 robots together are considered and prediction principle is undertaken.*

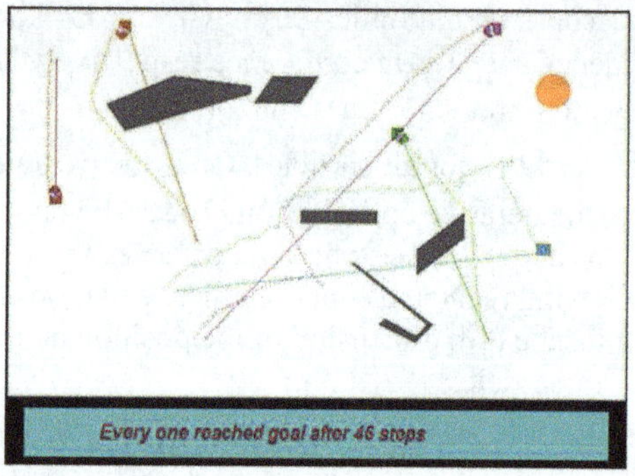

Figure 9. *Screenshot showing the complete path of the robots in the world map after the robots reached their goals and the dynamic obstacle has passed.*

6. Experiment Using Khepera-Ii Robot

Khepera II as shown in figure 10 is a miniature robot of diameter 7 cm and equipped with 8 in-built infrared proximity sensors, and 2 relatively accurate encoders for the two motors. The range sensors are positioned at fixed angles and have limited range detection capabilities[6][7]. The sensors are numbered from 0 to 7, with the leftmost sensor designated by 0, and the rightmost by 7. The sensor numbered 7 and 8 are on the back side of Khepera robot and not used in this experiment. The Khepera model we used is a table-top robot, connected to a workstation through a wired serial link.

Figure 10. *The Khepera II Robot.*

Now during online execution of the robot path traversal, both the robots are placed in their initial positions. They have to reach their goal avoiding all static and dynamic obstacles present in the world-map. The dynamic obstacles are pink in colour and move in the world map randomly. Sensors and camera are the main source of information. These data acquisition tools are already embedded in the robot.

The optimal path to the goal is decided by the PSO algorithm and static obstacles are detected by the sensory value. Here we have taken account of the first six sensors of the Khepera-II Mobile Robot. So the sensory data set is given by R={$R0$,$R1$,$R2$,$R3$,$R4$,$R5$}.The robots are programmed in such a way that whenever a static obstacle is in the robot's vicinity the reflected sensory value is more than the predefined threshold which denotes presence of obstacle. So the robots deviate from their path to avoid the obstacle and then again come back to its optimal path thus avoiding collision.

The dynamic obstacle as well as the kin is detected by its colour. The vision system generates information corresponding to the result of the processed image captured by the camera of the functional robot. After the online capture of images by the Khepera robot the set of images are sent to the PC for colour recognition. This is done so that the co-operating agent can recognize its kin as well as dynamic obstacle. The colour is expressed as an RGB triplet (r,g,b) each component of which can vary from 0 to a defined maximum value.

In this problem, the dynamic obstacle is pink in color and is identified through color image processing using Vector Distance approach. The images of.bmp format being of pink colour, the individual components of the RGB triplet will result in a value which is more than the predefined threshold for pink colour. Now, while searching for pink the program in the PC connected to the co-operating robot sets a tolerance or a threshold so that the images or colour close to pink can be recognized. The same procedure is applied for kin recognition which is done by identifying black colour of khepera robot. The only difference is that the images of.bmp format being of black colour, the individual components of the RGB triplet will result in zero value ideally.

That is the system works for the following situations in case of kin recognition:

$$\forall I \in P \cup P_{close\_to}$$

where I represents the set of images captured and B denotes pink set and P_{close_to} near to pink.

In practical world, the nearness factor is considered to remove the constraint of ideality of any particular colour.

The Euclidian distance between those two vectors is computed as follows:

$$D = \sqrt{(R-t)^2 + (G-t)^2 + (B-t)^2} \quad (9)$$

where t is the tolerance limit considered for near to pink factor.

The pseudo code for dynamic obstacle recognition is given below.

Pseudo-code :

Begin

{

For each pixel

{

Compute the distance between that pixel and the reference color (D)(Pink and black here) //calculation of distance vector//

If D < Threshold //Thresholding operation//

Then Current pixel is accepted as pink or black //selection of pink and black colour

else

Current pixel is not accepted

}

End

The real time experiment of multi-robot motion planning was supposed to be done with varying number of robots from two to twelve as done in case of simulation, but here for the sake of simplicity and ease of experimental set-up only two robots are considered. The white colored objects are the static obstacles, the pink colored ball being the dynamic obstacle is moved randomly in the world map. Figure 11 to 14 shows the actual image of the complete world map during the time of path planning in real environment.

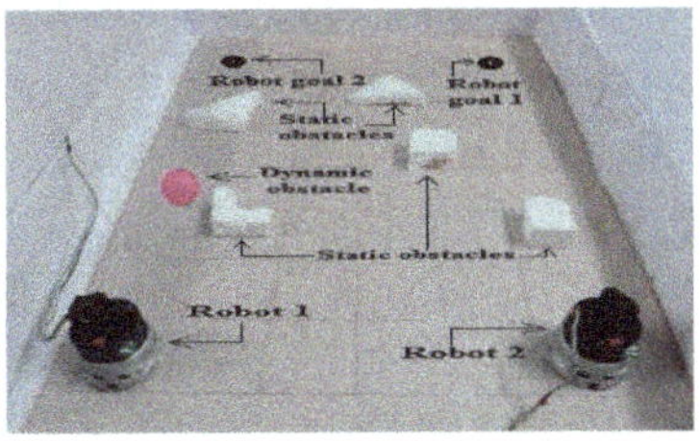

Figure 11. *The complete world-map showing the initial position of the Robots (R), the static obstacles and the dynamic obstacle, the goal (G) position of the respective robots and the boundary.*

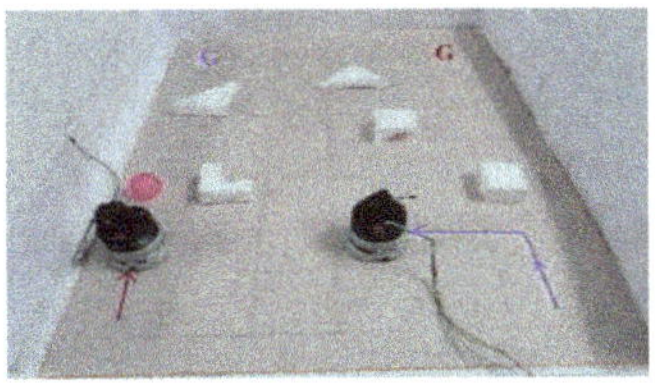

Figure 12. *The complete world-map showing the Robots (R) during online motion, here the robot-1 detects the dynamic obstacle in front of it as a result it will turn 90° to its right and continue its motion.*

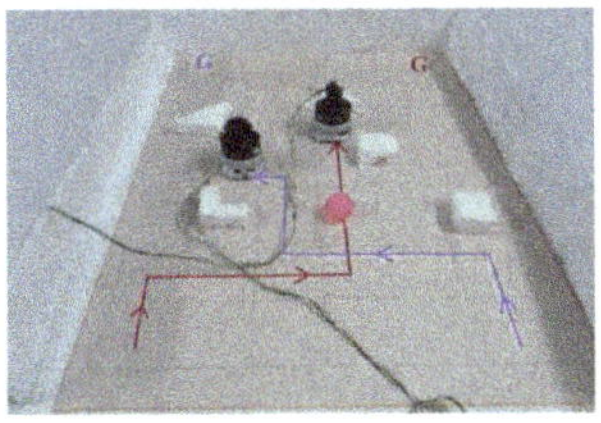

Figure 13. *The complete world-map showing the Robots (R) during online motion, its trajectory of motion of the robot-1 isshown is red line and that of of robot-2 is shown in blue line,the obstacles, the goal (G) and the boundary*

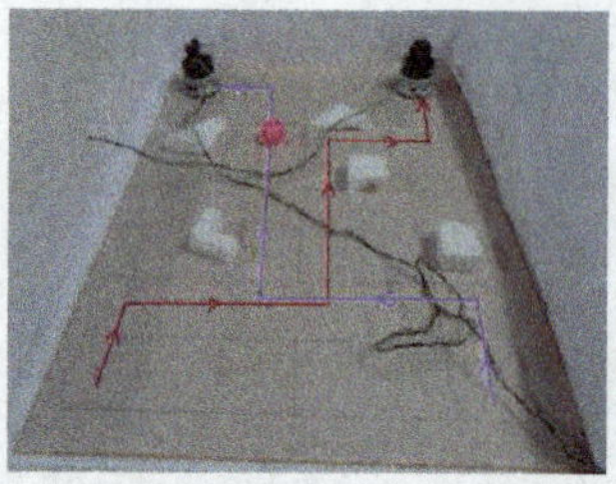

Figure 14. The complete world-map showing the Robot(R) has reached its goal (G) successfully. The trajectory of the path is also shown.

7. Performance Analysis

The performance of the prediction based system can be evaluated in many ways. Performance evaluation is a wide topic, and covers many techniques to measure the quality of the system. Different performance evaluation metrics covers different areas, which include how algorithms cope in different physical conditions in the scene, i.e. changing number of robots and obstacles, changing velocity of the robots, change in position of the obstacles etc. Two metrics average path deviation and average uncovered target distance have been considered for evaluating the performance of the system.

Average Uncovered Target Distance (AUTD)

Given a goal position G_i and the current position C_i of a robot on a 2-dimensional workspace, where G_i and C_i are 2-dimensional vectors, the uncovered distance for robot i is $\| G_i - C_i \|$, where $\| . \|$ denotes Euclidean norm. Initially when the robots start from its starting position the uncovered Target Distance is maximum and it decreases as the robots trends towards their respective goals. As soon as the robots reach their goals UTD becomes zero. For n robots, uncovered target distance (UTD) is the sum of $\|G_i - C_i\|$ i.e. UTD = $\sum_{i=1}^{n} \|G_i - C_i\|$.

Now, for k runs of the program, the average of UTDs is evaluated and it is called the average uncovered target distance (AUTD). For all experiments conducted in this study, k = 10 is considered.

Average Path Deviation (APD)

When no obstacles are present in the world map then the robots will follow the minimum path decided by the PSO algorithm. Whenever static and dynamic obstacles are introduced in the world map the robots deviate from their optimal path in order to avoid collision. Average path deviation is the difference between these two paths. For more number of robots the average path deviation will be the sum of individual APD for separate robots. The average path deviation should be as low as possible for better performance.

Let the average time taken by the robots to reach their goal in presence of no obstacle be t. Now, if the optimal path traversed by the robot be $D_{traversed}$ and the actual path between the robot and the goal in the presence of static and dynamic obstacles be $D_{static} + D_{dynamic}$, the Average Path Deviation is given by.

$$APD = | D_{traversed} - (D_{static} + D_{dynamic}) |$$

Figure.15 and figure.16 reflect that the uncovered target distance increases with the increasing number of robots. This happens due to the congestion in the path of the robots at the initiation of the iteration with more number of robots. The initial position of the dynamic obstacle is different in figure.15 from figure.16. So the predicted position of the dynamic obstacle is also different in each case. As a result the trajectories of the robots are also different in each case.

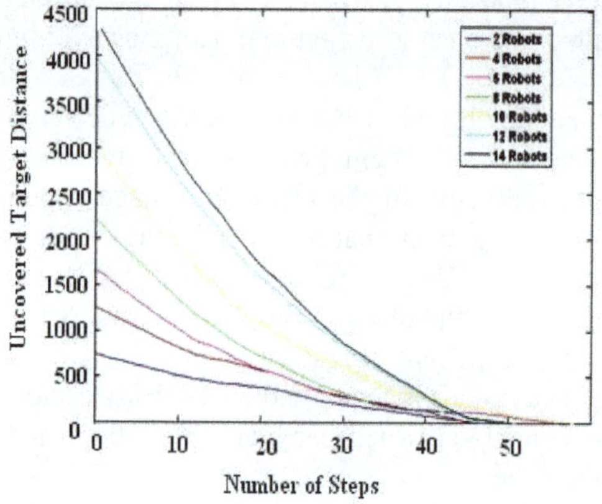

Figure 15. Plot of Uncovered Target Distance(UTD) vs Steps with number of robots as parameter and initial co-ordinate of the dynamic obstacle (200,300).

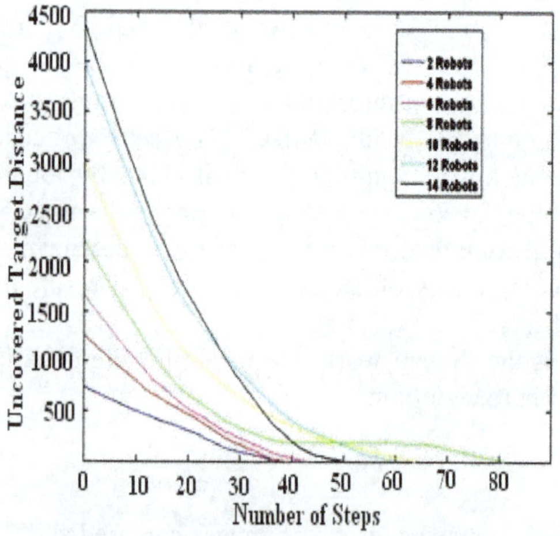

Figure 16. Plot of Uncovered Target Distance (UTD) vs Steps with number of robots as parameter and initial co-ordinate of the dynamic obstacle (50,300).

For a fixed uncovered target distance the number of steps undertaken by the robots decreases with increasing velocity of the robots, which is quite evident from the figure 17 and figure 18. This happens because robot with more speed reaches the goal fast.

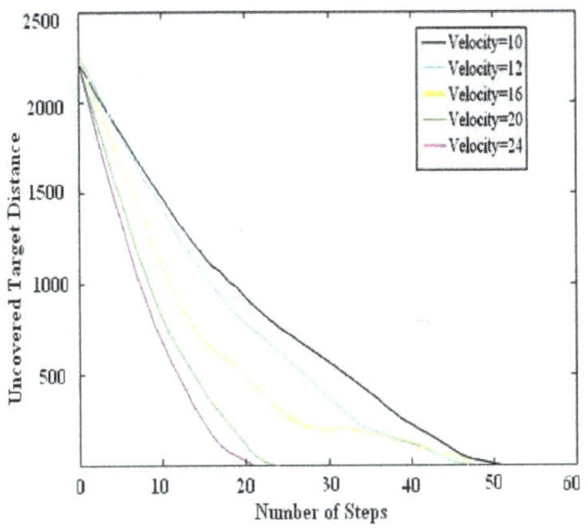

Figure 17. *Plot of Uncovered Target Distance(UTD) vs Steps with robot velocity as parameter and the initial co-ordinate of the dynamic obstacle (200,300).*

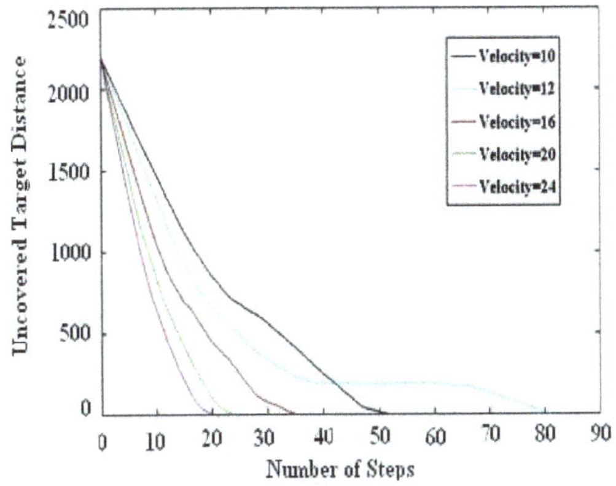

Figure 18. *Plot of Uncovered Target Distance vs Steps with robot velocity as parameter and the initial co-ordinate of the dynamic obstacle (50,300).*

The bar chart in figure.19 clearly shows that the number of steps covered by the robot to reach their goal in prediction based approach is much less than the approach without prediction.

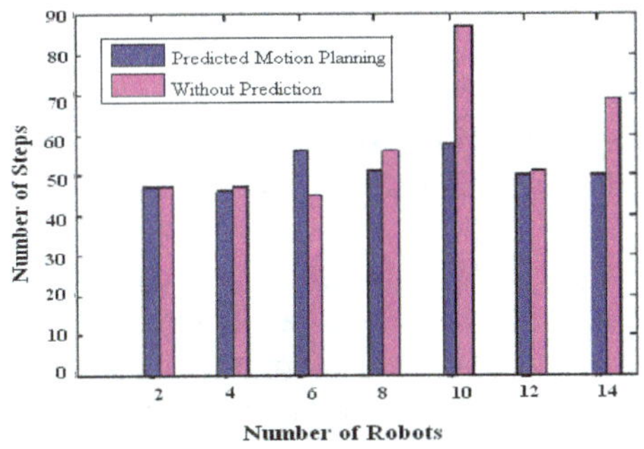

Figure 19. *Bar chart showing performance evaluation of robots with and without prediction.*

This proves the excellence of our work thereby increasing the efficiency of the system of dynamic obstacle avoidance thereby reducing the risk factor of collision.

8. Conclusion

We have introduced the concept of prediction to determine the next position of the dynamic obstacle. This prediction principle is also employed in online path planning using Khepera robot as well as in simulation apporach.In both the cases, the robots reach their goal without hitting any obstacle. The experimental results are in conformation with the fact that the prediction logic helps in minimizing the steps of motion of the robots, which encounter the dynamic obstacle within its trajectory compared to the non-prediction based approach. The prediction logic helps the robot to determine the location of the dynamic obstacle and plan its path accordingly minimizing the time of traversal. Thus both the number of steps as well as the execution time is minimized in the above problem. This is where our approach on multi-agent path planning amidst both static and dynamic obstacles supersedes the other works on multi-agent systems already existing in this domain.

References

[1] J. Kennedy, R. Eberhart, "Particle swarm optimization", In Proceedings of IEEE International conference on Neural Networks. (1995) 1942-1948.

[2] Jayasree Chakroborty, Amit Konar, Aruna Chakroborty, "Multi-robot co-operation by Swarm and Evolutionary Algorithms".

[3] M. Ryan, "Graph Decomposition for Efficient Multi-robot Path planning,"in Proceedings of the 20th International Joint Conference on Artificial Intelligence, pp. 2003-2008, Jan. 2007.

[4] A.Fujimori and S. Tani, 'A navigation of mobile robots with collision avoidance for moving obstacles', in Proc. IEEE International Conference on Industrial Technology, Bangkok, Thailand, Dec. 2002, pp. 16.

[5] F. van den Bergh and A.P. Engelbrecht, "Cooperative learning in neural networks using particle swarm optimizers," South African Computer Journal, 26:84-90, 2000.

[6] T. Tsubouchi and M. Rude, "Motion planning for mobile robots in a time-varying environment", J. of Robotics and Mechatronics, Vol. 8, No. 1, pp. 15-24, 1996.

[7] S. Ishikawa, "A method of indoor mobile robot navigation by using fuzzy control", in Proc. IEEE/RSJ Int. Workshop on Intelligent Robots and Systems, pp. 1013-1018, 1991.

[8] F. Kunwar, F. Wong, R. Ben Mrad, B. Benhabib, "Guidance-based online robot motion planning for the interception of mobile targets in dynamic environments", Journal of Intelligent and Robotic Systems, Vol. 47, Issue 4, pp. 341-360, 2006.

[9] J. Minura, H. Uozumi, and Y. Shirai, "Mobile robot motion planning considering the motion uncertainty of moving obstacles", in Proc. IEEE Int. Conf. on Systems, Man, and Cybernetics, Tokyo, pp. 692-698, 1999.

7

Multiplexed spatiotemporal communication model in artificial neural networks

Shinichi Tamura[1, 2, *], Yoshi Nishitani[2], Takuya Kamimura[3], Yasushi Yagi[3], Chie Hosokawa[4], Tomomitsu Miyoshi[2], Hajime Sawai[2], Yuko Mizuno-Matsumoto[5], Yen-Wei Chen[6]

[1]NBL Technovator Co., Ltd, 631 Shindachimakino, Sennan City, 590-0522 Japan
[2]Graduate School of Medicine, Osaka University, Suita, 565-0871 Japan
[3]ISIR, Osaka University, 8-1 Mihogaoka, Ibaraki City, Osaka, 567-0047 Japan
[4]AIST Kansai, 1-8-31 Midorigaoka, Ikeda 563-8577 Japan
[5]Graduate School of Applied Informatics, University of Hyogo, Kobe, 650-0047 Japan
[6]Graduate School of Sci. and Eng., Ritsumeikan University, Kusatsu, 525-8577 Japan

Email address:
tamuras@ nblmt.jp(S. Tamura)

Abstract: It is well known that there is intercommunication among the different areas of the brain. However, till date, the rules of communication have not been successfully analyzed. The spike trains from neuronal cells have been simply treated as density-modulated waves with an activation level of the corresponding neuronal cells, or, at most, they have been analyzed using traditional metrics between sequences. The spike trains from neuronal cells have a random-like pattern that provides few clues regarding a coding rule. Here in a randomly generated artificial 3 × 3 multiplexed spatiotemporal communication neural network composed of threshold elements, we showed that pseudorandom sequences were generated during the simulation, similar to the random sequences generated by the cultured neural network of the rat brain. The transiently generated sequence patterns in the simulation were regarded as reflecting the circuit structure. These randomly shaped circuits generated pseudorandom sequences that functioned as codes for multiplexing communication. Although the circuit weights are randomly generated at present, it will be possible to extend this approach to determine the network weights by learning. This paper provides simulation results that support findings on cultured neural network.

Keywords: M-Sequence, Neural Network, Pseudo Random Sequence, Spatiotemporal Communication, Spike Train

1. Introduction

We have developed a time-shift diagram method [1] for visualizing the propagation of brain waves. Figure 1 shows an example of a time-shift diagram in which the transmissions of magnetoencephalography (MEG) waves for a number counting task are shown with propagation times of less than 5 [ms] (in red; mainly within each hemisphere) and more than 10 [ms] (in blue; mainly across the callosum) [2]. Propagation times of 5–10 [ms] are indicated in green. When compared with the MRI dipole diagram method, which shows only a small number of major flows, our method follows an even smaller flow of signals. Questions arise as to how neuronal cells find their target cells and how the target cells obtain the necessary signals from the source neurons even if they are located at remote positions. Such multi-access communication requires codes. This issue served as the motivation for our research. However, till date, the rules of communication in the brain have not been successfully analyzed. The spike trains from neuronal cells have been treated simply as density-modulated waves with an activation level of the corresponding neuronal cells, and, at best, they have been analyzed using traditional metrics between sequences and from the viewpoint of spatial independent information.

1.1. Research on Spike Coding

To analyze spike trains, metrics between spike trains have been proposed via an alignment of distances and convolution metrics, including traditional rate coding [3]. However, the coding scheme of neurons has not been solved.

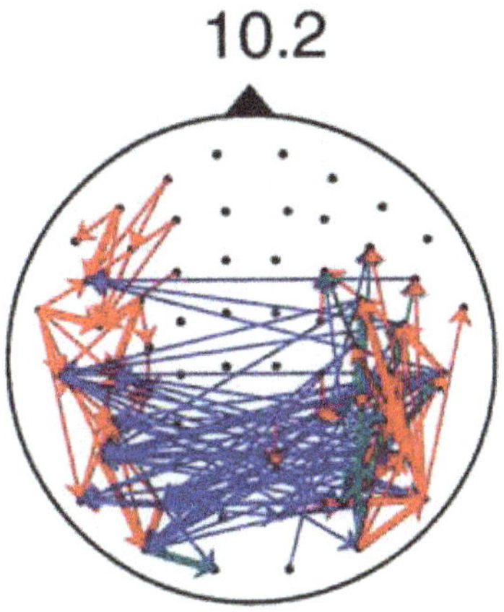

Figure. 1. *Time-shift diagram of 10.2 Hz MEG for a number-counting task; lag time < 5 ms (red) was primarily within each hemisphere and lag time > 10 ms (blue) was across the callosum. Green: between 5–10 ms.*

1.2. Spatiotemporal Coding

The extension of signals in a multidimensional manner permits dealing with many spatiotemporal patterns in artificial and natural neural networks [4-7]. In the visual system in particular, directional receptive fields, as seen in mammalian simple cells, emerge by a minimum information criterion [8] and an independent component analysis [9] for natural and facial images, i.e., spatially independent basis functions are derived by self-organization. Figure 2 shows how the receptive fields of the visual system are obtained by self-organization of the neural circuit with mutual inhibition to output only independent components [10]. Thus, it is reasonable to seek the temporally independent components of information representation in the brain as a pair of spatially independent components or seek the spatiotemporal information representation and communication coding scheme.

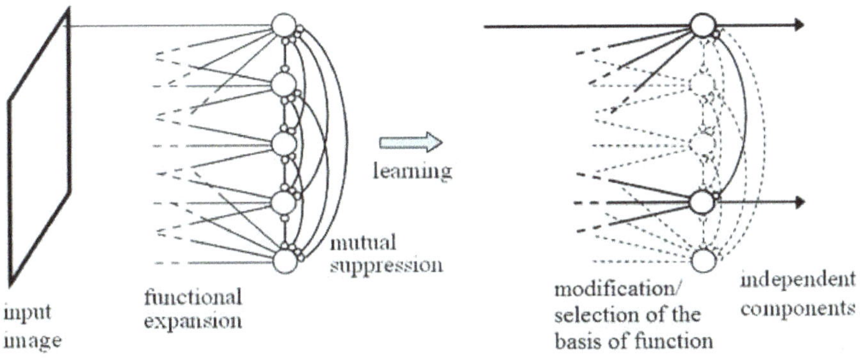

Figure. 2. *Independent component extraction by mutual suppression in the visual system (dotted lines represent those that disappear after learning)*

1.3. Pseudorandom Codes from Cultured Neural Networks

We have been analyzing the spike train structure of cultured neural networks to clarify intelligent processing in the brain [11-13]. We have decoded the spike trains of several samples of neural networks cultured on 8 × 8 multi-electrodes. From these, we observed significantly more M-sequences than observed from interval shuffled trains, which are representative pseudorandom sequences.

The question as to why neuronal spike sequences have a white-noise like pattern, such as M-sequences, then arises. Although many researchers have been tackling this problem, it has not yet been solved. The objective of this study is to support these *in vivo* data via simulation.

The remainder of this paper is organized as follows: After presenting some background information in Section 2 (as well as in the Appendix), we propose a 3 × 3 spatiotemporal communication neural network model in Section 3 and present a discussion and conclusions in Section 4.

2. M-Sequence

An M-sequence is an important basis of communication theory and systems [14-17]. The electrical 3-cell linear feedback shift register (LFSR) shown in Fig. 3 cyclically generates the M-sequence "0010111" of length 7. The operation in the figure is performed by exclusive OR (xOR) according to the standard theory. Although there are some exceptional xOR neurons [18], this may be equivalently realized by the combination of threshold elements as a standard neuronal model. One such example is shown in the Appendix.

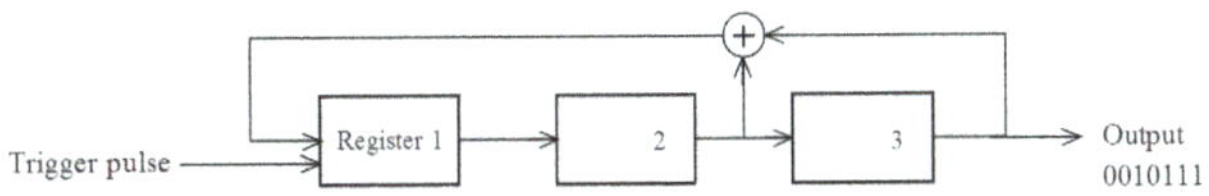

Figure 3. *Electrical linear feedback shift register with three logical elements (registers; cells) with an M-sequence output of length 7 (⊕ shows an Exclusive OR operation)*

The length of the M-sequence generated from n-cell LFSR is $2^n - 1$. In the case of a 3-cell LFSR, only one type of M-sequence exists, with the exception of mirror order and rotationally shifted sequences. We refer to this as M3, which is shown in Table 1. Table 1 also shows a 4-cell case (M4).

Table 1. M3 and M4 M-sequences

No. of Cells n	Regular /Reversal	Number and Comment	M-Sequence
3	Reversal (Rev)	(1)	1101000
		(2) mirror of (1)	1011000
	Regular (Non-Rev)	(3)	0010111
		(4) mirror of (3)	0100111
4	Reversal (Rev)	(5)	111011001010000
		(6) mirror of (5)	101001101110000
	Regular (Non-Rev)	(7)	000100110101111
		(8) mirror of (7)	010110010001111

Loop-shaped circuits, such as a LFSR, may become components of a large network or provide insights into larger loops in the intelligent network [19-22]. Furthermore, M-sequences are often used as codes in real communication systems, such as CDMA mobile phones, and may also become temporally independent components of information representation in the brain [10]. Therefore, we pay special attention to M-sequences as typical markers in the following sections and present some simulation results in the Appendix.

3. Communication Model in a Neural Network

3.1. Simplest Model

We can simulate brain neural networks as threshold-element (cell) networks. Figure 4 shows the simplest model of 2 × 2 spatiotemporal multiplexed communication. Each neuronal cell denoted by ○ works as one of these threshold elements in a synchronous mode, such that if the sum of the weighted inputs to the element is more than 0, it outputs "1," otherwise it outputs "0." Therefore, input "1" to n_1 is transmitted only to the opposite destination cell n_3, and input "1" to n_2 is transmitted to n_4.

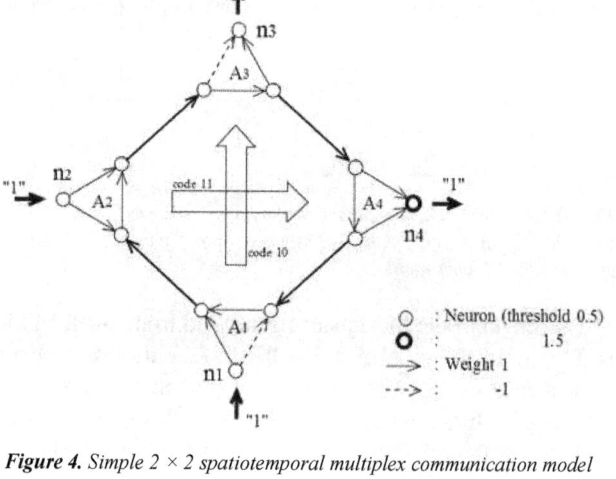

Figure 4. *Simple 2 × 2 spatiotemporal multiplex communication model*

3.2. 3 × 3 Spatiotemporal Multiplex Communication

As shown above, we can simulate brain neural networks using threshold element networks. In Fig. 5, we show a simple 3 × 3 communication model. With regard to the general behavior of synchronous threshold element networks, please refer to the Appendix.

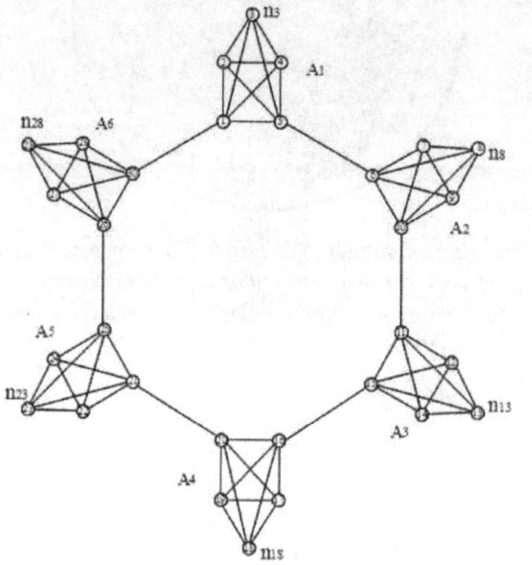

Figure 5. *A 3 × 3 communication model. Pulse "1" sent from neuron n_3 is supposed to be received only by n_{18}, that sent from n_8 is supposed to be received by n_{23}, and that sent by n_{13} is supposed to be received by n_{28}.*

Here, signal "1" sent from cell n_3 is not supposed to be received by n_8, n_{13}, n_{23}, and n_{28}. Signal "1" is to be received only by n_{18}. Similarly, the n_8 signal is supposed to be received by n_{23}, while that of n_{13} is supposed to be received by n_{28}. Each neuronal assembly of A_1, A_2, ..., A_6 is set to be able to work to encode or decode the signal. Main loops $\{n_1\text{-}n_5\text{-}n_6\text{-}n_{10}\text{-}n_{11}\text{-}n_{15}\text{-}n_{16}\text{-}n_{20}\text{-}n_{21}\text{-}n_{25}\text{-}n_{26}\text{-}n_{30}\text{-}n_1\}$ are set to give not one but two paths to any communication channels of $n_3 \rightarrow n_{18}$, $n_8 \rightarrow n_{23}$, and $n_{13} \rightarrow n_{28}$, i.e., each assembly works to encode/decode the signal and to pass such signals through that are not for its output neuron. Then, the signal can be spatiotemporally encoded.

3.2.1. Wide Time Gate

We generated networks with random weights (∈ {+1, 0, −1}) and selected those that satisfied the requirements mentioned above. We generated 2.012×10^8 networks in which each weight on the main loop was bidirectionally fixed to +1, i.e., weights from n_5 to n_6 and n_6 to n_5 were +1, and so on. Other weights were randomly fixed to +1, 0, and −1, with probabilities of 1/3 for each.

All cells in the network were synchronously driven. A single pulse "1" was given to the n_3 cell at time 1, and the output number of "1" was counted between time 1 and 16 at the n_{18} cell. Effectively, the count was between times 8 and 16 at the destination cell because the pulse arrived at time 8. If the number of output "1" is the largest among $\{n_8, n_{13}, n_{18}, n_{23}, n_{28}\}$, successful communication is achieved for the test of the $n_3 \rightarrow n_{18}$ channel. If the three communication channels of $n_3 \rightarrow n_{18}$, $n_8 \rightarrow n_{23}$, and $n_{13} \rightarrow n_{28}$ are all successful in the same network, we classified the network as being successful in the 3 × 3 multiplex communication.

We obtained 141 successful networks with the desired function, i.e., the success rate was 7.01 × 10^{-7} or, in other words, one network per 1.43 × 10^6 randomly generated networks exhibited 3 × 3 spatiotemporal multiplex communication.

Figure 6 shows an example of a successful network. Figure 7 shows the flow of the codes that worked as markers of the information flow; almost all sequences generated from a cell were transient within the short time of the communication and were too long to analyze. Thus, to visualize the information flow, we selected several remarkable short codes with length ≦7 from the observed propagating wavefront sequence. We marked sequences "1011," "10101," "11111," and "0010111." "1011" is a core part of the reversal M-sequence "1011000;" "10101" is a typical alternating sequence of "0" and "1;" "11111" is a representative of long continuous "1" sequences; and "0010111" is a conventional non-reversal M-sequence.

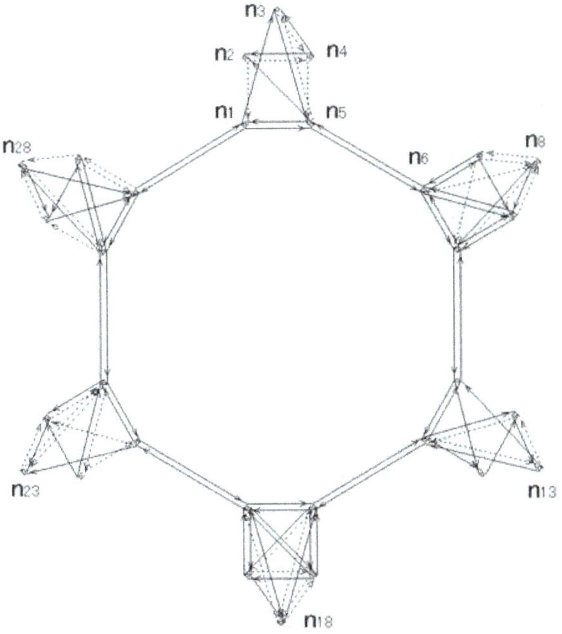

Figure 6.. Example of a network realizing 3 × 3 spatiotemporal multiplex communication (solid lines indicate a weight of +1 and dotted lines indicate a weight of −1)

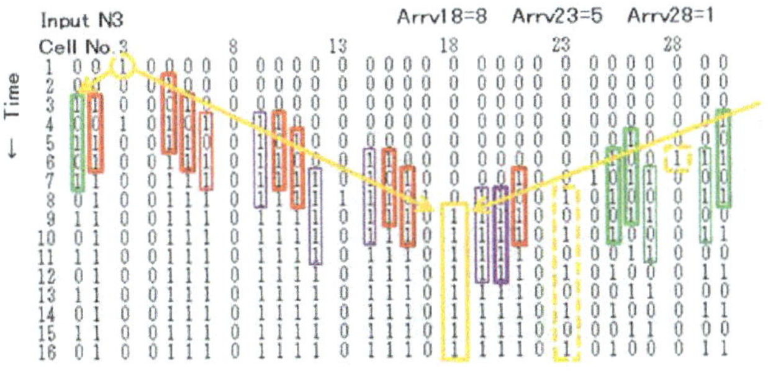

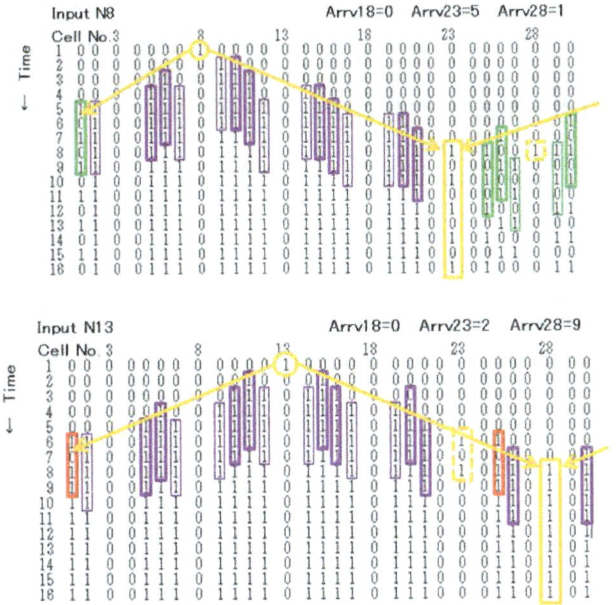

Figure 7. State transition diagram of the multiplex communication network shown in Fig. 6 with a wide time gate, where "1" is the input to n_3 (top), n_8 (middle), and n_{13} (bottom). It can be considered that the source input "1" is encoded to "1011," "11111," and "10101" and finally gives the maximum output at the destination n_{18} (top), n_{23} (middle), and n_{28} (bottom), respectively.

Input stimulation to one of $\{n_3, n_8, n_{13}\}$ was encoded by the corresponding cell assembly $\{A_1, A_2, A_3\}$ and spreads in both directions to the right and left of the loop. Sometimes they were transformed to another code at the passing cell assembly, decoded at the destination assembly, and transmitted to the destination cell. If the coming sequence is not for its assembly, the assembly does not take it in but passes it to the next assembly.

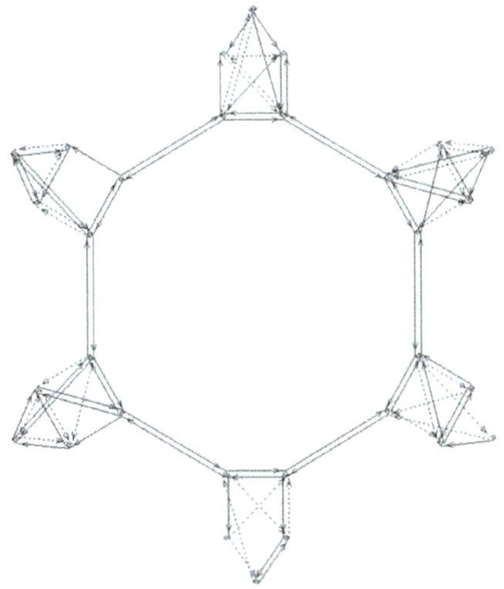

Figure 8. Another example of network weight realizing a 3 × 3 communication

Figures 8 and 9 show another example. Here, a conventional non-reversal M-sequence, "0010111," and reversal M-sequences, "1010001" and "0100011," were generated by the initial source input pulse "1" to the n_{13} cell.

However, only the first code contributed to the output via an anticlockwise rotation route, whereas latter codes via a clockwise rotation route did not contribute. The maximum output was given from the destination cell n_{28}.

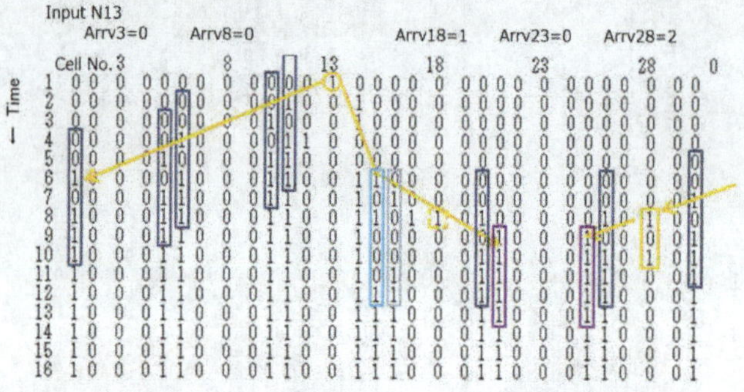

Figure 9. State-transition diagram of the communication network shown in Fig. 8. "1" is given to n_{13}, which is internally encoded to the M-sequences "0010111," "1010001," and "0100011."

The temporal response function (time gate) at the receiving cell is shown in Fig. 10. In the case of a wide time gate, the number of "1s" at the destination cell is effectively counted after time 8, which represents the fastest arrival time of the pulse.

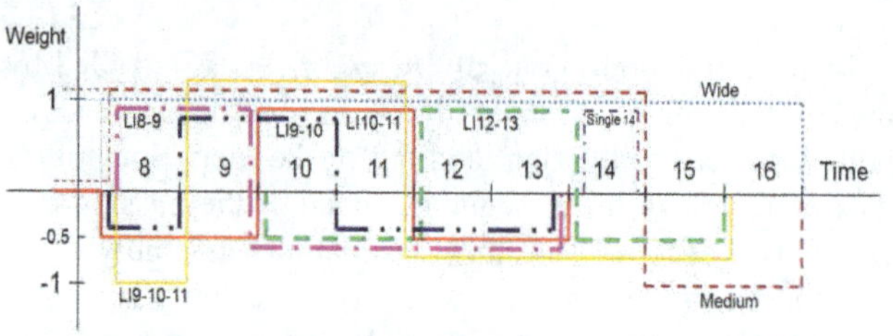

Figure 10. Time gate weights at a receiving cell. Time 8 is the fastest arrival time of the pulse at the destination cell. In the case of a single time gate, only time 14 is shown. Double and triple time gates, which are the central part of LI-type gates, are not shown.

3.2.2. Medium Time Gate

In the network described in Section ***3.2.1***, the gate only counted the number of pulses that arrived at the destination cell during the observation period from time 1 to 16. In the medium time gate, we restricted the arrival period at the receiving cell, i.e., the "1s" that arrived at times 15 and 16 were counted as penalties, and their number was subtracted from the number of "1s" that arrived between times 1 and 14 (roughly counting pulses between time 8 and 12; see Fig. 10 Medium). We obtained 222 successful networks from the 2.295×10^8 generated candidate networks. The success rate was 9.67×10^{-7}, or, in other words, one network per 1.03×10^6 generated candidate networks met the required communication function. This figure was higher than that of the wide time gate. However, we consider that such communication tasks are still difficult. Figure 11 shows the number of codes of "1011," "11111," "10101," and "1101" obtained at the pulse wavefront on the main loop. This result shows that rate distribution was influenced by the task of the network. There was a definite difference between the appearance average rates of "1011" of the wide time gate network and those of the medium time gate as an ensemble. However, because the appearance rates had large standard deviations, the average difference was not particularly important. In other words, there are many ways to realize a given task, as shown by the large standard deviations depicted in Fig.12, and the appearance rate reflects the structural difference only as an average.

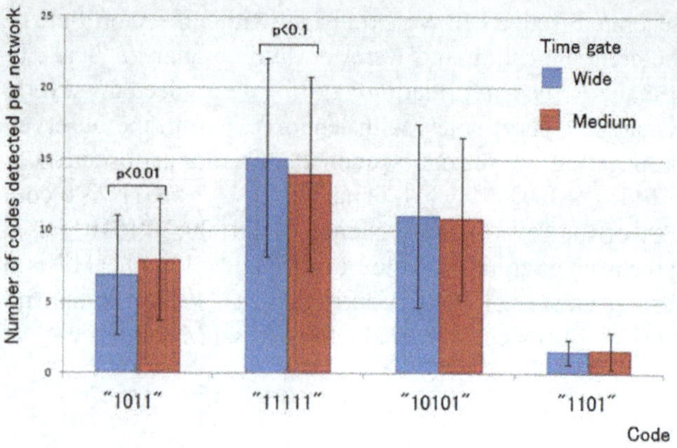

Figure 11. Major code spectrum of wide and medium time gate communication networks

3.2.3. Single Time Gate

In this case, we imposed a restriction of the time gate to specific single times of 8, 9, 10, 13, and 15. The corresponding success rates are shown in Fig. 12.

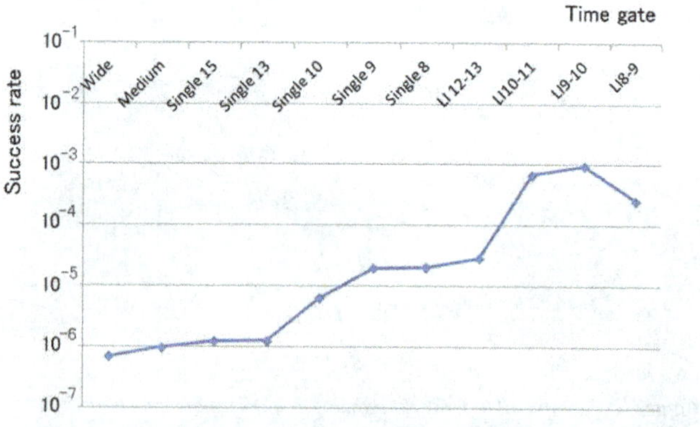

Figure 12. Success rate of a random network for the communication tasks

3.2.4. Double and Triple Time Gates

In case of a double time gate, we combined two single time gates, such as D8–9, D9–10, D10–11, and D12–13. The corresponding success rates are shown in Fig. 13; the success rates of the double time gate are close to the average of the corresponding single time gates. Furthermore, we combined three single time gates, such as T9–10–11. In this case, the success rate was higher than that of each of the corresponding single time gates.

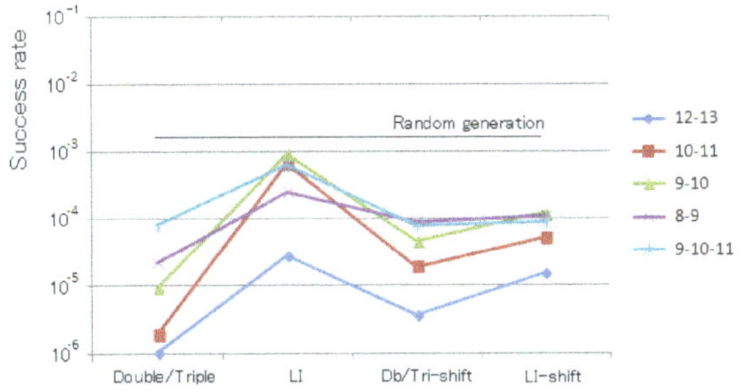

Figure 13. *Success rates of a randomly generated network for a 3 × 3 communication task with double/triple, LI gates, and their time shift gates. "Random generation" is a theoretical upper limit of a randomly generated network that can attain the given task by chance.*

3.2.5. Lateral Inhibition (LI) Type Time Gate

Because the LI-type response is universal in natural neural networks, we applied it to the time response of the receiving cells.

The central peak was fixed to weight 1 and width 2. The central positions were taken at times 8–9, 9–10, 10–11, and 12–13. Four negative bases on both sides were set at a weight of −0.5. However, because the pulse arrived first at the destination cell at time 8, the front negative bases before time 8 were moved to the tail, as shown in Fig. 10. Each weight from time 8 to 16 was:

LI8–9 = (1, 1, −0.5, −0.5, −0.5, −0.5, 0, 0, 0)
LI9–10 = (−0.5, 1, 1, −0.5, −0.5, −0.5, 0, 0, 0)
LI10–11 = (−0.5, −0.5, 1, 1, −0.5, −0.5, 0, 0, 0)
LI12–13 = (0, 0, −0.5, −0.5, 1, 1, −0.5, −0.5, 0).

The code spectrum obtained is shown in Fig. 14. Some average differences existed between these LIs. However, the standard deviations were close to these averages; therefore, we could only observe almost random pulse sequences. The situation was the same in the cases of a single time gate described in *3.2.3* in which the spectra were almost the same as those shown in Fig. 14.

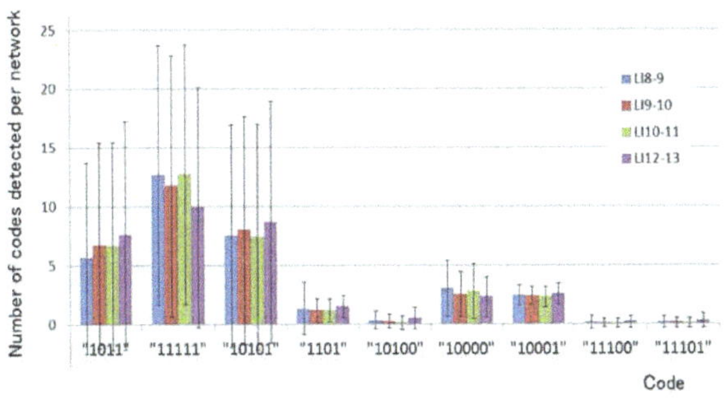

Figure 14. *Code spectrum with standard deviation on the main loop of LI-type time gate communication networks*

The weight of the LI-type triple time gate was set as LI9–10–11 = (−1, 1, 1, 1, −0.5, −0.5, −0.5, −0.5, 0).

3.2.6. Time Shift Gate

The time-gate settings described above may create a bias to give the destination cell the maximum output, while potentially giving the non-destination cell located nearest to the source cell a lower output for pulses arriving through the shorter main loop route. To compensate for this bias, we shifted the center of the time gate of the halfway cells by 2, 4, and 6 according to the distance from the source cell. However, this remains incomplete at the point at which the pulses arriving through the longer main loop route are neglected.

3.3. Success Rate for Communication

We randomly generated networks, and if the network randomly gave the communication pair, the success rate ideally should become 5^{-3} = 1.6 × 10^{-3}. However, the success rate of these networks with regard to communication was significantly different according to the type of time gate, as is shown in Figs. 12 and 13. The success rate of the wide time gate was the worst. In this time gate, there was no restriction on the arrival time of pulses, and only the pulses arriving at the receiving cell were counted. Single time gates, double time gates, and triple time gates had a better success rate. Among these, the faster time gate was relatively good. The LI-type time gates had the best success rate. In cases of LI9–10, LI10–11, and LI9–10–11, the success rate was approximately 10^{-3}. The improvement in detectability according to the time-gate shape had two effects on the success rate. One is the direct effect of raising the success rate. The other is lowering the success rate by raising the detectability of non-destination cells.

Figure 13 shows the success rates of time shift gates for double-type, triple-type, and LI-type time gates. LI-type time gates seemed to be sufficiently powerful for detecting the proper sequence. The time shift gates raised the detectability of non-destination cells, thus reducing the success rates. This shows that a bias of the time gate adjusted for destination cells helped detection at the destination cell. Figure 15 shows an example of a state-transition diagram of an LI-type time shift gate case.

The success rates of all time gates were lower than 5^{-3} of the ideal random channel selection. The reasons for this observation may be that the random lines are sometimes disconnected, the network outputs the same levels, and the three directional tasks are not independent. However, LI-type reception is commonly effective in spatiotemporal communication with high success rates close to the ideal rate as well as in other general neural networks.

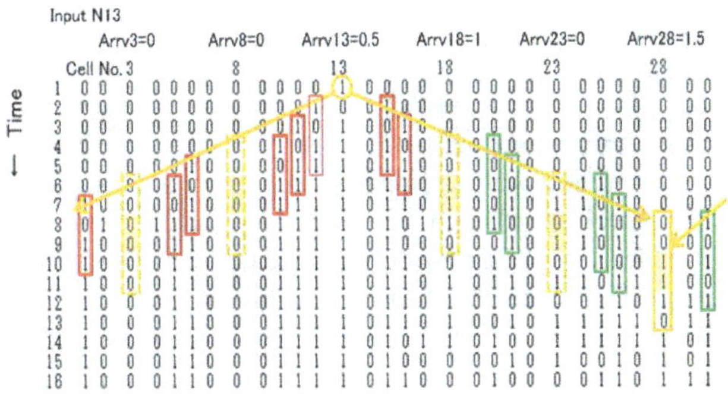

Figure 15. *Example of a state-transition diagram for a stimulation where "1" is given to n_{13} and an LI-type time shift gate is used*

4. Discussion and Conclusions

We believe that the communication function between neuronal cells provides a basis for the intelligence functions of the brain, such as memory, association, and abstraction. However, to date, the neuronal spike trains have been treated as random noise-like signals, and the coding scheme of the natural neural network has not been completely elucidated. We have been tackling this issue from a communication engineering viewpoint and have proposed a model of spatiotemporal multiplex communication. In this model, each cell works as a transmitting/receiving cell and as a relay cell, the roles of which are I/O and intermediate communication, respectively. In this sense, the network works like a multi-hop ad hoc communication network.

Although only the shape of the network is given by 3 × 3 spatiotemporal multiple communication loop types, various network shapes can be considered; for example, a homogeneous network without a pre-assigned loop. Moreover, there are various ways to determine the task of the communication, i.e., various possible network shapes can have additional routes to the destination and timings. This represents a problem of balance between space and time.

We have shown that the simulated pulse sequence from each cell of the threshold element network resembled our experimental data, including the code spectrum. However, though not shown here, our experimental data were obtained in an unsorted state, including a single neuronal cell, and several neuronal cells and cell assemblies, such as synfire chains [23,24], which may cause synchronous spikes, including codes. Codes may be composed from these cell "groups," which represents an open problem.

We obtained target networks via random generation and selection. Although this method is useful to demonstrate the feasibility of such communication tasks, it is ineffective in real situations because of necessary computational capacity. In the brain, target networks are considered to be formed by learning. Therefore, we are now developing such networks based on learning. In addition, noise immunity, interference effects between communication channels and successive transmissions, and stability should be investigated. However, the objective of this paper is to demonstrate the feasibility of spatiotemporal multiplex communication in neural networks, and we believe we were able to show its feasibility using a minimum-size model.

The content of this paper can be summarized as follows:

1) A neural network with a spatiotemporal communication function was proposed.

2) Each neuronal cell worked as a transmitting and receiving cell as well as a relay cell.

3) Pulses running in the network seemed to be like noise, similar to the code observed in naturally cultured neuronal networks.

Appendix

Here, we present some results of an output analysis, particularly for the M-sequences included in the state transition output of each threshold element of the network in general.

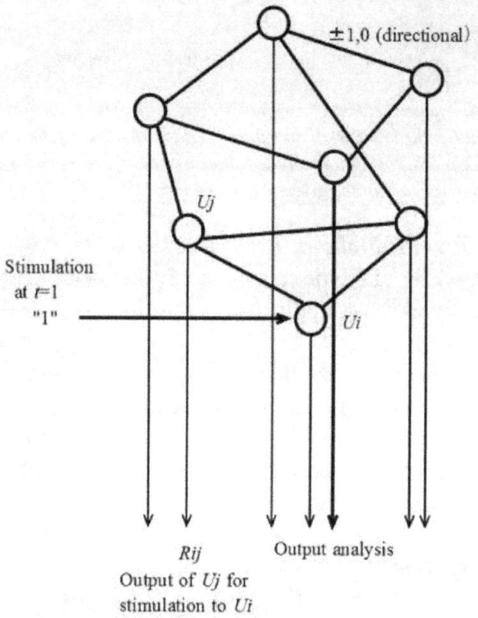

Figure A1. *Output analysis of threshold cell network*

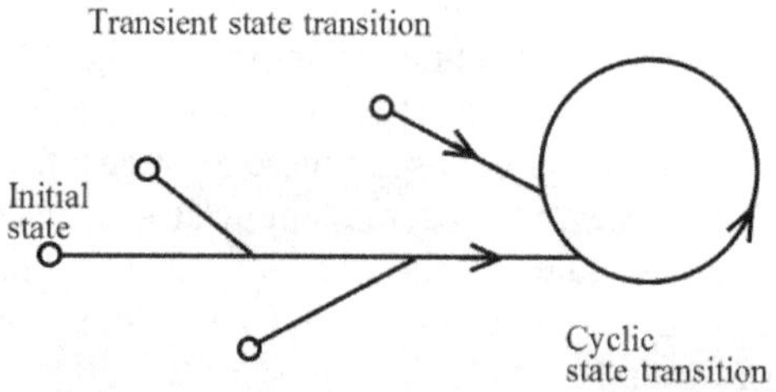

Figure A2. *State transition of the network*

Figure A1 shows the method used to analyze the output of the network. At time t = 1, a single stimulation "1" is given to cell U_i; $i \in \{1, 2, \ldots, n\}$, where n is the number of cells. Cell U_j; $j \in \{1, 2, \ldots, n\}$ gives an output of "0" if the sum of the weighted input given to U_j is $\leqq 0$ and "1" if this sum is > 0 synchronously in the network. The threshold is 0 unless otherwise specified. Each weight is +1, 0, or −1. Generally, the state expressed by the combination of all cell states ("0" or "1") changes with time, as shown in Fig. A2.

A1. Four-Cell Network

In this case, all possible 3^{16} = 43,046,721 networks could be generated non-randomly as a treatable maximum number. Although we did not use this approach, this number may be reduced by using the symmetric characteristic of the weights. Instead, stimulation was given only to a fixed cell, e.g., U_1.

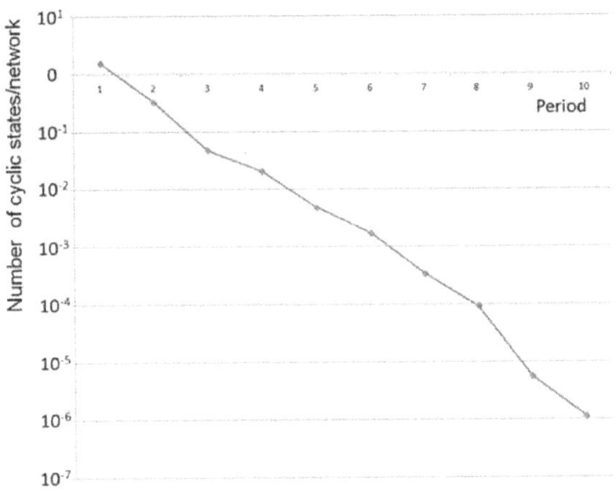

Figure A3. *Period distribution of cycles of state transition in a 4-threshold-cell network*

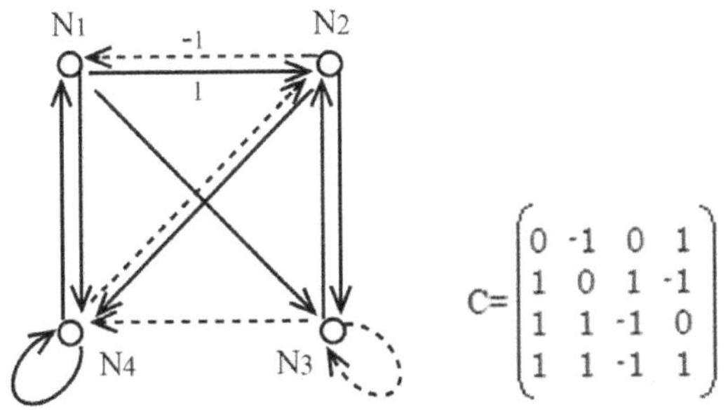

Figure A4. *Threshold cell network with weights of ±1 and 0, which generates the M-sequence "1001110" from N_1 for stimulation of "1" to N_1*

The average number of cycles of state transition per network is shown in Fig. A3. Period 1 means that the state did become stationary. Some networks generated an M-sequence (see Fig. A4 for an example). The *i*, *j* entry of matrix ***C*** shows the connection weight from N_j to N_i. The output sequence from N_1 is "1001110," which is an inverse-order, rotationally-shifted, and non-reversed version of the M-sequence "1101000" listed in Table 1.

A2. Number of M-Sequences Detected

The output from each cell was checked among 32 clock times after stimulation. Then, the total number of 1.34×10^{10} positions on the sequences was checked. The detected M-sequences are shown in Table A1, in which (1)–(4) correspond to those listed in Table 1. In this case, the rate of Non-Rev was larger than that of Rev.

A3. Twelve-Cell Network

In this case, we were no longer able to check all networks. Thus, we generated networks with various weights {+1, 0, −1}, as shown in Table A1. By changing the stimulation cell, the output from each cell was checked among 64 (practically 52 time positions) clock times after the stimulation.

Table A1. *Rate of M-sequences detected*

No. of Cells n	Possible Networks	Rate of Weights +1, 0, −1	Generated Networks	Sequence Positions Checked $\times 10^6$	Rate of M3 (length 7) Rev [(1), (2)] $\times 10^{-3}$	Rate of M3 (length 7) Non-Rev [(3), (4)] $\times 10^{-3}$
4	3^{16}	(All Checked) Equivalently 1/3, 1/3, 1/3	3^{16}	13,400	6.61 (4.47, 2.14) $\times 10^{-5}$	16.6 (5.00, 11.6) $\times 10^{-5}$
		0.4, 0.2, 0.4	6000	44.9	12.0 (6.08, 5.89)	9.95 (4.88, 5.07)
		0.3, 0.4, 0.3	6000	44.9	11.1 (5.59, 5.46)	8.44 (4.22, 4.22)
12	3^{144}	0.1, 0.8, 0.1	3000	22.4	1.53 (0.785, 0.749)	0.912 (0.456, 0.456)
		0.15, 0.8, 0.05	23,000	172	3.32 (1.56, 1.76)	2.94 (1.69, 1.25)
		0.06, 0.92, 0.02	9,000	67.4	0.117 (0.0472, 0.0702)	0.0802 (0.0561, 0.0241)

Figure A5 shows a time flow for the case of the weight rate of (0.1, 0.8, 0.1). Here, "Theor" represents the theoretical values calculated for cases in which the observed sequence was white random (no correlation within the sequence) and the observed "0" and "1" rates were used at each time position. The observed rates of the M-sequence code were rather low, with the exception of the initial time area. This means that the states at the initial time area are transient and close to the random state. In addition, after some time, the states shift to steady states that include final cyclic states, most of which are not M-sequence cycles, but are shorter (such as "1111111," "0000000," or "101010101").

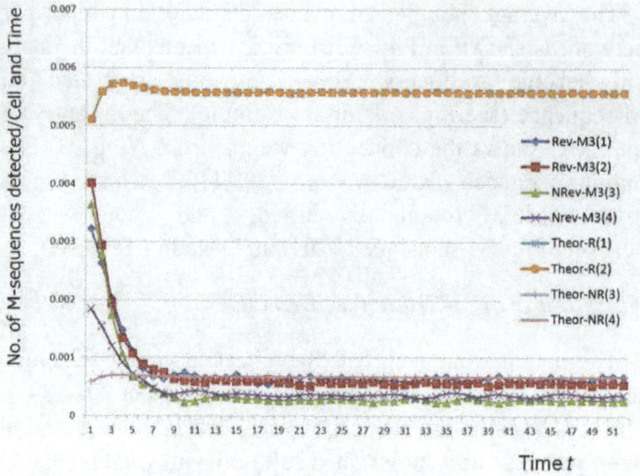

Figure A5. *Number of M-sequences detected per cell and time. Symbols correspond to those listed in Table 1. "Theor" refers to the theoretical values under the assumption of a random sequence. Curves of Theor-(1) and (2) almost overlap. Curves of Theor-(3) and (4) also almost overlap.*

As a result, we can say that there was a tendency for the rate of Rev M-sequences, including those that are fragmental, to be greater than that of Non-Rev sequences. In addition, information from the source cell was included in the early transient period. This explains why we analyzed only the wavefront in this study.

Acknowledgements

The authors are grateful to Dr. T. Shimosakon of Osaka Institute of Technology for his support and advice.

References

[1] Y. Mizuno-Matsumoto, K. Okazaki, A. Kato, T. Yoshimine, Y. Sato, S. Tamura, and T. Hayakawa., "Visualization of epileptogenic phenomena using cross-correlation analysis: Localization of epileptic foci and propagation of epileptiform discharges," IEEE Trans. Biomed. Eng. 46, pp.271-279, 1999.

[2] Y. Mizuno-Matsumoto, M. Ishijima, K. Shinosaki, T. Nishikawa, S. Ukai, Y. Ikejiri, Y. Nakagawa, R. Ishii, H. Tokunaga, S. Tamura, S. Date, T. Inouye, S. Shimojo, and M. Takeda, "Transient Global Amnesia (TGA) in an MEG Study," Brain Topography 13, pp.269-274, 2001.

[3] B. Cessac, H. Paugam-Moisy, T. Viéville, "Overview of facts and issues about neural coding by spikes," J. Physiol.-Paris 104, (1-2), pp.5-18, 2010.

[4] O. Kliper, D. Horn, B. Quenet, and G. Dror, "Analysis of spatiotemporal patterns in a model of olfaction," Neurocomputing 58-60, pp.1027-1032, 2004.

[5] K. Fujita, Y. Kashimori, and T. Kambara, "Spatiotemporal burst coding for extracting features of spatiotemporally varying stimuli," Biol. Cybern., 97, pp.293-305, 2007. DOI 10.1007/s00422-007-0175-z

[6] I. Tyukin, T. Tyukina, and C. van Leeuwen, "Invariant template matching in systems with spatiotemporal coding: A matter of instability," Neural Networks 22, pp.425-449, 2009.

[7] A. Mohemmed, S. Schliebs, S. Matsuda, and N. Kasabov, "Training spiking neural networks to associate spatio-temporal input–output spike patterns," Neurocomputing 107, pp.3-10, 2013.

[8] B.A. Olshausen and D.J. Field, "Emergence of simple-cell receptive field properties by learning a sparse code for natural images," Nature 381, pp.607-609, 1996.

[9] A. Bell and T. Sejnowski, "The independent components of natural scenes are edge filters," Vision Research 37, pp.3327-3338, 1997.

[10] S. Tamura, Y. Mizuno-Matsumoto, Y-W. Chen, and K. Nakamura, "Association and abstraction on neural circuit loop and coding," The Fifth International Conference on Intelligent Information Hiding and Multimedia Signal Processing (IIHMSP2009) A10-07(No.546) , Kyoto, September 12-14, 2009. Available: IEEE XPlore

[11] Y. Nishitani, C. Hosokawa, Y. Mizuno-Matsumoto, T. Miyoshi, H. Sawai, and S. Tamura "Detection of M-Sequences from Spike Sequence in Neuronal Networks," Computational Intelligence and Neuroscience 2012, Article ID 862579, 9 pages, 2012. doi:10.1155/2012/862579.

[12] S. Tamura, Y. Nishitani, C. Hosokawa, Y. Mizuno-Matsumoto, T. Kamimura, Y-W. Chen, T. Miyoshi, and H. Sawai, "M-sequence family from cultured neural circuits," The 3rd Int'l Workshop on Computational Intelligence for Bio-Medical Science and Engineering (CIMSE-2012), Taipei, October 23-25, 2012.

[13] S. Tamura, Y. Nishitani, C. Hosokawa, Y. Mizuno-Matsumoto, T. Kamimura, Y-W. Chen, T. Miyoshi, and H. Sawai, "Pseudo random sequences from neural circuits," IFMIA 2012, Daejeon, November 16-17, 2012.

[14] R.C. Dixon, Spread Spectrum Systems, John Wiley & Sons Inc., 1976.

[15] D.V. Sarwate and M. B. Pursley, "Crosscorrelation properties of pseudorandom and related sequences," Proc. IEEE 68, pp.593-619, 1980.

[16] S. Tamura, S. Nakano, and K. Okazaki, "Optical code-multiplex transmission by Gold-sequences," IEEE/OSA J. Lightwave Tech. 1, No.3, pp.121-127, 1985.

[17] S.W. Golomb and G. Gong, Signal Design for Good Correlation: For Wireless Communication, Cryptography, and Rader, Cambridge University Press, 2005.

[18] P. Fromherz and V. Gaede, "Exclusive-OR function of single arborized neuron," Biol Cybern 69, pp.337-344, 1993.

[19] C. Lecerf, "The double loop as a model of a learning neural system," Cybernetics and Informatics 1, pp.587-594, 1998.

[20] Y. Choe, "Analogical cascade: A theory on the role of the thalamo-cortical loop in brain function," Neurocomputing 52-54, pp.713-719, 2003.

[21] T. Kamimura, K. Nakamura, K. Yoneda, Y-W. Chen, Y. Mizuno-Matsumoto, T. Miyoshi, H. Sawai, and S. Tamura, "Information communication in brain based on memory loop neural circuit," ICIS2010 & SEDM2010, Chengdu, June 23-25, 2010. Available: IEEE Xplore

[22] T. Kamimura, Y-W. Chen, Y. Yagi, and S. Tamura, "Learning of Loop Neural Circuit for Memory," CIMSE2011, ICCIT2011, Jeju, November 29-December 1, 2011. Available: IEEE Xplore

[23] M. Abeles, Local Cortical Circuits: An Electrophysiological Study, Springer, Berlin, 1982.

[24] M. Abeles, "Synfire chains," Scholarpedia, 4(7), pp.1441, 2009.

Load frequency control for interconnected power system using different controllers

Atul Ikhe, Anant Kulkarni

P. G. Department, College of Engineering Ambajogai, Dist. Beed, Maharashtra, India

Email address:
atulikhe1@gmail.com(A. Ikhe), anantkulkarni2002@gmail.com(A. Kulkarni)

Abstract: This paper explores the potential of using soft computing methodologies in controllers and their advantages over conventional methods. PID controller, being the most widely used controller in industrial applications, needs efficient methods to control the different parameters of the plant. As reported by several researchers, the conventional approach of PID controller is not very efficient due to the presence of non-linearity in the system of the plant. Also, the output of the conventional PID system has a quite high overshoot and settling time. The main focus of this work is on the controller to obtain good output frequency responses. The tuning of PID controller is necessary to get an output with better dynamic and static performance. The application of PID controller imparts it the ability of tuning itself automatically in an on-line process while the application. The output response of PID-tuning is compared with I, PI and conventional PID controller and found reasonably good over these conventional controllers.

Keywords: Conventional Controller, Interconnected Power System, Load Frequency Control (LFC), PID Tuning, Tie-Line

1. Introduction

The problem of controlling the real power output of generating units in response to changes in system frequency and tie-line power interchange within specified limits is known as load frequency control (LFC) [1]. The Objectives of LFC are to provide zero steady-state errors of frequency and tie-line exchange variations, high damping of frequency oscillations and decreasing overshoot of the disturbance so that the system is not too far from the stability [2]. The interconnected power system is typically divided into control areas, with each consisting of one or more power utility companies. Sufficient supply for generation of each connected area to meet the load demand of its customers.

The above mentioned objectives are carried successfully in previous works by different authors using PI and PID controllers [4] & [5].

In this paper PID-tune controller is used for better frequency responses. This type of controller is used in power system so reducing the steady state error. System load is never steady using this controller these can be controlled. When uncontrolled case more oscillation, negative overshoot be observed but while comparing to conventional type controller PID and propose work result gives better performances of dynamic responses.

2. PID Controller

There are many types of controller such like proportional, integral, derivative and combinational of these (PI, PID).

2.1. PID Controller

The block diagram of Proportional Integrative Deri vative (PID) controller is shown in Fig.1

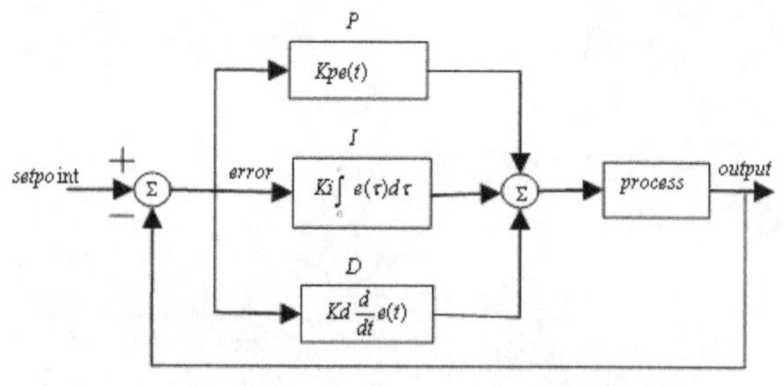

***Figure 1:** Block diagram of a PID controller.*

The PID controller improves the transient response so as to reduce error amplitude with each oscillation and then output is eventually settled to a final desired value. Better margin of stability is ensured with PID controllers. The mathematical equation for the PID controller is given as [4]

& [9].

$$y(t)= Kpe(t) + Ki\int_{0}^{t} e(\tau)d\tau + Kd\frac{d}{dt}e(t) \quad (1)$$

Where y (t) is the controller output and u (t) is the error signal. K_p, K_i and K_d are proportional, integral and derivative gains of the controller. The limitation conventional PI and PID controllers are slow and lack of efficiency in handling system non-linearity. Generally these gains are tuned with help of different optimizing methods such as Ziegler Nicholas method, Genetic algorithm, etc., The optimum gain values once obtained is fixed for the controller. But in the case deregulated environment large uncertainties in load and change in system parameters is often occurred. The optimum controller gains calculated previously may not be suitable for new conditions, which results in improper working of controller. So to avoid such situations the gains must be tuned continuously.

The tuning parameters are

2.1.1. Proportional Gain (K_p)

Larger values typically mean faster response since the larger the error, the larger the Proportional term compensation. An excessively large proportional gain will lead to process instability and oscillation.

2.1.2. Integral Gain (K_i)

Larger values imply steady state errors are eliminated more quickly. The trade-off is larger overshoot: any negative error integrated during transient response must be integrated away by positive error before we reach steady state.

2.1.3. Derivative Gain (K_d)

Larger values decrease overshoot, but slows down transient response and may lead to instability due to signal noise amplification in the differentiation of the error.

2.2. Advantages of PID Controller

They can perform poorly in some applications. PID controllers, when used alone, can give poor performance when the PID loop gains must be reduced so that the control system does not overshoot, oscillate or hunt about the control set point value. A problem with the Derivative term is that small amounts of measurement or process noise can cause large amounts of change in the output.

3. Model of Two Area Power System

Each area is assumed to have only one equivalent Generator and is equipped with governor- turbine system. They are the control signals from the controllers A two area model is adapted in the work is shown in Figure.2 [2] & [11].

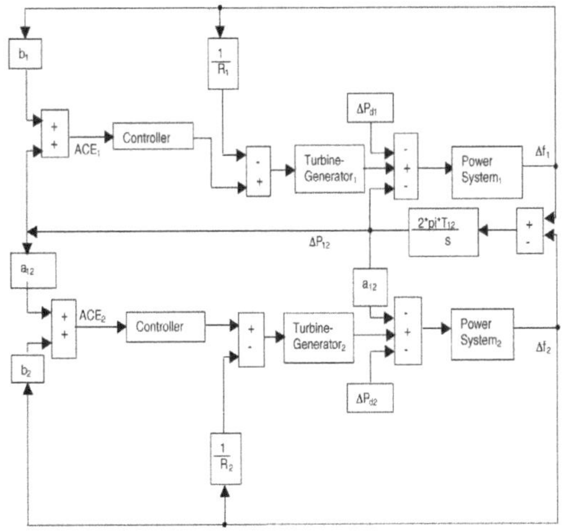

Figure 2: Block diagram of two area power system.

The terms showed in the Figure 2 are termed given below:

fi :Nominal system frequency of ith area. [HZ]

ΔfI :Incremental frequency deviation of ith area. [HZ pu]

Tsi : Speed governor time constant of i th area [sec.]

Kgi : Gain of speed governor of i th area

Ri :Governor Speed regulation of the of ith area [ZH /pu.MW]

Tti : Governor Speed regulation of the of ith area [ZH /pu.MW]

Kti : Gain of turbine of ith area

Kpi :Gain of power system (generator load) of i th area. [ZH /pu.MW]

Kpi = 1/D

Tpi Gain of power system (generator load) of i th area. [ZH /pu.MW]

Tpi = 2Hi /Difi

Hi : Inertia constant of i th area . [MW-sec/MVA]

ΔPGi : Incremental generator power output change of ith area .[pu MW]

ΔPti : Incremental turbine power output change of i th area. [pu MW]

Ki : Gain of controller of ith area.

The plant for a power system with a non-reheated turbine consists of three parts:

• Governor with dynamics:

$$G\,g(s)=\frac{1}{TGs+1}. \quad (2)$$

• Turbine with dynamics:

$$Gt(s)=\frac{1}{TTs+1}. \quad (3)$$

• Load and machine with dynamics:

$$Gp(s)=\frac{Kp}{TPs+1}. \quad (4)$$

Now the open-loop transfer function without droop characteristic for load frequency control is

$$\widetilde{P} = GpGtGg = \frac{Kp}{(TPs+1)(TTs+1)(TGs+1)}. \quad (5)$$

4. Matlab Simulink Model

4.1. Power system Model Using Different Controllers

In two area system, two single area systems are interconnected via tie-line. Inter connections established increases the overall system reliability. Even if some generating units in one area fail, the generating units in the other area can compensate to meet the load demand. The basic block diagram of five area interconnected power system is shown in Fig.2. A conventional integral controller is used on a power system model. The PID controller improves steady state error simultaneously allowing a transient response with little or no overshoot. As long as error remains, the integral output will increase causing the speed changer position, attains a constant value only when the frequency error has reduced to zero. The SIMULINK model of a two area interconnected power system using PID controller is shown in Figure 3[6].

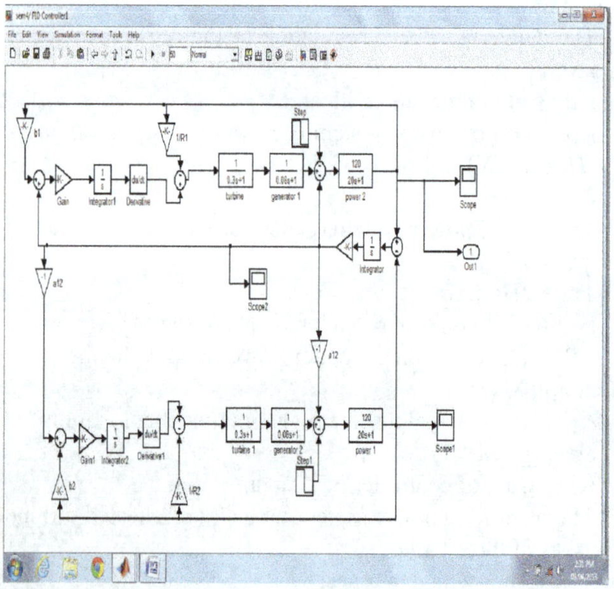

Figure 3: *Simulink model of two area power system using PID controller.*

The output response is shown in Fig.4,which having the comparison results between simple integral(I), proportional integral (PI) ,Proportional integral derivative (PID).

The output frequency response using PID is better than I and PI.

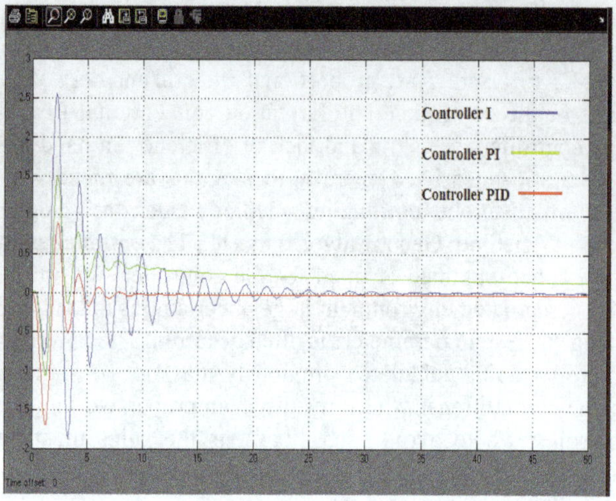

Figure 4: *Output frequency response using different controller.*

The gain value of different types of controller using in two areas power system is given in Table 1.

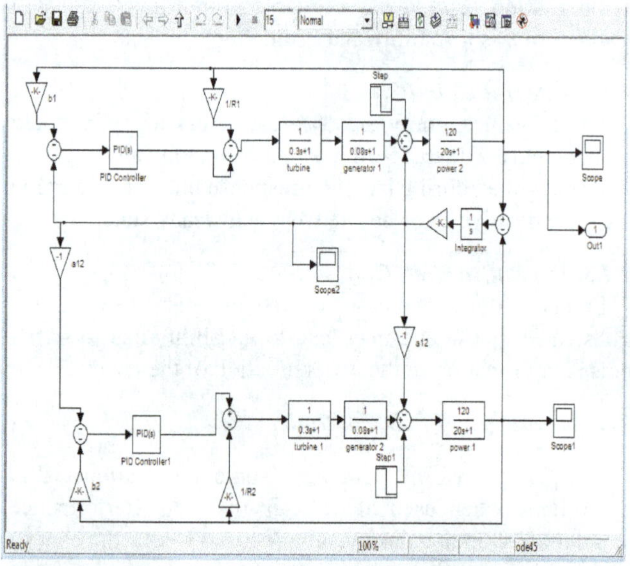

Figure 5: *Simulink model of two area power system using PID tuning controller.*

Table 1: *Different values of gain for controllers [citation].*

Controller	Kp Area1	Kp Area2	Ki Area1	Ki Area 2	Kd Area 1	Kd Area 2	Settling time (sec.)
I	-	-	0.2742	0.4680	-	-	35
PI	0.1109	0.0121	0.2742	0.2019	-	-	25
PID	0.1109	0.0121	0.2742	0.2019	0.1110	0.003	10

It shows that for different controllers getting different settling time value. The settling time of PID controller is less than I, PI controller. We can control oscillations, rise time and settling time of PID controller using PID tuning method.

4.2. Model of PID Tuning

The gain value of controller is automatically fixed when we select PID tuning controller. The MATLAB Simulink diagram is shown in Figure 5.

The output response of PID tuning method for area1, area 2 and Tie-line is shown in Figure 6(a), 6(b),6(c) respectively.

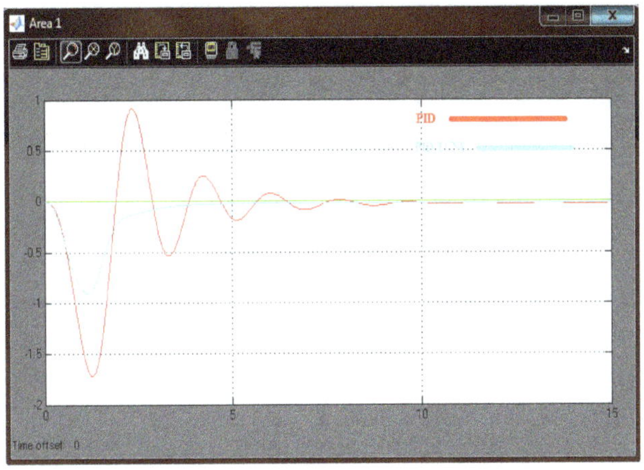

Figure 6(a): *Output response of area 1.*

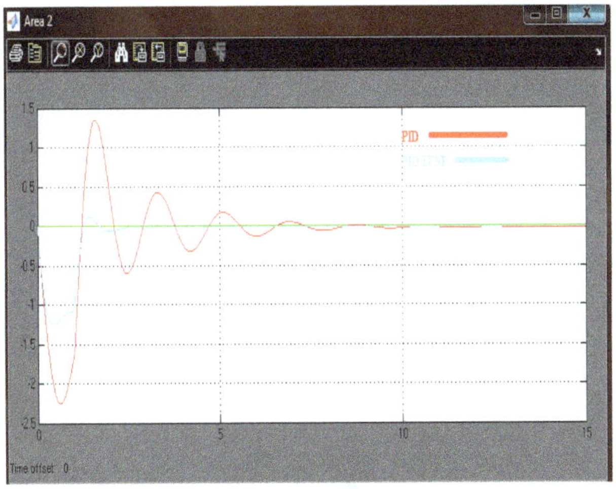

Figure 6(b): *Output response of area 2.*

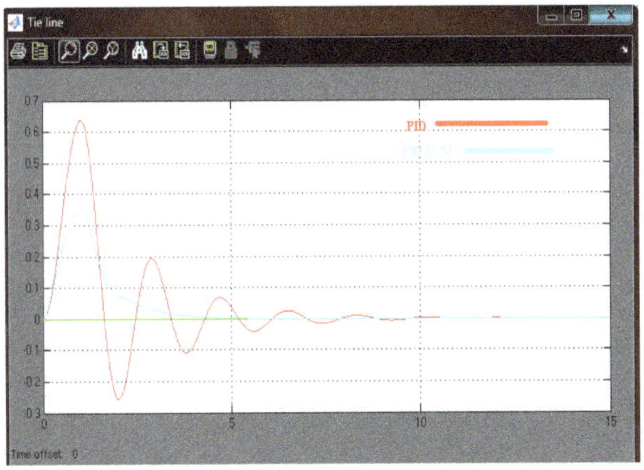

Figure 6(c): *Output response of tie-line of power system.*

For better dynamic responses using PID tuning method, we reduce settling time, oscillation. The response of power system also varies according to rated power capacity of any system.

5. Conclusions

A tuning of PID controller used for load frequency controller of two area interconnected power system has been presented. It can be implemented in four area power system and controlled by using advanced controller systems. The system performance was observed on the basis of dynamic parameters i.e. settling time, overshoot and undershoot. The system performance characteristics reveals that the performance of PID tuning method better than other controllers. As a further study, the proposed method can be applied to multi area power system load frequency control (ALFC) and also optimum values can be obtained by Fuzzy Logic Controller (FLC), Genetic Algorithm and Neural networks.

References

[1] Wen Tan, "Unified tuning of PID load frequency controller for power system via IMC",*IEEE Trans. Power Systems*, vol. 25, no. 1, pp. 341-350, 2010.

[2] G. Raj Goutham,Dr. B. Subramanyam, "IMC Tuning of PID Load Frequency Controllerand Comparing Different Configurations for Two Area Power System",*Internationa lJournal of Engineering Research and Applications*, Vol. 2, Issue 3, May-Jun 2012,pp.1144-1150.

[3] Emre Ozkop, Ismail H. Altas, Adel M. Sharaf, "Load Frequency Control in Four Area Power Systems Using Fuzzy Logic PI Controller",*16th National Power Systems Conference*, 15th-17th Dec., 2010.

[4] K. P. Singh Parmar, S. Majhi, D. P. Kothari, "Optimal Load Frequency Control of an Interconnected Power System", MIT *International Journal of Electrical and Instrumentation Engineering*, vol. 1, No. 1, pp 1-5, Jan 2011.

[5] Mohammad Soroush Soheilirad, Mohammad Ali Jan Ghasab, Seyed mohamm-dhossein Sefidgar, Aminmohammad Saberian,"Tuning of PID Controller for Multi AreaLoad Frequency Controlby Using Imperialist Competitive Algorithm", J. Basic. Appl. Sci.Res., 2(4)3461-3469, 2012.

[6] Akanksha Sharma, K.P. Singh Parmar and Dr. S.K. Gupta, "Automatic Generation Controlof Multi Area Power System using ANN Controller", International Journal of ComputerScience and Telecommunications [vol. 3, Issue 3, March 2012]

[7] S.Ganapathy, S.Velusami, "Design of MOEA based Decentralized Load-Frequency Controllers for Interconnected Power Systems with AC-DC Parallel Tie-lines", International Journal of Recent Trends in Engineering, Vol 2, No. 5, November 2009.

[8] R.Francis, Dr. I. A.Chidambaram , "Automatic Generation Control for an Interconnected Reheat Thermal Power Systems Using Wavelet Neural Network Controller", International Journal of Emerging Technology and Advanced Engineering Website: www.ijetae.com (ISSN 2250-2459, Volume 2, Issue 4, April 2012)

[9] K.RamaSudha,V.S.Vakula, R.VijayaShanthi, "PSO based Design of Robust Controller for Two Area Load Frequency Control with Non linarites", *International Journal of Engineering Science and Technology*, Vol. 2(5), 2010, 1311-1324

[10] G.Karthikeyan, S.Ramya, Dr. S.Chandrasekar, "Load Frequency Control for Three Area System with Time Delays Using Fuzzy Logic Controller", International Journal of Engineering Science & Advanced Technology, ISSN: 2250–3676 Volume-2, Issue-3, 612 – 618.

[11] Kanika Wadhwa, Sourav Choubey, Pardeep Nain, "Study of Automatic Generation Control Of two area thermal-thermal system with GRC and without GRC", First National conference on Power System Engineering (PSEC'12) Paper code PS1015.

[12] K. S. S. Ramakrishna1," Automatic generation control of interconnected power system with diverse sources of power generation," International Journal of Engineering, Science and Technology, Vol. 2, No. 5, 2010, pp. 51-65

[13] GaddamMallesham, Akula Rajani, "Automatic Generation Control Using Fuzzy Logic",8th International Conference on development and application systems Suceava, Romani May 25 – 27, 2006.

Matlab simulation of temperature control of heat exchanger using different controllers

Neeraj Srivastava, Deoraj Kumar Tanti, Md Akram Ahmad

Electrical Engineering Department, BIT Sindri, Dhanbad, India

Email address:
nee2k8@gmail.com (N. Srivastava), dktanti@yahoo.com (D. K. Tanti), akram14407@gmail.com (Md A. Ahmad)

Abstract: Heat exchanger system is widely used in chemical plants because it can sustain wide range of temperature and pressure. The main purpose of a heat exchanger system is to transfer heat from a hot fluid to a cooler fluid, so temperature control of outlet fluid is of prime importance. To control the temperature of outlet fluid of the heat exchanger system a conventional PID controller can be used. Due to inherent disadvantages of conventional control techniques, Fuzzy logic controller is employed to control the temperature of outlet fluid of the heat exchanger system. The designed controller regulates the temperature of the outgoing fluid to a desired set point in the shortest possible time irrespective of load and process disturbances, equipment saturation and nonlinearity.

Keywords: PID Controller, FLC, Heat Exchanger

1. Introduction

In practice, all chemical process involves production or absorption of energy in the form of heat. Heat exchanger is commonly used in a chemical process to transfer heat from the hot fluid through a solid wall to a cooler fluid. There are different types of heat exchanger used in the industry but most of the industry use shell and tube type heat exchanger system. Shell and tube heat exchangers are probably the most common type of heat exchangers applicable for a wide range of operating temperatures and pressures. In shell and tube heat exchanger one fluid flows through the tubes and a second fluid flows within the space between the tubes and the shell[1]. The outlet temperature of the shell and tube heat exchanger system has to be kept at a desired set point according to the process requirement. Firstly a classical PID controller is implemented in a feedback control loop so as to achieve the control objectives. PID controller exhibits high overshoots which is undesirable. To minimize the overshoot Fuzzy logic controller is implemented. Fuzzy logic has become one of the most successful of today's technologies for developing sophisticated control systems. The reason is very simple: Fuzzy logic addresses applications perfectly as it resembles human decision making with an ability to generate precise solutions from certain or approximate information.

The paper is organized as follows: section 2 gives a brief introduction of Heat exchanger system. Assumptions made and sources of disturbances are also described along with mathematical modeling of system. Section 3 describes PID Controller and its tuning method. Section 4 describes Fuzzy logic controller, its membership function and rule base. Section 5 shows simulation model and its resultant graphs. Section 6 and 7 shows result, discussions and conclusion.

2. Heat Exchanger System

A typical interacting chemical process for heating consists of a chemical reactor and a shell and tube heat exchanger system. The super-heated steam comes from the boiler and flows through the tubes. Whereas, the process fluid flows through the shells of the shell and tube heat exchanger system. The process fluid which is the output of the chemical reactor is stored in the storage tank. The storage tank supplies the fluid to the heat exchanger system. The heat exchanger heats up the fluid to a desired set point using super-heated steam supplied from the boiler. The storage tank supplies the process fluid to a heat exchanger system using a pump and a non returning valve. There is also a path of non condensed steam to go out of the shell and tube heat exchanger system in order to avoid the blocking of the heat exchanger.

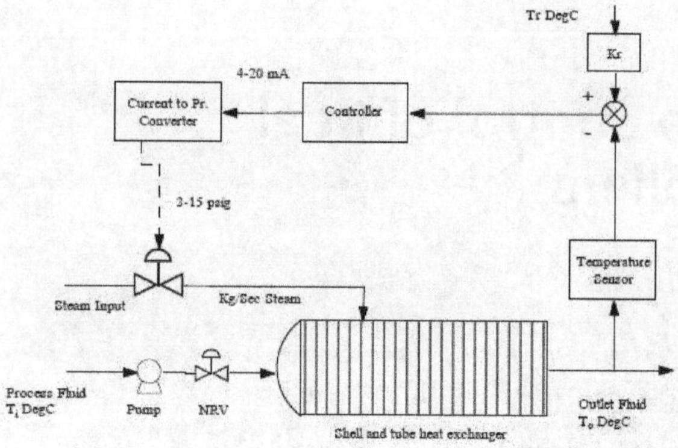

Fig 1. *Shell and tube heat exchanger system control scheme.*

2.1. Assumptions

Different assumptions have been considered in this research paper. (i) Inflow and the outflow rate of fluid are same, so that the fluid level is maintained constant in the heat exchanger. (ii) The heat storage capacity of the insulating wall is negligible.

A thermocouple is used as the sensing element which is implemented in the feedback path of the control architecture. The temperature of the outgoing fluid is measured by the thermocouple and the output of the thermocouple is sent to the transmitter unit, which eventually converts the thermocouple output to a standardized signal in the range of 4-20 mA. This output of the transmitter unit is given to the controller unit. The controller implements the control algorithm, compares the output with the set point and then gives necessary command to the final control element via the actuator unit. The actuator unit is a current to pressure converter and the final control unit is an air to open valve. The actuator unit takes the controller output in the range of 4-20 mA and converts it in to a standardized pressure signal in the range of 3-15 psig. The valve actuates according to the controller decisions. Fig 1 shows the control scheme adopted in heat exchanger system.

2.2. Sources of Disturbances

There can be two types of disturbances in this process. (i) the flow variation of input fluid (ii) the temperature variation of input fluid.

2.3. Mathematical Modeling of Heat Exchanger System

In this section the heat exchanger system, actuator, valve, sensor are mathematically modeled using the available experimental data. The experimental process data's are summarized below[2].

Exchanger response to the steam flow gain $= 50°C/kg/sec$

Time constants $= 30\ sec$

Exchanger response to variation of process fluid flow gain $= 1°C/kg/sec$

Exchanger response to variation of process temperature gain $= 3°C/°C$

Control valve capacity for steam $= 1.6\ kg/sec$

Time constant of control valve $= 3\ sec$

The range of temperature sensor $= 50°C\ to\ 150°C$

Time constant of temperature sensor $= 10\ sec$

From the experimental data, transfer functions and the gains are obtained as below.

Transfer function of process $= \frac{50e^{-s}}{30s+1}$

Gain of valve $= 0.13$

Transfer function of valve $= \frac{0.13}{3s+1}$

Gain of current to pressure converter $= 0.75$

Transfer function of disturbance variables

(i) Flow$= \frac{1}{30s+1}$ (dominant). (ii) Temperature $= \frac{3}{30s+1}$

Transfer function of thermocouple $= \frac{0.16}{10s+1}$

3. Proportional-Integral-Derivative (PID) Controller

The mnemonic PID refers to the first letters of the names of the individual terms that make up the standard three-term controller. These are P for the proportional term, I for the integral term and D for the derivative term in the controller. PID controllers are probably the most widely used industrial controller. Even complex industrial control systems may comprise a control network whose main control building block is a PID control module. In PID controller Proportional (P) control is not able to remove steady state error or offset error in step response. This offset can be eliminated by Integral (I) control action. Output of I controller at any instant is the area under actuating error signal curve up to that instant. I control removes offset, but may lead to oscillatory response of slowly decreasing amplitude or even increasing amplitude, both of which are undesirable. Derivative (D) control action has high sensitivity. It anticipates actuating error, initiates an early correction action and tends to increase stability of system[2].

Ideal PID controller in continuous time is given as

$$y(t) = K_p\left(e(t) + \frac{1}{T_i\int_0^t e(t)dt} + T_d\frac{de(t)}{dt}\right) \quad (1)$$

Laplace domain representation of ideal PID controller is

$$Gc(s) = \frac{Y(s)}{E(s)} = K_p(1 + \frac{1}{T_i s} + T_d s) \quad (2)$$

3.1. Tuning of PID Controller

Ziegler and Nichols proposed rules for determining values of $K_p, T_i\ and\ T_d$ based on the transient response characteristics of a given plant. Closed loop oscillation based PID tuning method is a popular method of tuning PID controller. In this kind of tuning method, a critical gain K_c is induced in the forward path of the control system. The high value of the gain takes the system to the verge of instability. It creates oscillation and from the oscillations, the value of frequency and time are calculated. Table 1 gives experimental tuning rules based on closed loop oscillation method[3,4].

Table 1. *Closed loop oscillation based tuning methods.*

Type of Controller	K_p	T_i	T_d
P	$0.5K_c$	∞	0
PI	$0.45K_c$	0.83T	0
PID	$0.6K_c$	0.5T	0.125T

The characteristic equation $1 + G(s)H(s) = 0$ in this case is obtained as below

$$900s^3 + 420s^2 + 43s + 0.78K_c + 1 = 0 \quad (3)$$

Applying Routh stability criterion in above eq gives $K_c = 24.44$

Auxiliary equation is $420s^2 + 0.78K_c + 1 = 0$ (4)

Substituting $s = j\omega$ gives $\omega = 0.218$ and $T = 28.82$

For the PID controller the values of parameters obtained using Ziegler Nichols closed loop oscillation based tuning methods are

$$K_p = 14.66 \quad T_i = 14.41 \quad T_d = 3.60$$

Usually, initial design values of PID controller obtained by all means needs to be adjusted repeatedly through computer simulations until the closed loop system performs or compromises as desired. These adjustments are done in MATLAB simulation.

4. Fuzzy Logic Controller (FLC)

The design of fuzzy logic controller is attempted in heat exchanger. The fuzzy controllers are designed with two input variables, error and rate of error and one output variable (i.e.) the hot water flow rate to the shell side. The mamdani based fuzzy inference system uses linear membership function for both inputs and outputs. For the fuzzy logic controller the input variables are error $(e)$and rate of error (Δe), and the output variable is controller output(Δy). Triangular membership functions are used for input variables and the output variable. The universe of discourse of error, rate of error and output are [-13, 13], [-4, 4] and [-5, 5] respectively. The rule base framed for shell and tube heat exchangers are tabulated in Table 3[5,7].

The structure of the rule base provides negative feedback control in order to maintain stability under any condition. For the evaluation of the rules, the fuzzy reasoning unit of the FLC has been developed using the Max-Min fuzzy inference method[8]. In the particular FLC, the centroid defuzzification method is used. Linguistic variables for error, rate of error and controller output are tabulated in table 2.[6]

Fig 2, 3 and 4 shows membership functions of different variables implemented in FIS editor in MATLAB toolbox and fig 5 shows surface view of all variables in 3 dimension[9].

Table 2. *Linguistic variables.*

VBN	Very big negative	PS	Small positive
NB	Big negative	PM	Medium positive
NM	Medium negative	PB	Big positive
NS	Small negative	VBP	Very big positive
Z	Zero		

Table 3. *Rule base for fuzzy logic controller.*

$e \rightarrow$ $\Delta e \downarrow$	VBN	NB	NM	NS	Z	PS	PM	PB	VBP
VBN	VBN	VBN	VBN	VBN	VBN	NB	NM	NS	Z
NB	VBN	VBN	VBN	VBN	NB	NM	NS	Z	PS
NM	VBN	VBN	VBN	NB	NM	NS	Z	PS	PM
NS	VBN	VBN	NB	NM	NS	Z	PS	PM	PB
Z	VBN	NB	NM	NS	Z	PS	PM	PB	VBP
PS	NB	NM	NS	Z	PS	PM	PB	VBP	VBP
PM	NM	NS	Z	PS	PM	PB	VBP	VBP	VBP
PB	NS	Z	PS	PM	PB	VBP	VBP	VBP	VBP
VBP	Z	PS	PM	PB	VBP	VBP	VBP	VBP	VBP

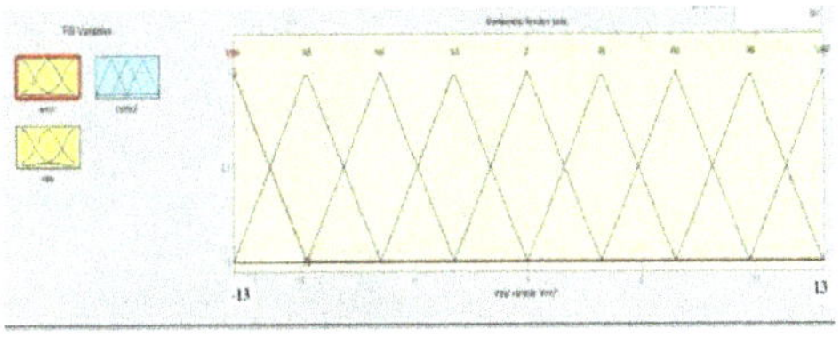

Fig 2. *Membership function for error.*

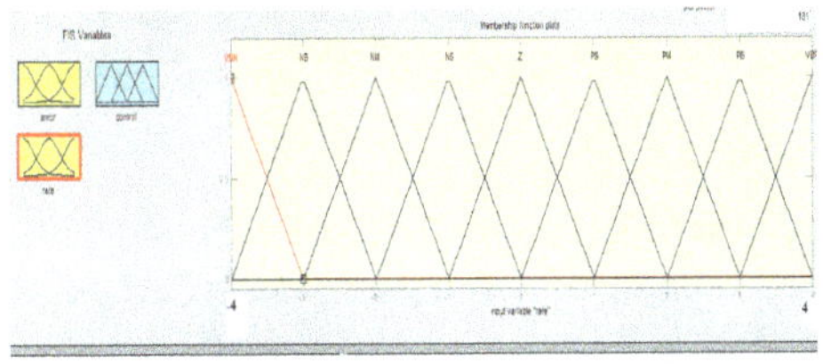

Fig 3. *Membership function for rate of error.*

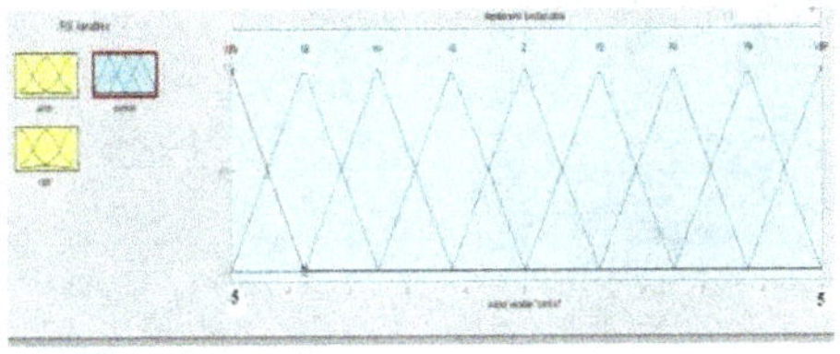

Fig 4. *Membership function for control output.*

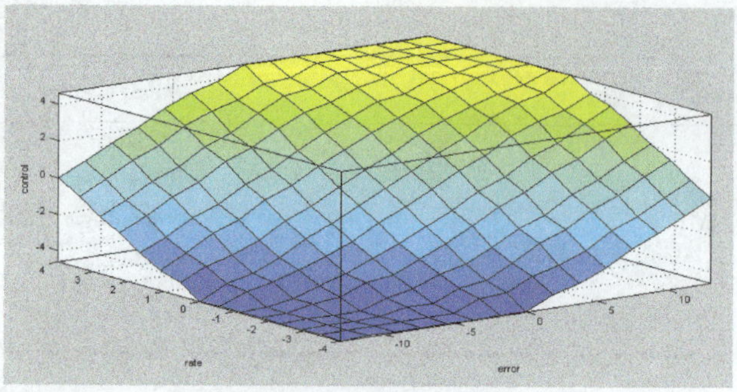

Fig 5. 3D Surface view.

5. Simulation

The simulation for different control mechanism discussed above were carried out in Simulink in MATLAB and simulation results have been obtained. Fig 6 and 7 shows the PID controller and FLC system block diagram which is simulated in matlab.

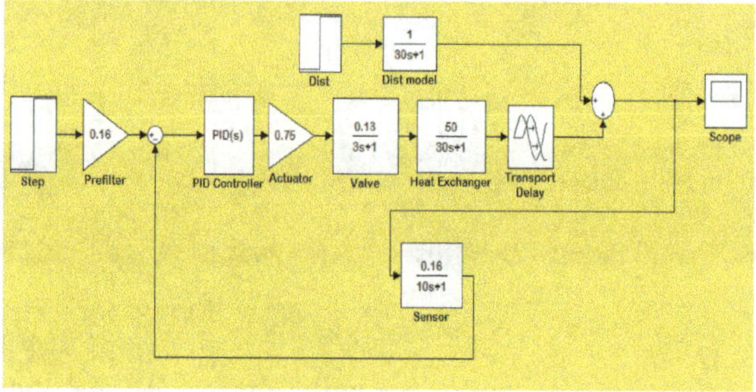

Fig 6. PID Controller.

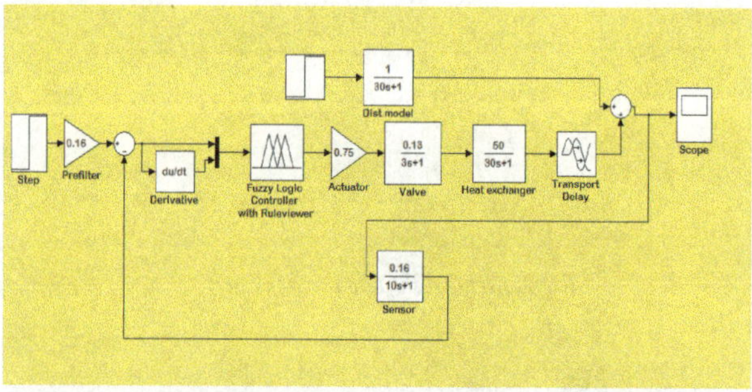

Fig 7. Fuzzy logic controller.

Fig 8 and 9 shows step response of PID and FLC system where x axis denote time and y axis denote set point value.

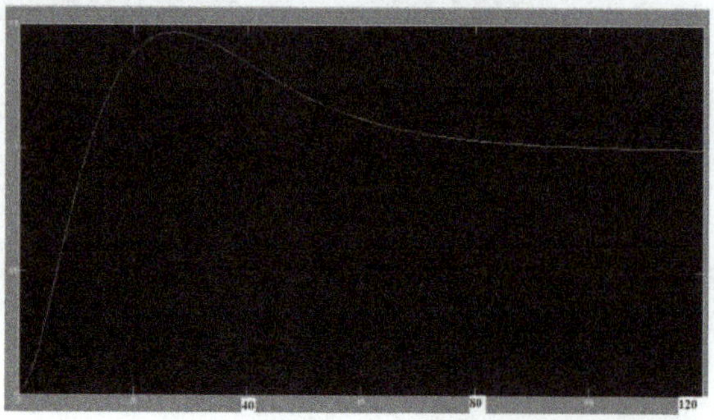

Fig 8. PID Step response.

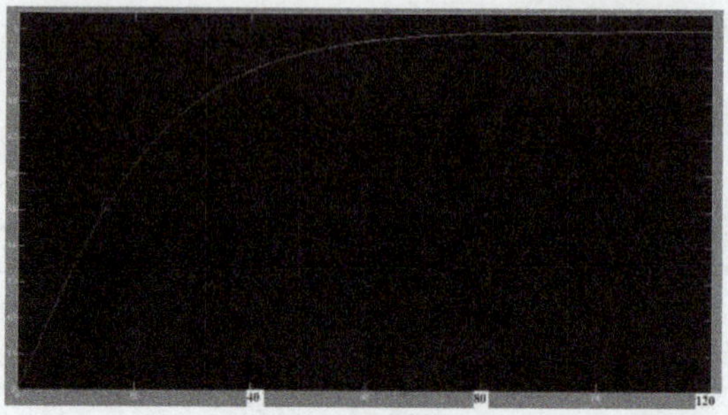

Fig 9. FLC Step response.

6. Simulation Result and Discussion

To evaluate the performance of the different controllers this paper has considered two vital parameters of the step response of the system. The first parameter is the maximum overshoot and the second parameter is the settling time.

Peak Overshoot: It indicates maximum positive deviation of output with respect to its desired value. It is defined as[8] $\%M_p = \frac{c(t_p)-c(\infty)}{c(\infty)} \times 100\ \%$

Settling Time: It is the time required for the response to reach and stay within a specified tolerance band of its final value. The tolerance band is taken randomly as 2%.

In this paper control of temperature of heat exchanger is done by 2 different controllers. In PID controller we set the parameters by using Ziegler Nichols closed loop method. After this we simulated the model in MATLAB and tuned the parameters until the response is satisfactory. In step response, we found overshoot and large settling time both of which are undesirable. Moreover there are three tuning parameters which simultaneously be adjusted to get desired result.

Then a Fuzzy logic controller is developed with 9 membership functions for each variable. There were total 81 rules generated. The response is smooth as well as fastest as compared to PID controllers. So FLC is recommended because it is easy to implement, low cost and no need to know exact plant parameters. Different parameters are tabulated in table 4.

7. Conclusion

In this paper, a comparative study of performance of conventional (PID) and intelligent (FLC) controllers is studied. The aim of the proposed controller is to regulate the temperature of the outgoing fluid of a shell and tube heat exchanger system to a desired temperature in the shortest possible time and minimum or no overshoot

After comparing results for different controllers, we obtain that fuzzy logic controller is the one which gives quick response without any oscillations. It is easy to implement fuzzy logic as it is computer oriented. PID controller, though good for industrial process but due to oscillatory response and large settling time is now going to be replaced by new technology like Fuzzy logic and Neural network.

Table 4. *Comparison of different parameters.*

Control System	Maximum Overshoot (%)	Settling Time (sec)
Feedback PID Controller	47.2	88
FLC	0	65

Nomenclature

K_p	Proportional gain
T_i	Integral time
T_d	Derivative time
K_c	Critical gain
T	Time period of oscillation
ω	Angular frequency of oscillation
$Gc(s)$	Controller transfer function
e	Error between desired output and actual output
Δe	Rate of error

References

[1] *Anton Sodja et.al, "Some Aspects of Modeling of Tube-and-Shell Heat-Exchangers," in Proc of 7th Modelica Corif-, Italy, pp. 716-721, Sep 2009.*

[2] *Subhransu Padhee, Yuvraj Bhushan Khare, Yaduvir Singh "Internal Model Based PID Control of Shell and Tube Heat Exchanger System," IEEE, JAN 2011.*

[3] *Katsuhiko Ogata, "Modern Control Engineering". 5th edition 2010.*

[4] *Kiam Heong Ang, Gregory Chong and Yun Li, "PID Control System Analysis, Design, and Technology," IEEE Trans., Control Syst. Technol., vol. 13, no. 4, pp. 559-576, Jul2005.*

[5] *Mridul Pandey, K. Ramkumar & V. Alagesan "Design of Fuzzy Logic Controller for a Cross Flow Shell and Tube Heat-Exchanger," IEEE, Mar 2012*

[6] *Zadeh L.A., Fuzzy relation Equations and Applications to Knowledge Engineering, Kluwer Academic Publishers, Holland, 1989*

[7] *Larsen P.M., Industrial application of Fuzzy Logic Control, academic press, inc., may, 1979.*

[8] *BS Manke, "Linear Control System". 9th edition 2010*

[9] *Fuzzy Logic Toolbox Help file in MATLAB version 7.11*

Ant colony optimization with re-initialization

Matej Ciba, Ivan Sekaj

Institute of Control and Industrial Informatics, Bratislava, Slovakia

Email address:

bigmato@centrum.sk(M. Ciba), ivan.sekaj@stuba.sk(I. Sekaj)

Abstract: This contribution introduces an Ant Colony Optimization (ACO) algorithm with re-initialization mechanism. The whole search process is broken by re-initialization into shorter semi-independent steps called "macro cycles". The length of macro cycle depends on pheromone accumulation and can be adjusted by a user parameter. It is shown that re-initialization mechanism prevents ACO algorithm from pheromone saturation and consecutive stagnation. This approach avoids overhead caused by algorithm run with excessive pheromone values where further exploration is hardly possible. The solution offers lower CPU cost of the search process and enables automation of heuristic search especially in changing environments like dynamic networks. The efficiency of proposed method is demonstrated on a path minimization problem on 50 node graph.

Keywords: Ant Colony Optimization, Re-Initialization, Pheromone Saturation, Minimal Path Search

1. Introduction

Wide range of problems like Routing problem, Assignment problem, Scheduling problem and others can be transformed into graph representation. Exact algorithms for instance Dijkstra or Bellman-Ford appear to be slow and inefficient on large scale graphs. It is not always necessary to find the best possible solution. Instead good quality solution in reasonable time is often preferred. This is the reason why heuristic methods become popular. Among the well-known graph search algorithms that utilizes heuristic are A* search [1] and ACO algorithm.

Ant colony optimization represents an efficient tool for optimization and design of graph oriented problems. It is a multi-agent meta-heuristic approach and was first purposed in 1991 by Dorigo at al. as Ant system (AS) algorithm.

Ants are using indirect form of communication via trailing pheromone. As ants are passing the terrain (graph) they mark used routes (arcs of the graph) by chemical substance called pheromone. Other ants can sense the substance and follow the same track.

During the search process each ant sets off from the ant colony (start position) and moves to search food (destination). The aim is to find the shortest path. On their way back they use the same way from which abundant loops have been removed. The amount of pheromone (1) $\Delta\tau^{k}_{ij}(t)$ the k-th ant produces is inversely proportional to the tour length $L^{k}(t)$.

$$\Delta\tau_{ij}^{k}(t) = \begin{cases} Q/L^{k}(t) & if\,(i,j) \in T^{k}(t) \\ 0 & if\,(i,j) \notin T^{k}(t) \end{cases} \qquad (1)$$

$T^{k}(t)$ is the tour generated by ant k, Q is a constant and tuple (i, j) denotes beginning and termination node of an arc. All pheromone tracks (2) are preserved by arcs of the graph

$$\tau_{ij}(t+1) = (1-\rho)\tau_{ij}(t) + \sum_{k=1}^{m} \Delta\tau_{ij}^{k}(t) \qquad (2).$$

Where $\rho \in (0,1)$ is the pheromone persistence ($1 - \rho$ is evaporation rate) and m is the number of ants. Evaporation rate is a user adjusted parameter and affects pheromone durability; i.e. how long the acquired information will be available. Too high values causes random search (quick evaporation), too low values get algorithm stock in local optimum.

An ant in each node has to make a decision which arc to take. This probabilistic choice is called *random proportional rule* (3). Ant chooses the next node only from its neighborhood N^{k}_{i} except the node it visited previously.

$$p_{ij}^{k}(t)=\frac{\tau_{ij}^{\alpha}(t)+\eta_{ij}^{\beta}}{\sum_{j\in N_i^k}\tau_{ij}^{\alpha}(t)+\eta_{ij}^{\beta}} \quad (3)$$

The probability $p_{ij}(t)$ of choosing the particular arc (i, j) depends on pheromone $\tau_{ij}(t)$ and the heuristic η_{ij} values associated with the arc. Symbols α and β are weight parameters and represents balance between ants' gathered knowledge and user preferred area. Heuristic values η_{ij} affect probability only at the beginning when pheromone values are low. The more pheromone is located on particular arc, the more attractive it appears.

2. ACO Algorithms with Diversity Control

Parameters for setting the balance between exploration and exploitation belong to the most important variables for majority of heuristic algorithms.

Even author of ACO realized the need for diversity control and introduced Ant Colony System (ACS) [3] which differs from original Ant System in three main aspects: (i) *pseudo-random proportional* rule, (ii) global and (iii) local pheromone update rule.

Pseudo-random proportional rule uses random uniformly distributed variable $q \in (0,1)$ which is compared with a tunable parameter $q_0 \in \langle 0,1 \rangle$ (4).

$$j=\begin{cases}\arg\max_{j\in N_i^k}\{\tau_{ij}(t)\eta_{ij}^{\beta}\} & if \quad q\le q_0\\ J & otherwise\end{cases} \quad (4)$$

where J is a random variable selected according to the probability distribution given by (3). Low q_0 values prefer balanced exploration of new paths.

Since in a *global pheromone update rule* only the *best-so-far* ant is allowed to add pheromone the pheromone value is weighted by pheromone persistence parameter ρ to prevent rapid pheromone accumulation (5).

$$\tau_{ij}(t+1)=(1-\rho)\tau_{ij}(t)+\rho\Delta\tau_{ij}^{bs}(t);\forall(i,j)\in T^{bs} \quad (5)$$

According to *local pheromone update rule* each ant immediately after cross the arc reduces the pheromone value (6). This supports further exploration since the crossed arc becomes less desirable for the following ants.

$$\tau_{ij}(t+1)=(1-\xi)\tau_{ij}(t)+\xi\tau_0 \quad (6)$$

Nakamichi and Arita [4] chose another approach to diversity control. Instead of control diversity in depositing pheromone they control diversity in finding tours. Authors introduced a mechanism of *random selection* in addition to the probabilistic selection. Random selection is a simple operation which selects next node from the list of unvisited neighbors with the equal probability. *Random selection rate* r is probability with which random selection operates each time an ant has to choose the next node and represents a user parameter which adjusts the balance between exploration and exploitation.

Kumar, et al. [5] enriched AS with (i) prevention of quick convergence and (ii) stagnation avoidance mechanisms.

The mechanism for prevention of quick convergence (i) is based on pseudo-random proportional rule (4), but the tunable parameter q_0 is dependent on algorithm iteration (7).

$$q_0=\frac{\log_e(NC)}{\log_e(N_{\max})} \quad (7)$$

N_{max} is the maximum number of iterations and NC is the iteration counter. At the beginning of the search process when q_0 values are low the bias exploration is preferred. In this way a very quick convergence of the algorithm into locally optimized solution is prevented.

The stagnation avoidance mechanism (ii) is based on comparison of random generated quantity $q \in (0,1)$ with probability $p^k_{ij}(t)$ of the selected arc. If $q < p^k_{ij}(t)$, then use the probability selection rule $max(p_{ij}^k(t))$, if $q \geq p^k_{ij}(t)$, then choose the next node randomly. This occurs in later stages of the search process when pheromone values on the most selected arcs tend to be high and thus the chance of further exploration is low. Randomized selection at the later stages of the search process decreases the change of stagnation in local optimum.

Stützle and Hoos [6, 7] applied pheromone limit values in their MAX-MIN Ant system (MMAS). MMAS uses similar update rule to ACS [3] where only one ant is allowed to update the pheromone. But instead best-so-far solution only also iteration-best solution is used in an alternate way. Since the elitist strategy in update rule leads to stagnation in suboptimal solution, MMAS limits possible range of pheromone values to the interval $[\tau_{\min}, \tau_{\max}]$. Pheromone trail limits have the effect of limiting the probability values p_{ij} of selecting arc which favors exploration over exploitation and consecutive stagnation.

Furthermore, the pheromone trails are initialized to upper pheromone limit values, which, together with a small evaporation rate increases the exploration of new paths from the beginning of the search process. In the later phases when the system approaches stagnation (i.e. no improved has been generated for a certain number of consecutive iterations), pheromone values are re-initialized.

3. Pheromone Value Limits

The overall pheromone from all the ants which is accumulated on the graph is described by (2). Let suppose an ideal progress of the algorithm, i.e., from the first iteration all the ants will use the same path. The amount of pheromone accumulated on the arcs of the path is given by (1) and (2):

$$\tau_{ij}(t+1) = (1-\rho)\tau_{ij}(t) + m\frac{Q}{L} \tag{7}$$

Pheromone accumulation over iterations will be:

$$\begin{aligned}
&\tau_{ij}(0) = \tau_0 \\
&\tau_{ij}(1) = (1-\rho)\tau_0 + m\frac{Q}{L} \\
&\tau_{ij}(2) = (1-\rho)^2\tau_0 + (1-\rho)m\frac{Q}{L} + m\frac{Q}{L} \\
&\ldots \\
&\tau_{ij}(t) = (1-\rho)^t\tau_0 + \sum_{e=0}^{t-1}(1-\rho)^e m\frac{Q}{L}
\end{aligned} \tag{8}$$

The pheromone limit value for arc which is never used by any ant (*m=0*) is zero

$$\lim_{t\to\infty}\tau_{ij}(t) = \lim_{t\to\infty}(1-\rho)^t\tau_0 = 0;\ (1-\rho) \prec 1 \tag{9}$$

and the pheromone limit value for arc that is always used by all the ants is infinitive geometric series

$$\begin{aligned}
&\lim_{t\to\infty}\tau_{ij}(t) = \lim_{t\to\infty}\left((1-\rho)^t\tau_0 + \sum_{e=0}^{t-1}(1-\rho)^e m\frac{Q}{L}\right) = \\
&= m\frac{Q}{L}\sum_{e=0}^{\infty}(1-\rho)^e
\end{aligned} \tag{10}$$

of which convergence test is the ratio

$$\left|\frac{a_{n+1}}{a_n}\right| = \frac{(1-\rho)^1}{(1-\rho)^0} = (1-\rho) \prec 1 \tag{11}$$

$$\tau_{ij}(\infty) = m\frac{Q}{L\rho} \tag{12}$$

Simulation of (2) and its limit value is on Figure (1). Three phases of the search process can be distinguished: preliminary (0<t<40), saturation (40<t<80) and stagnation (t<80) phase.

Preliminary phase is characteristic by rapid growth of overall pheromone value. In saturation phase pheromone evaporation takes effect. Stagnation phase represents excessive pheromone values accumulated in the most attractive part of the search area given by local or global optimum.

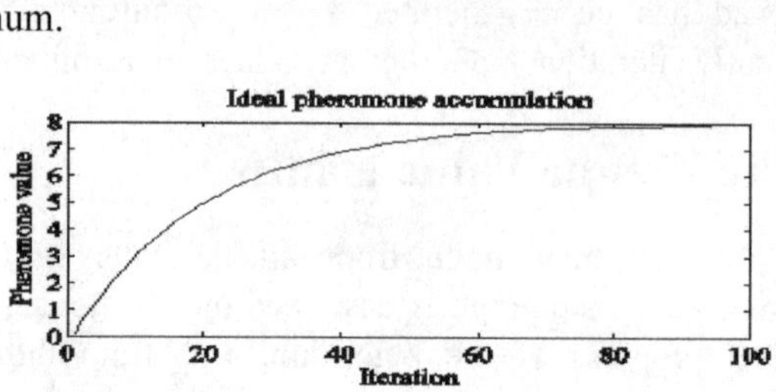

Figure 1. *Simulation of pheromone accumulation (2) for m=2, L=5, Q = 1 and ρ = 0.05.*

4. Re-Initialization

In the stagnation phase pheromone evaporation is in balance with accumulation. Due to pseudo-random proportional rule (4) not all the ants will use the same path. Majority of the ants will be attracted by the same search area with excessive pheromone values and further exploration is hardly possible. For this reason re-initialization is proposed. It is not able to estimate the pheromone limit value by using (12). However, pheromone derivation can be used instead to monitor the search process.

In this work it is proposed to re-start the search process in user adjustable point *x* where pheromone derivation equals pheromone value (13).

$$x\bar{\tau}(t) = \frac{d\bar{\tau}(t)}{dt} \tag{13}$$

At the re-initialization point, excessive pheromone values are decreased with the following requirements: deteriorate the difference between individual pheromone values and pheromone arithmetic mean (i), reduction will be directly proportional to its size (ii) and overall pheromone value will be reduced (iii). In this case a non-linear transformation is used (14).

$$\begin{aligned}
&\tau'_{ij}(t) = \left(s_{ij}(t)d_{ij}(t) + \bar{\tau}(t)\right)rc \\
&s_{ij}(t) = sign\left(\tau_{ij}(t) - \bar{\tau}(t)\right) \\
&d_{ij}(t) = \log_{10}\left(1 + \left|\tau_{ij}(t) - \bar{\tau}(t)\right|\right)
\end{aligned} \tag{14}$$

The overall pheromone value $\bar{\tau}$ should be decreased with respect for the information lost. The aim is to escape excessive pheromone values but preserve acquired knowledge about the problem. For that purpose reducing coefficient *rc* reflects the speed of pheromone accumulation (15).

$$rc = \frac{m\bar{\tau}(t)}{|A|(t-t_r)} \tag{15}$$

$|A|$ is a number of arcs of the graph, t is current iteration and t_r is iteration of the last re-initialization or zero in the beginning. Let call a section between two re-initializations a macro cycle.

The search process is divided into macro cycles. The number of macro cycles as user defined parameter is easier to estimate than number of iterations. It has been shown that low number of macro cycles is sufficient [8].

4.1. Stagnation Avoidance Mechanism

A *pseudo-random proportional* rule with utilization of q_0 variable parameter is used, but instead of the whole search process (7) it should be applied within each single macro cycle.

The simple approach is to create a rank of equally distributed vales for q_0 (16), one for each macro cycle [8]. Since q_0 is compared with probability of choosing particular arc $p^k_{ij}(t)$, the maximum interval span $\langle p_{\min}, p_{maz} \rangle$ for q_0 values falls in range $\langle 0,1 \rangle$. Provided the arc with the highest probability $p^k_{ij}(t)$ should not be always chosen during the first macro cycle (probabilistic selection (3)) and never chosen during the last macro cycle (pseudo-random selection rule (4)), the interval have be reduced according to user defined values.

$$q_0 \in \left(\begin{array}{l} q_{01}, q_{02}, q_{03} \ldots \mid q_{0i+1} - q_{0i} = \Delta; \\ \Delta = \dfrac{p_{\max} - p_{\min}}{N_{mc} - 1} \end{array} \right) \quad (16)$$

N_{mc} is the total number of macro cycles and N is current macro cycle.

In this article it is proposed to apply the similar approach for stagnation avoidance mechanism to original paper [5]. The parameter q_0 will vary within the range in each single macro cycle. The solution (17) is based on the difference (13) and will be close to zero at the beginning and close to one at the end of each macro cycle.

$$q_0 = 1 - \frac{d\bar{\tau}(t)/dt - x\bar{\tau}(t)}{d\bar{\tau}(t_r)/dt} \quad (17)$$

The name of the algorithm is derived from variable parameter q_0 as ACO with variable macro cycle (ACO_{VMC}).

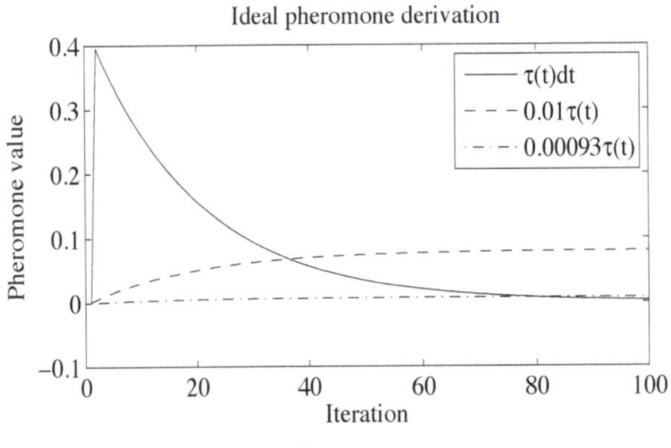

Figure 2. Ideal pheromone derivation

5. Test Case

The aim of the test case is to show the impact of (i) macro cycle length and (ii) q_0 value on algorithm performance as well as comparison ACO with re-initialization with one of the best performing ACO algorithm ACO_{KTS} [5].

Common ACO parameters for both algorithms were set in accordance with [9] and are listed in the Table 1. Other variable parameters like number of ants and number of macro cycles are listed in the result Table 2.

Limits for variable parameter q_0 for pseudo-random proportional rule (16) were set within the interval $\langle 0.1, 0.9 \rangle$.

Variable parameter for macro cycle length x (13) was chosen to cover (i) preliminary phase and (ii) preliminary and saturation phase. The values were estimated as (i) 0.0093 and (ii) 0.00093 respectively (Figure 2).

For each setting 500 trials were performed. At first ACO_{VMC} was tested and mean value of termination cycle $\bar{c}_t$ was determined for each setting. Then $\bar{c}_t$ was considered as number of iterations for ACO_{KTS}.

Test graph is random generated symmetrical multi-graph with 50 nodes and 200 arcs (Figure 3). Node coordinates x,y fall in range $\langle 0,1 \rangle$ and arc values c_{ij} are equal to Euclidean distances between arc nodes (i,j). Arc lengths c_{ij} have been considered as heuristic values η_{ij}. This setting simulates common traffic optimization problem. The task is to find the shortest path between two given nodes - the start node $n_s=2$ and the end node $n_e=31$.

For test evaluation only probability of finding the global optimum ($T = [2\ 26\ 47\ 22\ 18\ 24\ 31]$) was used.

Table 1. *Common ACO parameters settings*

Parameter name	Value
Initial pheromone value $\tau_{ij}(0)$	0.1
Weight of pheromone information α	0.5
Heuristic values η_{ij}	0.1
Weight of heuristic information β	0.1
Pheromone persistence ρ	0.05
Number of ants m	10
Number of cycles	200

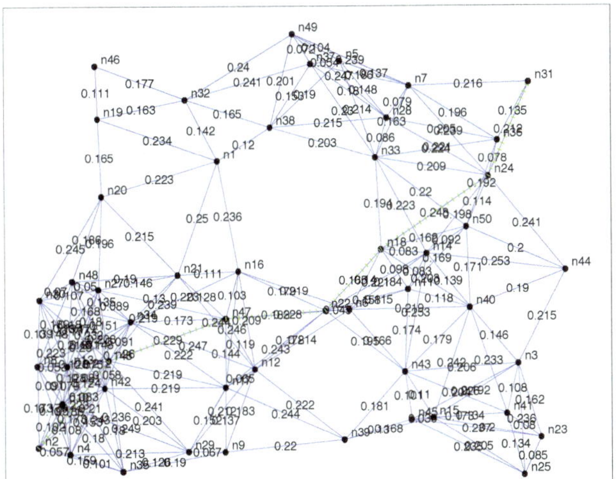

Figure 3. 50 node graph with the shortest path in green.

6. Results

The more resources (ants or macro cycles) are available the higher the probability of finding a global optimum is. Resource allocation between ants and macro cycles depends on graph complexity. The algorithm with macro cycles benefits more from additional ants than from more macro cycles (Table 2) which similar to other ACO algorithms and complies with [9].

6.1. Macro Cycle Length

Extended macro cycle which covers saturation phase has more than twice probability of finding the global optimum in most cases compared to short macro cycle which includes only preliminary phase (Table 2, columns 3, 4). Since most of the pheromone is already accumulated at the end of the preliminary phase (Figure 1), mainly re-distribution of trailing pheromone occurs in the later stages of the search process. Pheromone concentration into particular area of the search space is caused by further utilization of acquired knowledge in stagnation phase.

Depending on parameter x (13) setting for macro cycle length, extended macro cycle can be twice longer (Figure 1). This results in twice longer overall search process. Hoverer, ACO_{VMC} results comparison four short macro cycles with two long macro cycles reveals higher performance of long macro cycles in each case.

6.2. q_0 Parameter

Results show better performance of ACO with variable macro cycles (ACO_{VMC}) over constant macro cycles (ACO_{MC}) (Table 2, columns 4, 5). This can be explained by graduate values of q_0 parameter (17) in stagnation avoidance mechanism which disturbs autocatalytic process mainly during the saturation phase. This results weaker selection pressure which favors exploration and better local optimization during saturation phase.

6.3. ACO_{VMC} and ACO_{KTS}

Comparison between ACO with variable macro cycles and ACO_{KTS} (Table2, columns 4, 6) on sample graph reveals that results are varying and fully comparable. In general, test data shows slightly better performance of ACO_{VMC} for more resources (ants or macro cycles).

Table 2. Simulation results on 50 node graph.

ants	macro cycles	Probability of finding the global optimum [%]			
		ACO_{VMC} x = 0.01	ACO_{VMC} x = 0.00093	ACO_{MC} x=0.00093	ACO_{KTS}
2	2	5.2	14.8	9.8	11.4
2	4	12.8	18.4	20.4	28.2
2	6	14.4	34.2	29.2	42.2
4	2	13.2	27.2	23.2	31.4
4	4	20.6	49.4	37.6	49.4
4	6	31.2	65.2	53.2	62.6
6	2	19.4	45.2	35.2	41.8
6	4	32.4	65.6	54.6	63
6	6	46.6	80.6	68.4	78.8

7. Conclusion

It has been proved on sample test that ACO with long variable macro cycles is fully comparable with one of the best performing ACO adaption of Kumar et al. (2003). The ACO with re-initialization benefits from (i) parameter representing number of macro cycles which is easier to estimate than number of iterations, (ii) reducing overhead caused by saturation phase of the search process and (iii) self-termination ability.

The above mentioned advantages enable implementation of heuristic algorithms for automatic optimization in manufacturing systems especially in dynamic environment where good solution has to be delivered in real time.

Acknowledgments

Thanks to Science Publishing reviewers for valuable feedback and provided comments which increased the paper quality.

References

[1] P. E. Hart, N. J. Nilsson and B. Raphael, A Formal Basis for the Heuristic Determination of Minimum Cost Paths. IEEE Transactions on Systems Science and Cybernetics SSC4 4(2), 1968, 100–107

[2] M. Dorigo, V. Maniezzo and A. Colorni, The Ant System: An autocatalytic optimizing process, Technical report 91-016 revised, Dipartimento di Elettronica, Politecnico di Milano, Milan, 1991

[3] M. Dorigo and L. M. Gambardella, Ant Colony System: A cooperative learning approach to the traveling salesman problem. IEEE Transactions on Evolutionary Computation, 1997, 1(1), 53–66,

[4] Y. Nakamichi and T. Arita, Diversity control in ant colony optimization, In Abbas HA (ed) Proceedings of the Inaugural Workshop on Artificial Life (AL'01), Adelaide, Australia, Dec 11, 2001, 70-78

[5] R. Kumar M. K. Tiwari and R. Shankar, Scheduling of flexible manufacturing systems: an ant colony optimization approach, proc. Instn. Mech. Engrs Vol. 217 Part B: J. Engineering Manufacture, 2003, 1443–1453

[6] T. Stützle and H. H. Hoos, Improving the Ant System: A detailed report on the MAX-MIN Ant System. Technical report AIDA-96-12, FG Intellektik, FB Informatik, TU Darmstadt, Germany, 1996

[7] T. Stützle and H. H. Hoos, MAX-MIN Ant System. Future Generation Computer Systems, 2000, 16(8), 889–914

[8] M. Ciba, ACO algorithm with macro cycles, Proceedings on 14th Conference of Doctorial Students on Elitech'12, Slovak Technical University of Bratislava, May 2012

[9] M. Becker and H. Szczerbicka, Parameters influencing the performance of ant algorithms applied to optimization of buffer size in manufacturing, IEMS Vol. 4, No. 2, December 2005, 184–191

11

Robust generalized minimum variance controller using neural network for civil engineering problems

L. Guenfaf[1,*], M. Djebiri[2], M. S. Boucherit[2], F. Boudjema[2]

[1]LSEI Laboratory, USTHB University BP 32 El Alia 16111, Bab Ezzouar, Algiers, Algeria
[2]LCP, Laboratory, ENP, Hassan Badi El harrach, Algiers, Algeria

Email address:
Lakhdar.guenfaf@yahoo.fr (L. Guenfaf)

Abstract: This paper presents a robustness of the proposed generalized minimum variance algorithm. The main idea is to use artificial neural network for generalization of the GMV. This will give a neural network-based control method wich can be applied to civil engineering structures. The neural network learns the control task from an already existing controller, which is the generalized minimum variance (GMV) controller. The objective is to take advantage of the generalization capabilities and the nonlinear behavior of neural networks in order to overcome the limitations of the existing controller and even to improve its performances. Simulation results demonstrate the robustness of this algorithm and its capability to compensate the structural parameter variations and seismic ground motion.

Keywords: Structural Control, Neural Networks, Generalized Minimum Variance Control

1. Introduction

Structural control is an expanding field of study since the increasing need to save human being lives and reduce structural damage [18]. For this purpose, many active control techniques have been investigated and applied to structural systems in order to reduce their dynamic response under earthquake excitation. But most of these techniques suffer one or more of the following disadvantages: (1) the control law is usually designed assuming a linear or a linearized behavior around a desired state of the structure being controlled, (2) structural parameters are supposed to remain constant along all the control time history, i.e., the structure suffers no damage or deterioration, (3) the control is designed for a specific type of earthquake excitation. If the above mentioned statements are verified, control efficiency will be ensured in most cases. However, in practical situations structures may exhibit a nonlinear behavior caused either by large displacements or material nonlinearity and damage. In other hand, structural model uncertainties and parameter variations, especially under severe earthquakes, are very probably to occur. These practical considerations will limit the effectiveness of the control system and performances of the control technique will be severely affected.

Faced with this kind of problems, civil engineers have turned to a more powerful tool which can perform control actions while compensating for unpredictable environmental changes. Neural networks have demonstrated their abilities to accomplish this task. They become a challenging alternative to solve specific structural problems, handling their attractive features such as nonlinearity, parallel processing, learning and generalization capabilities. The ability of neural networks to approximate arbitrary nonlinear mappings makes them of great deal in control of nonlinear systems.

In the context of structural control, neural networks have been used since the late 1980s for identification and control of structures [13]. Chassiakos et al [7] used neural networks to identify multi-degree-of-freedom systems with unknown parameters under earthquake excitation. In [11] a multilayer feedforward neural network is used and trained by Back-propagation to emulate seismic response of a two-story building. An indirect predictive learning control scheme for control of large space structures was investigated in [23]. Venini and Wen [21] used a neural network to approximate the inverse dynamics of a multi-degree-of-freedom structure. The trained network is used to control the structure by a single actuator through a hybrid control scheme. In [13] Ghaboussi et al presented a linearly trained neurocontroller for the control of linear structures when the structural response remained within the linearly elastic range. Recently, Bani-Hani et al [6] extended the method of Ghaboussi et al to nonlinear structural control problems. They demonstrated that the nonlinearly trained neuro-controller was able to reduce the structural damage more than the linearly trained neuro-controller, but globally the two controllers have comparable performances.

In [4] the authors are interested to minimize the norm of the nominal control sensivity transfer function in the condition of non minimum phase system. The stochastic polynomial C is chosen by the authors and equal to one in the case study of the paper. Also in an other paper [5] the controller is calculated using the variance of the output with and without tracking polynomial. The chosen parameter controller is done according to the norm of the output in optimal and suboptimal cases. Whereas Grimble [17] develop the non linear generalized minimum variance for MIMO systems in general case without specification on the parameter variations. Also nonlinear smith predictor is used for implementation in the case of open loop stable systems. The authors in [19] present the generalized minimum variance for noisy free systems and STR algorithm for AR systems. The polynomial C is then chosen by the authors. In our approach we are interested to maintain the stochastic behavior of the system charatirized by soil structure interaction. The closed loop polynomial is the same as that one of the exogen input to insure the optimal parameter of the controller. By using the neural network, we introduce the stochastic behavior in the non linear new model. This one can occurs in the generalized case by operating in the non training data.

In this paper, we first investigate the so-called generalized Minimum Variance (GMV) algorithm for buildings under earthquake excitation. The GMV algorithm attempts, by using a certain model of the system to be controlled and its perturbations, to minimize a generalized cost function including output variance and control effort. The GMV algorithm has been widely studied in literature [8,9,14] and has demonstrated good tracking and regulation performances. But these performances, as we will see, are closely related the knowledge of the ARMAX (Auto-Regressive Moving Average eXogen) model of the system to be controlled since the control strategy is derived in the base of this model. Therefore if the ARMAX model changes by changing the seismic excitation model or by structural parameter variations, control performances may be questionable. In the classical control theory we can use the adaptive approach witch need more time calculation and fast sampling time.

To overcome this problem, we will investigate, in this approach, the use of a neural network that learns to control a single-degree-of-freedom (SDOF) structural building from the GMV controller. The neural network controller will acquire control skills from the GMV controller in one hand, Iin the other hand it will use the generalization capabilities of neural networks to overcome the model-based restrictions of the GMV control technique mentioned above. We have tested the robustness of the resulted algorithm against parameter variation of the model. It has also been showed that the seismic exciataion has been taken in charge by the proposed algorithm with out any change in the parameter of the exogen model.

The approach is to avoid the development of robust controller with complicated calculation. The idea is to use neural ANN to have a profit of the generalization behviour and parameter variation. We will see that the presented controller can maintain the closed loop performances more that the classical GMV. The robustness is inside the ANN that gives us more unknow information that we have not taken into consideration in the development of the control law. But changing the excitation signal, we are changing the parameter of the polynomial C which affect the closed loop performances. The main result that we are not using an adaptive algorithm which take much calculation time against our algorithm.

2. Dynamical Model of the Structure

In this section, we are interested in formulating the dynamical equations of motion of a single-degree-of-freedom structure under seismic excitation. where mechanical actuators are active tendons. In order to establish the dynamical model, the following assumptions are considered [10]:

1. the structure is supposed to be a lumped mass m in the girder
2. the two vertical axes are weightless and inextensible in the vertical direction with spring constant k/2 each.

After calculation we can Easily derive the dynamical model of the structure[13]:

$$m\ddot{x}(t)+c\dot{x}(t)+kx(t)=u(t)-m\ddot{x}_g(t) \tag{1}$$

where u(t) is the external control force.
m is the structural mass;
c is the internal viscous damping of the structure;
k is the elastic stiffness;
x(t) is the relative displacement;
xt(t) is the absolute displacement defined as
xg(t) is the ground motion;
$\ddot{x}_g(t)$ represents the ground acceleration.

This model will be used for testing the proposed algorithm. The parameter of this one is presented in the next sections of this paper.

We can see that the variation of the mass affect directly the model of the system whereas the seismic excitation do not.

3. ARMAX Model of the Structure

To formulate an optimal control problem, it is necessary to specify the process dynamics and its environment. It is assumed that the influence of the environment on the process can be characterized by disturbances, which are stochastic process. As the system is linear, we can use the principle of superposition and represent all disturbances as a single disturbance acting on the output. It is assumed that this disturbance is a stationary Gaussian process with rational spectral density.

The calculation of the ARMAX model of the structure under seismic excitation can be done using equation of motion (1). After dividing this equation by m and introducing the notations

$$\omega_0^2 = \frac{k}{m} \text{ natural frequency}$$

$$\xi = \frac{c}{2m\omega_0} \text{ damping ratio}$$

Equation (1) becomes:

$$\ddot{x}(t) + 2\xi\omega_0\dot{x}(t) + \omega_0^2 x(t) = \frac{1}{m}u(t) - \ddot{x}_g(t) \quad (2)$$

Now applying Laplace transform to equation (2), we obtain

$$X(s) = \frac{1/m}{s^2 + 2\xi\omega_0 s + \omega_0^2}U(s) - \frac{1}{s^2 + 2\xi\omega_0 s + \omega_0^2}\ddot{X}_g(s) \quad (3)$$

where $X(s)$, $\ddot{X}_g(s)$ and $U(s)$ are the Laplace transform of $x(t)$, $\ddot{x}_g(t)$ and $u(t)$ respectively; s is the Laplace operator. Figure 1 shows the bloc diagram of the structural model

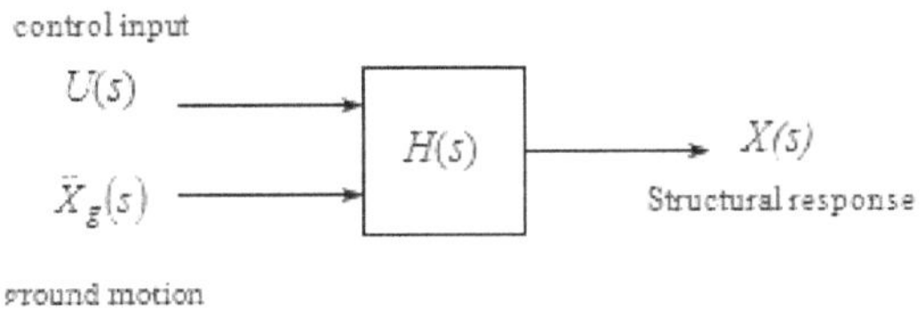

Figure 1. *Bloc diagram of the structural model.*

The ARMAX model of the structure is obtained by discretization of equation (3). We obtain [15,12]:

$$x(t) = \frac{B(q^{-1})}{A(q^{-1})}u(t) + \frac{C(q^{-1})}{A(q^{-1})}\ddot{x}_g(t) \quad (4)$$

where

$$A(q^{-1}) = 1 + a_1 q^{-1} + a_2 q^{-2}$$

$$B(q^{-1}) = b_1 q^{-1} + b_2 q^{-2}$$

$$C(q^{-1}) = c_1 q^{-1} + c_2 q^{-2}$$

q^{-1} shift operator defined as $q^{-1}y(t+1) = y(t)$

The ground acceleration is described by the Kanai-Tajimi model, i.e

$$\ddot{X}_g(s) = G_1(s)E(s) \quad (5)$$

Where

$$G_1(s) = \frac{G_{1N}(s)}{G_{1D}(s)} = \frac{2\xi_g\omega_g s + \omega_g^2}{s^2 + 2\xi_g\omega_g s + \omega_g^2}$$

E(s) is the Laplace transform of white noise.

Discretization of equation (5) gives the discrete ARMAX model of the structure under the Kanai-Tajimi ground acceleration. We can see that the parameter of the polynomial C depend directly on those of the structure. The main result here is the excitation signal in the model. So when we change the seismic ground motion, the final model will change as parameter variation. This affects directly the parameter of the exogen input represented by the polynomial C.

4. Generalized Minimum Variance Controller

The Generalized Minimum Variance (GMV) algorithm was introduced by Clarke [8,9] to control non-minimum phase systems. It is an extension of the Minimum Variance algorithm [1,2] which, by choosing a certain performance criterion, attempts to minimize the variance of the output.

The ARMAX (Auto-Regressive Moving Average eXogen) model of the system is used

$$A(q^{-1})y(t) = q^{-d}B(q^{-1})u(t) + C(q^{-1})e(t) \quad (6)$$

where

$$A(q^{-1}) = 1 + a_1 q^{-1} + \ldots + a_n q^{-n}$$

$$B(q^{-1}) = b_1 q^{-1} + \ldots + b_m q^{-m}$$

$$C(q^{-1}) = 1 + c_1 q^{-1} + \ldots + c_l q^{-l}$$

noted A, B and C

$d \geq 0$ is the time delay of the system

y(t) process output

u(t) control

e(t) white noise with zero mean and of variance σ^2.

The polynomial C is stable.

The performance index to be minimized is

$$J = E\left[(Py(t+d+1) - R_w w(t+d+1))^2 + (Q'u(t))^2\right] \quad (7)$$

where

E mathematical esperance

$w(t+d+1)$ reference signal

$P(q^{-1})$, $R_w(q^{-1})$ and $Q'(q^{-1})$ weighing polynomials with $P(q^{-1}) = P_N(q^{-1}) / P_D(q^{-1})$ and $Q'(q^{-1}) = Q'_N(q^{-1}) / Q'_D(q^{-1})$.

The degrees of P and Rw can be chosen arbitrarily. We remark in this criterion that w(t+d+1) is a disposable information, but y(t+d+1) is not. It is a future information that we must predict. After some calculations, we derive the GMV control strategy given by [15,16]:

$$u(t)=\frac{P_D C R_w w(t+d+1)-Ry(t)}{P_D(S+QC)} \tag{8}$$

Where

$$Q(q^{-1})=\frac{q'_{N_0} p_{D_0}}{q'_{D_0} p_{N_0} b_1} Q'(q^{-1})$$

q'_{N_0}, q'_{D_0}, p_{D_0} and p_{N_0} are the first coefficients of the polynomials Q'_N, Q'_D, P_D and P_N respectively.

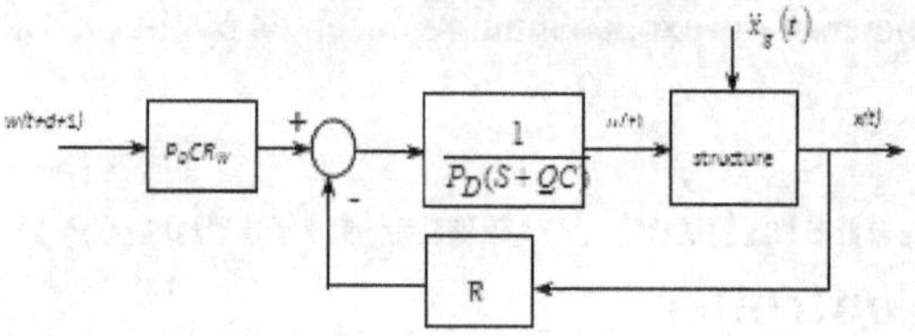

Figure. 2. Generalized Minimum Variance Control architecture.

It is seen from the previous equations that the performance of the GMV control is closely related to the accuracy of the ARMAX model of the system to be controlled. Therefore, if the ARMAX model used to derive the control law differs from the true ARMAX model of the system, control objectives will not be guaranteed. This can be the case in practical situations where structural parameters change if the building undergo damage or deterioration. Also, if the GMV algorithm is initially designed for a specific model and magnitude of the seismic excitation, its ability to reduce the structural response for a different excitation model is uncertain.

The calculation of the control depends on the polynomial C. This polynomial corresponds to the dynamical model as we can see in the Diophantine equation.

4.1. Closed Loop Analysis

To evaluate the regulator performances, we have to calculate the closed loop transfer function. From equations 8 and 6 we obtain

$$y(t)=\frac{P_D R_w}{P_N} w(t)+\frac{P_D S'}{P_N} e(t) \tag{9}$$

The difference between the output and the obtained model is a noise defined as

$$\xi_1(t)=\frac{P_D S'}{P_N} e(t) \tag{10}$$

Let

$$P'_D = P_D S' = P'_{D_0} + P'_{D_1} q^{-1} + \ldots + P'_{D_{nd+d}} q^{-(n_d+d)}.$$

Equation (10) becomes

$$\xi_1(t)=-\frac{1}{P_{N_0}}\left[\sum_{i=1}^{n_N} P_{N_i}\xi_1(t-i)+\sum_{i=0}^{n_D+d} P'_{D_i} e(t-i)\right] \tag{11}$$

then

$$E[\xi_1^2(t)] \le \left[1-\frac{1}{P_{N_0}^2}\left(\sum_{i=1}^{n_N} P_{N_i}^2\right)\right]^{-1}\left(\sum_{i=0}^{n_D+d} P'^2_{D_i}\sigma^2\right) \tag{12}$$

if

$$1-\frac{1}{P_{N_0}^2}\left(\sum_{i=1}^{n_N} P_{N_i}^2\right)>0$$

In this case, we have evaluated the maximum of the output variance, which is the closed loop performance (in the stochastic sense). The closed loop transfer function (between the output and the reference $w(t)$) is

$$F_{BF}(q^{-1})=\frac{P_D R_w}{P_N} \tag{13}$$

So the choice of weighing polynomials P and R_w define the reference model. The optimal control is achieved by choosing the closed loop dynamic as C polynome.

5. The Neural Network Controller

The Unlike the conventional control algorithms where the control task is formulated explicitly, in the neural network-based structural control methods the neurocontroller learns the control task. The neurocontroller acquires the knowledge of structural control from a set of training cases or an existing controller and stores it in the connection weights[23].

The aim of this section is to use a neural network that replaces the GMV control algorithm. The neural network learns control actions from data generated when the GMV algorithm is controlling the structural system. By using its learning and generalization capabilities the neural network will attempt to perform well and to overcome some of the restrictions of the GMV algorithm.

The overall control is implemented in two stages:

1. Training stage (1): in this stage, the neural network learns the function input/output of the GMV controller. When this later controls the structural system, training examples are generated. Then a supervised learning algorithm is used in order to train the neural network.
2. Control stage (2): after training, the network controls the structural system and the GMV controller is removed.

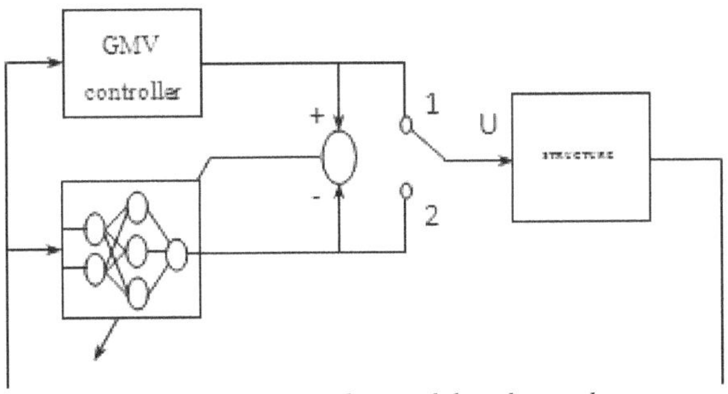

Figure 3. The neural network-based control.

5.1. Training Algorithm

The neural network training method used in this study is the well-known Backpropagation algorithm. It is based on the minimization of a quadratic cost function of the error between the desired output and the neural network output. This is done by continuously changing the neural network weights in the direction of the steepest descent of the error cost function.

The output of a unit in the output or hidden layer Oj is given by the following equation:

$$O_j = f(net_j) \\ net_j = \sum_i w_{ji} O_i \tag{14}$$

where

net_j is the weighted sum of the outputs of the previous layer

w_{ji} is the connection weight between the ith node of the previous layer and the jth node of the present layer

f(.) denotes the activation function of the nodes. In this study, we have used the tangent hyperbolic function

The error function to be minimized at each iteration of the Backpropagation algorithm is defined by

$$E_p = \frac{1}{2}\sum_k (y_{pk} - O_{pk})^2 \tag{15}$$

where

O_{pk} is the output of the network

y_{pk} is the desired output

the index k ranges over all the nodes of the output layer

the index p denotes the training pattern p from the set of training data.

The weight w_{ji}, which could belong to any layer of the network, is adjusted so as to minimize E_p, according to the following equation

$$w_{ji} = w_{ji} - \eta \frac{\partial E_p}{\partial w_{ji}} \tag{16}$$

where η is the learning rate.

5.2. Network Architecture

In order to implement the control method described above, we have used a three-layered feedforward neural network consisting of 8 inputs, 10 hidden units and 1 output. Figure 4 shows the architecture of the neural network.

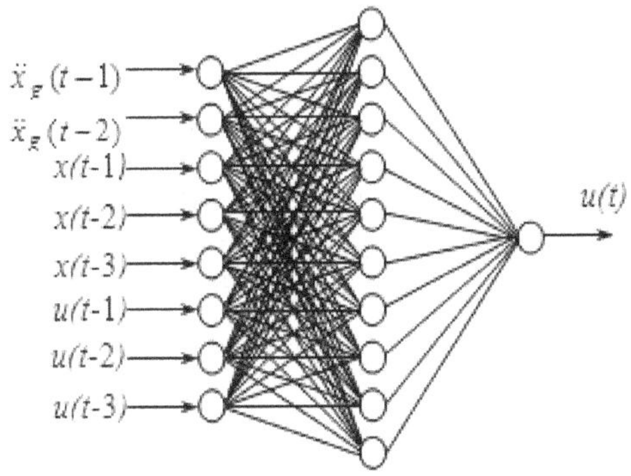

Figure 4. Neural network architecture.

The input units represent the ground acceleration at two past time steps, the relative displacement at three past time steps and the control signal at three past time steps. The output unit represents the control to be sent to the actuator. The number of past time steps for each variable included in the input layer is chosen so as to provide the neural network with sufficient amount of information to implement the GMV control law and to take into consideration environmental changes. This choice is generally based on intuition and trial and error [6]. There are no systematic methods for determining these numbers.

5.3. Generation of Training Data

The set of training data is generated from the GMV control of the structural system. The Kanai-Tajimi excitation model is used in the derivation of ARMAX model of the structure and the Kanai-Tajimi ground acceleration is used as an excitation in the base of the structure. 150 examples were selected from the 30 s control time history to train the network with Backpropagation algorithm.

We deduce that the training has been done on one stochatic model C. Any change of the stochastic model affect directly the regulator parameter for the classical algorithm.

6. Mathematical Model of Earthquake Ground Motion

The earthquake ground acceleration is modeled as a uniformly modulated non-stationary random process [10,22]

$$\ddot{x}_g(t) = \psi(t)\ddot{x}_s(t) \tag{17}$$

where $\psi(t)$ is a deterministic nonnegative envelope function and $\ddot{x}_s(t)$ is a stationary random process with zero mean and a Kanai-Tajimi power spectral density

$$\phi_g(\omega)=\left[\frac{1+4\xi_g^2\left(\frac{\omega}{\omega_g}\right)^2}{\left(1-\left(\frac{\omega}{\omega_g}\right)^2\right)^2+4\xi_g^2\left(\frac{\omega}{\omega_g}\right)^2}\right]S_0^2 \quad (18)$$

Where: ξ_g, ω_g are filter parameters and S_0 is the constant spectral density of the white noise.

However, it can be shown that the velocity and displacement spectra, which are derived from the acceleration spectra that are described by equation (17), have strong singularities at zero frequency. These singularities can be removed by using high-pass filter, as suggested by Clough-Penzien [10]. Using such a second high pass filter, the Kanai-Tajimi spectrum is modified as follows to obtain the Clough-Penzien spectrum

$$\phi_c(\omega)=\left[\frac{1+4\xi_g^2\left(\frac{\omega}{\omega_g}\right)^2}{\left(1-\left(\frac{\omega}{\omega_g}\right)^2\right)^2+4\xi_g^2\left(\frac{\omega}{\omega_g}\right)^2}\right]\left[\frac{\left(\frac{\omega}{\omega_c}\right)^4}{\left(1-\left(\frac{\omega}{\omega_c}\right)^2\right)^2+4\xi_c^2\left(\frac{\omega}{\omega_c}\right)^2}\right]S_0^2 \quad (19)$$

A particular envelope function $\psi(t)$ given in the following will be used

$$\psi(t)=\begin{cases}0 & \text{for} \quad t<0\\ \left(\frac{t}{t_1}\right)^2 & \text{for} \quad 0\le t\le t_1\\ 1 & \text{for} \quad t_1\le t\le t_2\\ \exp[-a(t-t_2)] & \text{for} \quad t\ge t_2\end{cases} \quad (20)$$

where t_1, t_2 and a are parameters that should be selected appropriately to reflect the shape and duration of the earthquake ground acceleration.

Numerical values of parameters are [15]:

t1=3s, t2=13s, a=0.26, ξg=0.65, ωg=19rad/s, ξc=0.6, ωc=2rad/s, S0=0.8 10-2m/s.

The Kanai-Tajimi and Clough-Penzien ground accelerations have been simulated for yhe excitation of the structure. According to the parameter of this model we can determine the discrete model. So we obtain one valid regulator for the chosen structure.

But using neural network algorithm, we can obtain an infinite kind of stochatic regulators from the ANN generalization.

7. Simulation Results

In order to demonstrate the efficiency of the neural network-based controller and show its superiority in comparison with the GMV controller, a single-degree-of-freedom structure with the following structural properties [20] m=2921Kg, k=1389kN/m, ξ=0.0124 is used. An active tendon controller is installed in the story unit and the angle of incline of the tendons with respect to the floor is 25°. Thus, the control force vector from the controller is u/cos25°. Thus we can suppose that the force is applied at the top of the structure and assumed to be activated externally by an independent power supply. Figure 5 shows the open response for the seismic excitation.

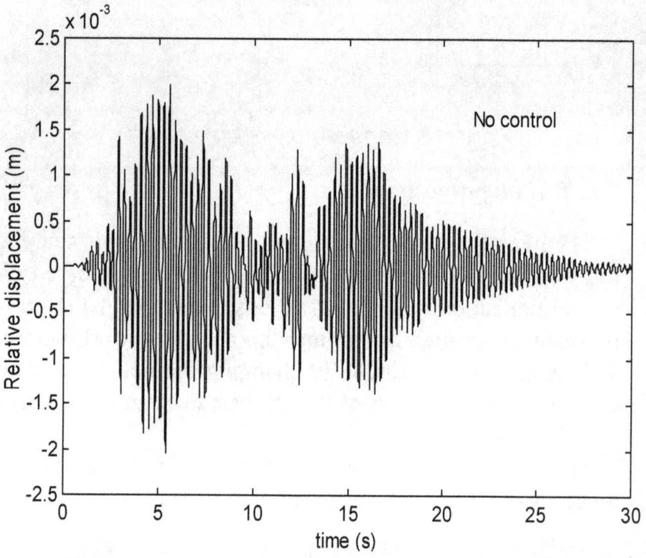

***Figure 5**. Open loop response for ground accelerations.*

To implement the GMV algorithm we have used a sampling period T_s=0.02s. Ponderation polynomials used in the GMV algorithm are:

$$P_D(q^{-1})=1,\ P_N(q^{-1}) = 1 - 0.5q^{-1},\ Q(q^{-1})= 10^{-8}$$

We have used to train the neural network a learning rate η of 0.2 for the first layer and 0.1 for the second layer. Weights were arbitrary initialized between –0.5 and 0.5. Training took around 2500 cycles to achieve an acceptable error.

The building was subjected to excitations of Kanai-Tajimi model and a parameter variation (diminution of the structural mass m of 20% at t=10s) for purposes of comparison. Figures 6 and 7 show the structural response for seismic escitation amplitude of 200%. Whereas in figure 8, we show that the classical algorithm can not maintain its performances. But in figure 9 we see the robustness of the proposed algorithm and the response is stable with the same calculated nework without any changes. Table I gives the output variance for each case.

***Table 1** Output variance for different cases.*

	Kanai-Tajimi model	200% of Kanai-Tajimi	Diminution of m
No control	$3,4016.10^{-7}$	$1,3606.10^{-6}$	$2,5232.10^{-7}$
GMV	$1,6286.10^{-10}$	$6,5143.10^{-10}$	Divergence
ANN	$1.7129.10^{-10}$	$7.0974.10^{-10}$	$1.7705.10^{-11}$

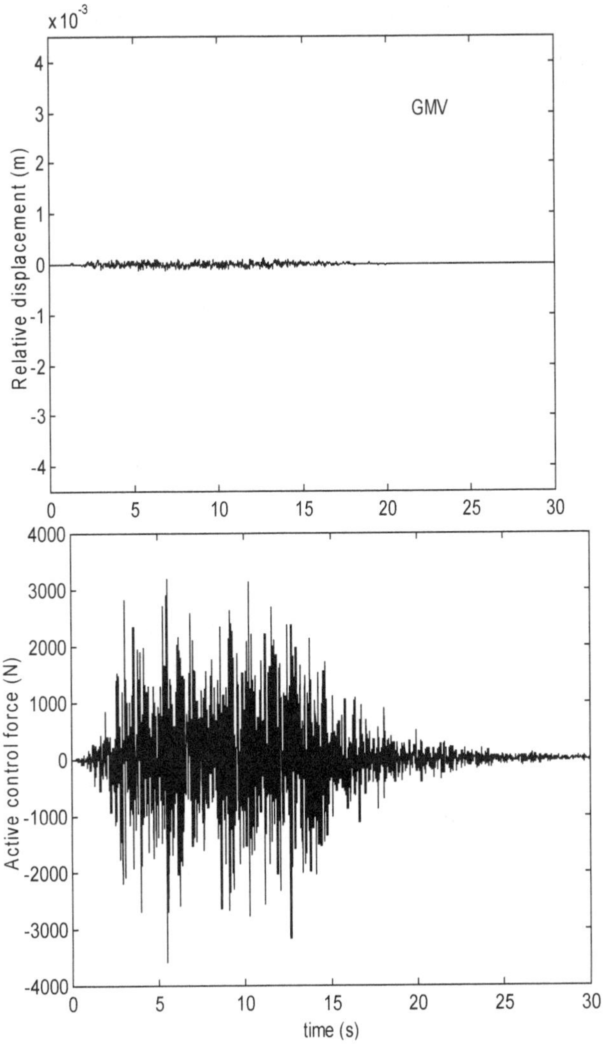

Figure 6. *Classical GMV Structural response to 200% of Kanai-Tajimi ground acceleration.*

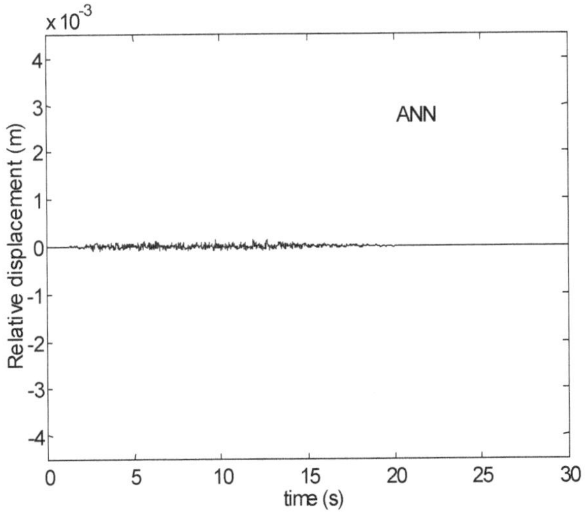

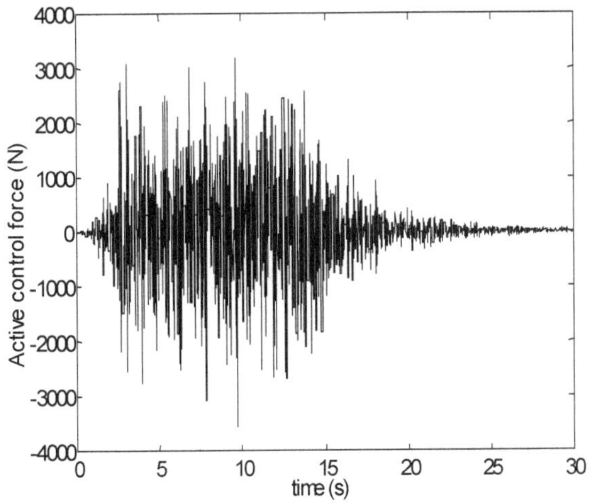

Figure 7. *NN GMV approach Structural response to 200% of Kanai-Tajimi ground acceleration*

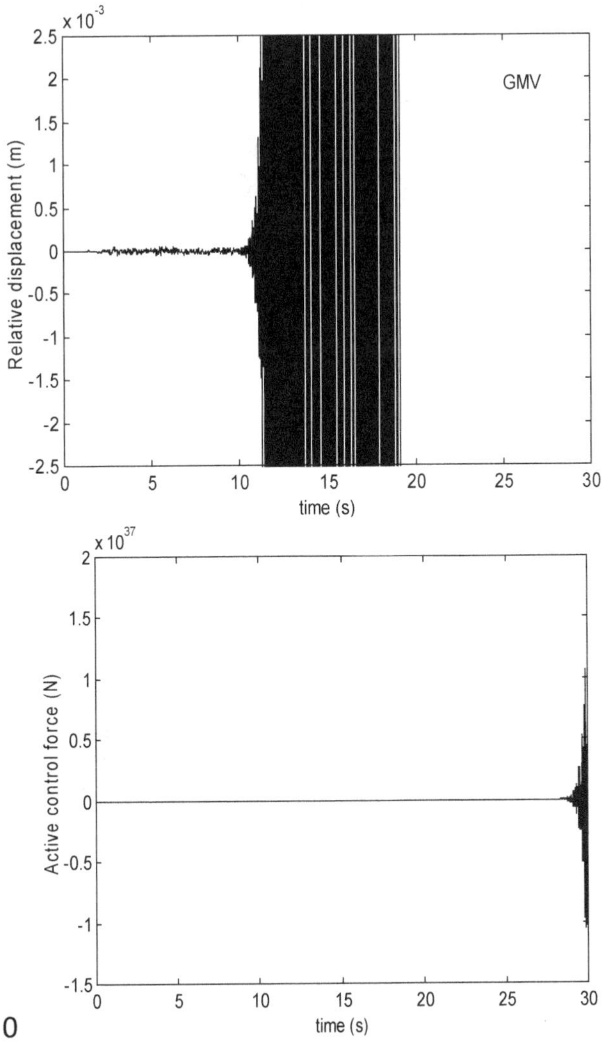

0

Figure 8. *Classical GMV Structural response to Kanai-Tajimi ground acceleration with structural parameter variation.*

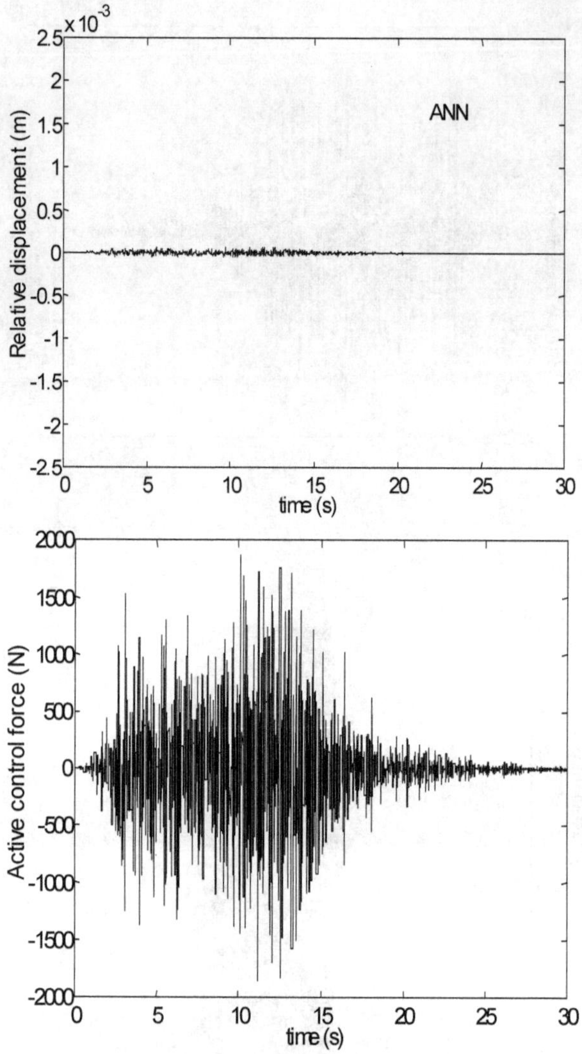

Figure 9. ANN GMV Structural response to Kanai-Tajimi ground acceleration with structural parameter variation.

It is easily seen that the neural controller has good performances comparable to those of the GMV controller. The neural controller was able to reduce the structural response with only a limited amount of control effort. But the important result here is that the neural controller could compensate for structural parameter variation whereas the GMV controller performances are severely degraded and the control showed to be instable in this case. This result shows the robustness of the proposed algorithm against this kind of variations.

8. Conclusions

A neural network-based controller was used in this paper to overcome limitations of the classical GMV algorithm. Also to ovoid the use of adaptive algorithm which needs more stability analysis and more time calculation. Despite that the neural network was trained on a specified data including only the Kanai-Tajimi earthquake excitation model, it was able to generalize to non trained data and showed to have good performances with another types of seismic excitation models. It has been shown also that the neural controller uses its generalization capabilities to compensate for structural parameter variation and maintains its good performances along all the control time history, whereas the GMV control performances become very bad since structural parameter variation will modify the ARMAX model of the structure.

As a result, we can say that the neural network acquired control skills from the GMV controller, while using its intrinsic capabilities to positively interact with the environmental changes. We obtain a robust neural network GMV algorithm witch give much better result than the classical one. The aim was to present a simple robust algorithm based on the stochastic control theory for many kind of seismic excitations. According to the seismic excitation amplitude, it is possible to develop many controller for each excitation and then to choose the corresponding one. The calculation of one ANN controller is sufficient to maintain the performance of the closed system without any other calculation or using adaptive control theory. In our case we can calculate the ANN parameter once and then to implement this one in parallel calculators as transputer or equivalent circuits. These kind of solution is more convenient in seismic application for civil engineering.

References

[1] K. J. Åström, "Introduction to stochastic control theory", Academic Press, 1970.

[2] K. J. Åström, "theory and applications of self tuning regulators", Automatica, 13(99), 457-476.

[3] Pergamon press, (1977).K. J. Åström, "Computer-controlled systems", Prentice-Hall, Inc., 1990.

[4] M.M. Al Imam and M. M. Mustapha,Robust minimum variance controller using over parametrized controller; ARPN Journal of Engineering and applied Sciences, vol 4; NO 10,pp11-18 December 2009.

[5] M.M. Al Imam and M. M. Mustapha, A Robust Controller for output variance reduction and minimum with application on permanent Field DC Motor; world Academy of science and technology, pp703-708, December 2009.

[6] K. Bani-Hani and J. Ghaboussi, "Nonlinear structural control using neural networks", J. eng. mech., ASCE 124(3), 319-327 (1998).

[7] A. G. Chassiakos and S. F. Masri, "Modelling unknown structural systems through the use of neural networks", Earthquake eng. struct. Dyn. 25, 117-128, (1996).

[8] D. W. Clarke, "Introduction to self tuning controller", IEE, Control Eng. Series 15, H. Nichelson and B. H. Swanick, "self tuning and adaptive control : theory and application", Peter pregrinus, 1981.

[9] D. W. Clarke, "Self tuning control of non-minimum phase systems", Automatica 20(5), 501-517, (1984).

[10] R. W. Clough and J. Penzien, "Dynamics of structures", McGraw-Hill, New York, 1993.

[11] J. P. Conte, A. J. Durrani and R. O. Shelton, "Response emulation of multistory buildings using neural networks", Proc. First World Conf. Struct. Control, WP1, 59-68, (1994).

[12] P. J. Gawthrop, "Some interpretations of self tuning controller", Proc. IEE 124(10), 889-894, (1977).

[13] J. Ghaboussi and A. Joghatie, "Active control of structures using neural networks", J. eng. mech., ASCE 121(4), 555-567, (1995).

[14] L. Guenfaf, N. Bali, R. Illoul and M. S. Boucherit, "Performances study of multirate generalized minimum variance algorithm applied to robotic manipulators", ITHURS'96, International conference, ENG'95 with IEEE.

[15] L.Guenfaf, "On the use of automatic control and neural networks in structural dynamics," Phd Thesis, 2001, LCP, ENP, Algiers, Algeria.

[16] L.Guenfaf and Al, "Generalized minimum variance control for buildings under seismic ground motion", Earthquake Eng. Strut. Dyn., Vol 30, Issue 7, pp945-960, 2001.

[17] M.J. Grimble Non-linear generalized minimum variance feedback, feedforward and tracking control Automatica 41 (2005) pp 957 – 969 Elsevier.

[18] G. W. Housner and Al, "Structural control: past, present and future", J. eng. mech., Special issue ASCE 123(9), 897-970, (1997).

[19] A. Patetea, K. Furuta, M. Tomizuka Self-tuning control based on generalized minimum variance criterion for auto-regressive models Automatica 44 (2008) pp1970–1975.

[20] H. A. Smith, W. -H. Wu and R. I. Borja, "Structural control considering soil-structure interaction effects'" Earthquake eng. struct. Dyn. 23, 609-626, (1994).

[21] P. Venini and Y.-K. Wen, "Hybrid vibration control of MDOF hysteretic structures with neural networks", Proc. First World Conf. Struct. Control, TA3, 53-6, (1994).

[22] J. N. Yang, A. Akbapour and P. ghaemmaghami, "New optimal control algorithms for structural control", J. eng. mech. ASCE 113(9), 1369-1386, (1987).

[23] G. G. Yen, "Reconfigurable learning control in large space structures", IEEE Trans. On Control Sys. Technol., 362-371, (1994).

A co-design for CAN-based networked control systems

Nguyen Trong Cac, Nguyen Van Khang

School of Electronics and Telecommunications, Hanoi University of Science and Technology, Hanoi, Vietnam

Email address:
cac.nguyentrong@hust.edu.vn (N. T. Cac), khang.nguyenvan1@hust.edu.vn (N. V. Khang)

Abstract: The goal of this paper is to consider a co-design approach between the controller of a process control application and the frame scheduling for CAN-Based Networked Control Systems in order to simultaneously improve the Quality of Control (QoC) of the process control and the Quality of Service (QoS) of the CAN-based network. First, we present a way to calculate the closed-loop communication time delay and we compensate this time delay using the pole-placement design method. Second, we propose a hybrid priority scheme for the message scheduling which allows to improve the QoS. Finally, we present a co-design of the communication time delay compensation and the message scheduling, which gives a more efficient Networked Control System.

Keywords: CAN Bus, Networked Control Systems, Message scheduling, Hybrid Priority Schemes, Communication Time Delay, Pole Placement Design, Co-Design

1. Introduction

The study and design of Networked Control Systems (NCSs) is a very important research area today because of its multidisciplinary aspect (Automatic Control, Computer Science, and Communication Network). The current objective of NCS design today is to consider a co-design in order to have an efficient control system [1, 2]. Several works [2-6] have considered the co-design problems by combining control and scheduling messages. The works [2, 3] have considered the pole-placement design for time delay compensation and Large Error First (LEF) scheduling algorithm for message scheduling. The value of error is encoded directly into the priority. The higher the error is, the higher the message priority is and vice versa. A limitation is that, we have a wide range of the error value and this is not bounded (for example when the system is unstable, the error is infinite). Mapping these error values in the definite number of priority bits is not an easy task. Paper [4] has considered a hybrid priority scheme for message scheduling using the control signal u. The identifier field of the message which represents the message priority is divided into 2 small fields, the first represents a static priority and the second represents a dynamic priority. However, the limitation of this work is to only use 4 bits for static priority field, so they can determine a maximum number of 16 data flows (or nodes) which is not enough to address all nodes in a NCS. The work [5] only compensates the time delay from the controller to the actuator, not the closed-loop time delay. The work [6] considers the issue of co-design combines scheduling messages (static priority scheduling) and communication delay compensation.

From the above analysis, we found the same problem co-design of NCS design by combining control system (considering delay compensation issues) and communication networks (considering message scheduling problems) is a new field many investors and should be studied more. Therefore, in this paper will present the co-design of closed-loop time delay compensation and the message scheduling to improve QoC and QoS simultaneously and to have a more efficient NCS.

This paper includes the following sections: the section 2 presents the general context of the study; the section 3 presents the proposal for computation and compensation of closed-loop time delay; the section 4 presents the proposal of a hybrid priority scheme for message scheduling; the section 5 presents the co-design; the conclusion is represented in the section 6.

2. Context of the Study

2.1. Inverted Pendulum Application

Structure diagram of an inverted pendulum mounted on a trolley is shown on Fig. 1.

The parameters are chosen as follows: the weight of the trolley $M = 0.94$ kg, the trolley has the weight of $m = 0.23$ kg and the length of $l = 0.3$m, free fall acceleration $g = 9.81$m/s^2, θ is the deviation angle of the pendulum, x is the position of the vehicle, u is the force putting into the trolley.

The purpose of the control problem is to move the trolley from position $x_0 = 0$ (initial position) to desired position $x_1 = 0.1$m while keeping the pendulum vertical $\theta(t) = 0$.

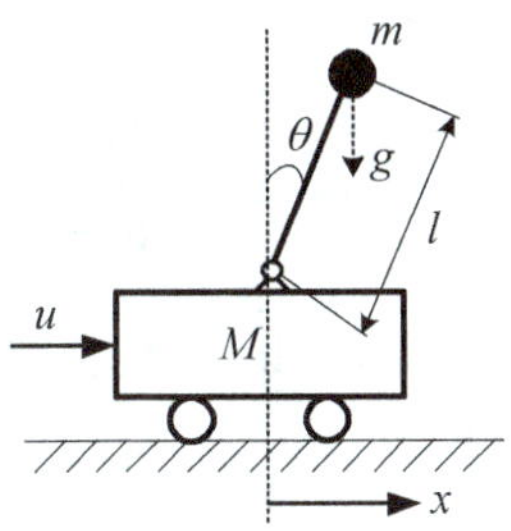

Fig 1. Structure diagram of an inverted pendulum mounted on a trolley.

The set point x_1 is a position echelon type applied at the time $t = 0.5$ s. The desired control parameters include: damping coefficient $\zeta = 0.707$, rise time $t_r = 600$ ms, natural pulsation $\omega_n = 1.8/t_r = 3$ rad/s.

State space model in continuous time [7]:

$$\begin{cases} \dot{x} = Ax(t) + Bu(t) \\ y(t) = Cx(t) + Du(t) \end{cases} \quad (1)$$

Where:

$$A = \begin{bmatrix} 0 & 1 & 0 & 0 \\ 0 & 0 & -\dfrac{mg}{M} & 0 \\ 0 & 0 & 0 & 1 \\ 0 & 0 & \dfrac{(M+m)g}{Ml} & 0 \end{bmatrix}, \quad B = \begin{bmatrix} 0 \\ \dfrac{1}{M} \\ 0 \\ -\dfrac{1}{Ml} \end{bmatrix}$$

$$C = \begin{bmatrix} 1 & 0 & 0 & 0 \end{bmatrix}, \quad D = 0$$

State-space model in discontinuous time domain with the sampling period h is described as follows [7]:

$$\begin{cases} x(kh+h) = \Phi x(kh) + \Gamma u(kh) \\ y(kh) = Cx(kh) + Du(kh) \end{cases} \quad (2)$$

Where:

$$\Phi = e^{Ah} \quad (3)$$

$$\Gamma = \int_0^h e^{As} ds B \quad (4)$$

The discrete controller is:

$$u(kh) = -K_d x(kh) \quad (5)$$

The sampling period h is defined by considering the following formula [7]:

$$0.1 \le \omega_n h \le 0.6 \quad (6)$$

Where ω_n is the natural pulsation (rad/s). We choose the sampling period $h = 50$ ms.

The state matrix Φ and Γ are calculated by Equation (3) and (4). The state feedback matrix K_d is calculated by using Ackerman function.

The performances of the discrete time system are as followed: the Overshoot of the angle $O = 5.07$ %, the setting time $t_s = 0.4$ s, and the time responses are shown on Fig. 2.

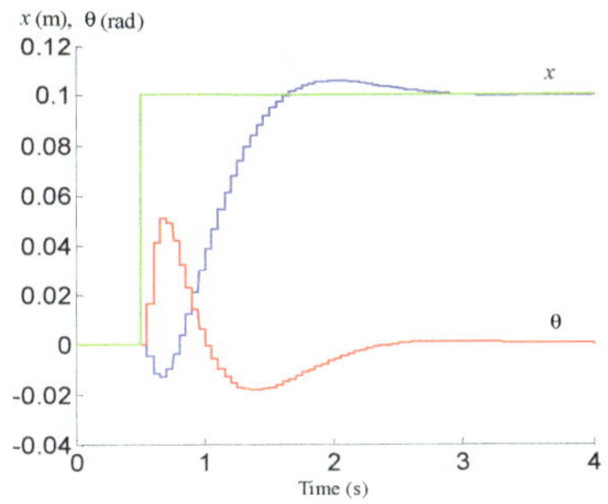

Fig 2. Time responses.

2.2. Implementation of a Process Control Application on a CAN Network

The general model of the implementation of a process control application through a network is shown on Fig. 3.

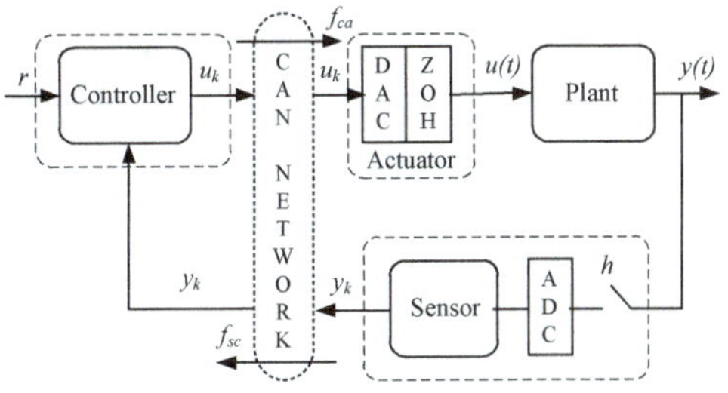

Fig 3. Implementation of a process control application on a CAN network.

We have three tasks: sensor task, controller task and actuator task. The sensor task is Time-Triggered while the controller task and the actuator tasks are Event-Triggered.

The sensor receives the sampled output y_k provided by the Analog Digital Converter (ADC) and sends a frame including y_k to the controller via the communication network. The controller receives y_k from the sensor, then calculates the control signal u_k and sends a frame including u_k to the actuator via the communication network. The actuator receives u_k, converts u_k into analog signal ($u(t)$) using the Digital Analog Converter (DAC) and then directly applies $u(t)$ to plant. The Zero Order Hold (ZOH) keeps the value $u(t)$ until the reception of the next value.

We have two flows of frames: the Sensor-Controller flow (f_{sc}) concerning the frames going from the sensor to the controller (denoted "f_{sc} frame"), and the Controller-Actuator flow (f_{ca}) concerning the frames going from the controller to the actuator (denoted "f_{ca} frame").

2.3. Communication Time Delay

Communication time delays in each sampling period consist of 2 components:

- The time delay is due to transferring f_{sc} frames (noted τ_{sc}) which is the duration between the sampling instant (t_k, $k = 0, 1, 2...$) and the reception instant of this frame by the controller. The time delay τ_{sc} includes the waiting time for medium access and the transmission time of a f_{sc} frame (noted D_{sc})
- The time delay due to transferring f_{ca} frames (noted τ_{ca}) elapsed from the ready-to-send instant of the f_{ca} frame till the reception instant of this frame by the actuator. The time delay τ_{ca} includes the waiting time for medium access and the transmission time of a f_{ca} frame (noted D_{ca})

Note that D_{sc} and D_{ca} can be easily calculated based on the frame length and the network bit rate.

Therefore, the communication time delay of a closed-loop control system is:

$$\tau = \tau_{sc} + \tau_{ca} \tag{3}$$

In this paper, we consider the following hypotheses:

- Communication time delay $\tau < h$.
- Computational time in the controller, sensors is neglected.
- There is no data loss.
- Clocks of the sensor and the controller are synchronized, *i.e.* the controller recognizes the sampling instant t_k.

2.4. Model of NCSs under Communication Time Delay

As shown in Fig. 3, a NCS has a continuous plant (Equation (8)) and a discrete controller (Equation (9)).

$$\begin{cases} \dot{x} = Ax(t) + Bu(t) \\ y(t) = Cx(t) + Du(t) \end{cases} \tag{8}$$

$$u(kh) = -K_d x(kh),\ k = 0,1,2... \tag{9}$$

Where $x(t)$ is the state vector, $u(t)$ is the input vector and $y(t)$ is the output vector, A is the system matrix, B is the input matrix, C is the output matrix, D is the connected matrix. And K_d is the state feedback gain matrix.

State-space model in discrete time domain with the communication time delay τ is described as follows:

$$\begin{cases} x(kh+h) = \Phi x(kh) + \Gamma_0(\tau)u(kh) + \Gamma_1(\tau)u(kh-h) \\ y(kh) = Cx(kh) + Du(kh) \end{cases} \tag{10}$$

Where Φ, $\Gamma_0(\tau)$, $\Gamma_1(\tau)$ are the state matrix defined as follows:

$$\Gamma_0(\tau) = \int_0^{h-\tau} e^{As} ds B \tag{4}$$

$$\Gamma_1(\tau) = e^{A(h-\tau)} \int_0^{\tau} e^{As} ds B \tag{12}$$

State-space model in Equation (10) can be re-written as follows:

$$\begin{bmatrix} x(kh+h) \\ u(kh) \end{bmatrix} = \begin{bmatrix} \Phi & \Gamma_1(\tau) \\ 0 & 0 \end{bmatrix} \begin{bmatrix} x(kh) \\ u(kh-h) \end{bmatrix} + \begin{bmatrix} \Gamma_0(\tau) \\ I \end{bmatrix} u(kh) \tag{5}$$

The discrete controller is:

$$u(kh) = -K_d(\tau) \begin{bmatrix} x(kh) \\ u(kh-h) \end{bmatrix} \tag{6}$$

Note that the parameters in Equation (10) and Equation (13) depend on h and τ.

2.5. Stability Analysis

Equation (13) can be rewritten as follows:

$$\begin{bmatrix} x(kh+h) \\ u(kh) \end{bmatrix} = \left(\begin{bmatrix} \Phi & \Gamma_1(\tau) \\ 0 & 0 \end{bmatrix} - \begin{bmatrix} \Gamma_0(\tau) \\ I \end{bmatrix} K_d(\tau) \right) \begin{bmatrix} x(kh) \\ u(kh-h) \end{bmatrix} \tag{7}$$

The matrix of the closed-loop control system is defined as follows:

$$\Phi_{cl} = \begin{bmatrix} \Phi & \Gamma_1(\tau) \\ 0 & 0 \end{bmatrix} - \begin{bmatrix} \Gamma_0(\tau) \\ I \end{bmatrix} \cdot K_d(\tau) \tag{8}$$

With each different sample period, we will find Φ_{cl}. We call k is the number of sampling period; we have the following cases:

$$k = 1 \Rightarrow x(h) = \Phi_{cl} x(0)$$

$$k = 2 \Rightarrow x(h+h) = \Phi_{cl} x(h) = \Phi_{cl}\Phi_{cl} x(0) = \Phi_{cl}^2 x(0)$$

$$k = 3 \Rightarrow x(2h+h) = \Phi_{cl} x(2h) = \Phi_{cl}\Phi_{cl}^2 x(0) = \Phi_{cl}^3 x(0)$$

$$k = 4 \Rightarrow x(3h+h) = \Phi_{cl} x(3h) = \Phi_{cl}\Phi_{cl}^3 x(0) = \Phi_{cl}^4 x(0)$$

$$\cdots$$

$$k \quad \Rightarrow x(kh+h) = \Phi_{cl} x(kh) = \Phi_{cl}\Phi_{cl}^{k-1} x(0) = \Phi_{cl}^k x(0)$$

Thus the closed-loop matrix will be the product of the matrix elements, which are calculated as follows:

$$\left\{ \prod_{k=1}^{\infty} \Phi_{cl_k} \right\} \tag{9}$$

Closed-loop matrix in Equation (17) is used to analyze the stability of feedback control systems.

The stability condition is that largest eigenvalue of a closed-loop matrix (Equation (17)) is smaller than 1 [8]:

$$\left| \lambda_{max} \left(\prod_{k=1}^{\infty} \Phi_{cl_k} \right) \right| < 1 \tag{10}$$

2.6. Global Control System which is Considered

We now present the system which will be analyzed in this work by means of the simulator TrueTime [9], a toolbox based on Matlab/Simulink which allows to simulate real-time distributed control systems.

2.6.1. General Considerations

We implement 8 identical process control applications (noted P_1, $P_{2...}$ P_8) through a CAN network (Fig. 4). So we have 24 different nodes connected through a CAN network and 16 data flows (8 f_{sc} and 8 f_{ca} flows) sharing the network. We consider other parameters and conditions as followed:

- Bit rate = 125 Kbit/s.
- Data field length of a frame = 8 bytes. Thus, $D_{sc} = D_{ca} =$ 150 bits [10].
- The sensor tasks are synchronous and have the same sampling period.

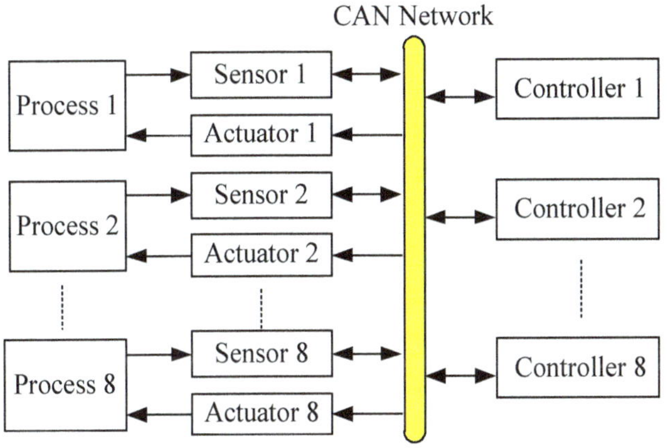

Fig 4. *Implementation of 8 process control applications on the CAN network.*

The scheduling of the frames is done in the MAC layer by means of priorities represented by the Identifier (ID) field of the frame.

Generally, the priorities are static priorities, *i.e.* each flow has a unique priority (specified *a priori* out of line) and all the frames of this flow have the same priority. Concerning the priorities, we will consider here either static priorities (*i.e.* each flow has a unique priority specified out of line) or hybrid priorities (as we said in the introduction) *i.e.* with two priority levels. One level represents the flow priority which is a static priority. The other level represents the frame transmission urgency. The urgency can be the same for all the frames of the flow and, in this case, the transmission urgency is also a static priority. The urgency can vary (for example, if the conditions of the application, which uses the flow, change) and, in this case, the transmission urgency is a dynamic priority. This concept has a great interest during the transient behavior of systems [12, 13].

The consideration of hybrid priorities requires structuring the field ID in two levels (Fig. 5) where the Level 1 represents the flow priority and the Level 2 represents the urgency priority [4].

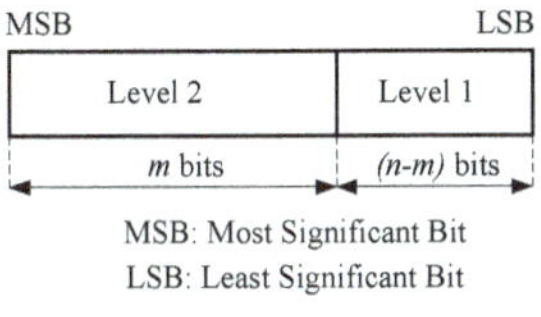

Fig 5. *Structure of the ID field.*

In the context of the competition based on these hybrid priorities, the frame scheduling is executed by comparing first the bits of the Level 2 (urgency predominance). If the urgencies are identical, the Level 1 (static priorities which have the uniqueness properties) resolves the competition.

2.6.2. Static Priorities Associated to the F_{sc} and F_{ca} Flows

Here we consider the conclusion shown in [11]: The priority of the f_{ca} flow should be higher than the priority of the f_{sc} flow in order to get the best results. Considering 8 process control applications (P_1, $P_{2...}$ P_8), each process has one f_{sc} flow and one f_{ca} flow. We call *Prio_*f_{ca}*_i* and *Prio_*f_{sc}*_i* the priorities of the f_{sc} and f_{ca} flows of the process P_i (i = 1, 2… 8), respectively. The priorities of 16 data flows are arranged in the following order:

*Prio_*f_{ca}*_1 > Prio_*f_{ca}*_2 > … > Prio_*f_{ca}*_8 > Prio_*f_{sc}*_1 > Prio_*f_{sc}*_2 > … > Prio_*f_{sc}*_8. i.e.* the controller has a higher priority than the sensor and during a sampling period, the order of medium access is arranged as P_1, P_2, …, P_8.

2.6.3. Criteria of the QoC Evaluation

We will consider the time response for the QoC evaluation.

2.6.4. Stability Analysis

As we know the lengths of frames of all processes and the order of medium access, we can calculate the communication time delay of each process. The time delays of P_1, P_2… P_8 are 2.4ms, 4.8ms, 7.2ms, 9.6ms, 12ms, 14.4ms, 16.8ms, 19.2ms respectively. We consider also a time delay of 23.55ms.

The largest eigenvalues of the closed-loop matrix corresponding to the time delay τ above are represented in Table 1. For the case of non-compensation, the higher the delay is, the higher the $|\lambda_{max}(\Phi_{cl})|$ is, which is logical. For the 8 processes, the $|\lambda_{max}(\Phi_{cl})|<1$, thus our system is stable. We see that if the time delay is 23.55ms, $|\lambda_{max}(\Phi_{cl})| = 1.009 > 1$, the system will be unstable.

Table 1. *Stability Analysis.*

τ (ms)	$\lvert\lambda_{max}(\Phi_{cl})\rvert$ **(non-compensation)**	$\lvert\lambda_{max}(\Phi_{cl})\rvert$ **(compensation)**
2.4	0.000598	0.000597
4.8	0.000623	0.000597
7.2	0.000635	0.000597
9.6	0.000647	0.000597
12	0.000658	0.000597
14.4	0.000669	0.000597
16.8	0.000679	0.000597
19.2	0.0028	0.000597
23.55	1.009	0.000597

For the case of Compensation, the $\left|\lambda_{\max}\left(\Phi_{cl}\right)\right|$ is equal for all time delays. It is due to the fact that we have maintained the dominant poles, hence the closed-loop matrix has not been changed, thus the eigenvalues of the closed-loop matrix have not been changed.

3. Proposal for Computation and Compensation of Time Delays

3.1. Ideas

The goal of this subsection is to propose a way to calculate the closed-loop communication time delay and we compensate this time delay using the pole placement design method in order to improve the QoC for CAN-based NCSs. We consider the implementation of several process control applications on a CAN network. Then we show the interest of the proposed method by comparing the QoC in cases of time delay compensation and of without time delay compensation.

3.2. Computation of Closed-Loop Communication Time Delays

The computation of closed-loop communication time delays is done by the controller in each sampling period.

Concerning the time delay τ_{sc}, as the controller has the knowledge of sampling instants t_k ($k = 0, 1 \ldots$), it can easily deduce the value of τ_{sc} by the time difference between the reception instant of f_{sc} frames and t_k.

Concerning the time delay τ_{ca}, it cannot be calculated because the f_{ca} frames have not been transmitted yet. However, due to the hypotheses in section 2.6 (*i.e.* the priority of the controller is higher than this of the sensor; there is no competition between the controllers; there is no data lost), the controller can immediately send its frame. *Therefore, τ_{ca} is equal to the duration of a frame transmission (D_{ca}).*

The closed-loop time delay will be computed by the controller as followed:

$$\tau = \tau_{sc} + D_{ca} \qquad (11)$$

3.3. Time Delay Compensation Steps

The compensation for time delays done by the controller in each sampling period has the following steps:

- Step 1: Identifying expected poles including the dominant poles and the other poles [7] which are selected equally to the real part of the dominant pole divided α, with $\alpha = 2 \div 10$.
- Step 2: Computing closed-loop communication time delay.
- Step 3: Computing the controller parameters according to the time delay value in order to maintain the position or the value of the expected poles.
- Step 4: Computing he control signal based on the new control parameters calculated in the previous step.

3.4. Considering the Implementation of the 8 Process Control Applications on CAN Network

We want to show here the interest of the pole-placement design method (which is based on an adaptive controller *i.e.* the parameters of the controller are modified according to the communication time delay) in comparison with the case where we without time delay compensation (*i.e.* we have a fixed controller).

The QoC is represented in Table 2 ($\Delta J/J_0$ %) and Fig. 6 and Fig. 7.

Table 2. *QoC ($\Delta J/J_0$ %).*

Process	Non-Compensation	Compensation
P_1	0.47	0.33
P_2	1.02	0.74
P_3	1.80	1.29
P_4	2.94	2.08
P_5	4.67	3.22
P_6	7.57	4.96
P_7	13.18	7.87
P_8	27.54	13.50

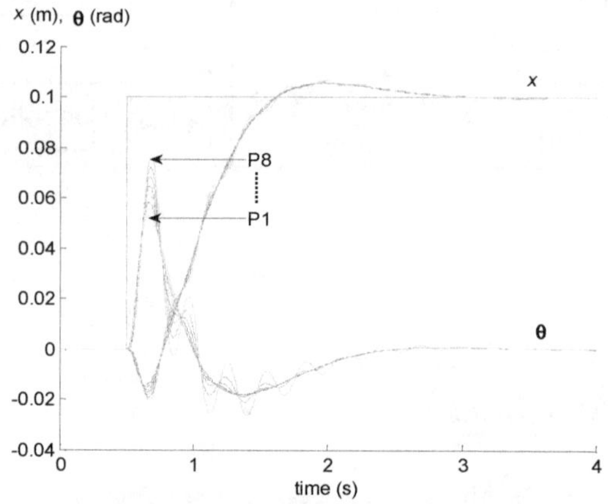

Fig 6. *Time responses when non-compensation.*

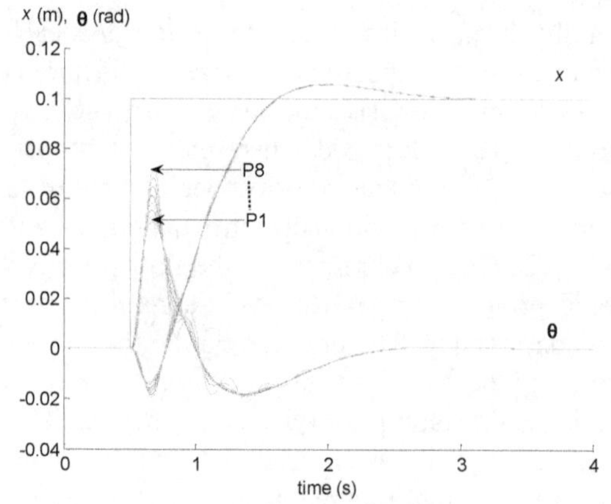

Fig 7. *Time responses when compensation.*

Remind that the priority of the process P_i is higher than this of the process P_j. We see in Table 2 for the both cases that the QoC follows the order of the priority, it is reasonable because the higher the priority is, the lower the time delay is and so the better the QoC is.

In the both Table 2 and Fig. 6 & 7, we can see the improvement of the QoC for the case of compensation compared with the case of non-compensation expressed through smaller values of $\Delta J/J_0$ % and less oscillatory time responses

4. Proposal of a Hybrid Priority Scheme for Message Scheduling

4.1. Limitations of Static Priorities

Considering static priorities, each node in the network has a unique and fixed priority. Hence, the application which has data flows of low priorities could not get the desired QoS and QoC. For example, we consider a control system with two nodes A and B, in which A has higher priority than B. At the instant *t*, both A and B have frames to transmit, B has a stronger transmission urgency in order to satisfy the QoS requirement (for example the deadline) but B cannot transmit its frames before the end of the frame transmission of A. That will make the system less efficient.

Other limitations of static priority can be found when we consider time response of process control applications which is characterized by two regimes: transient regime (because of an input change or a noise) and permanent regime (system is in the steady state). In transient regime, frames must be transmitted as soon as possible in order to obtain some QoC requirements such as response time, overshoot, rise time (*i.e.* frames have high transmission urgencies). Contrariwise, in permanent regime, frames are required to be transmitted quickly (*i.e.* frames have low transmission urgencies). It is clearly noticed that static priority cannot overcome this problem.

4.2. Hybrid Priorities and Related Works

4.2.1. Idea of Hybrid Priorities

The idea of hybrid priority results from the limitations of static priority presented in the previous section. The idea is that frame priorities can vary depending on transmission urgency. That can be done by restructuring the ID field into 2 small fields as represented in Fig. 5.

Level 2 (*m* high significant bits) represents transmission urgency which is called dynamic priority part with its value *ID_dyn*. The *ID_dyn* can be changed during system operation and several data flows can share the same *ID_dyn*. Level 1 (*n-m* bits) which is called static priority part represents the uniqueness of data flows as its value *ID_sta* is fixed, unique and specified before system running. The uniqueness means that there are no two or more nodes having the same *ID_sta*. The term "hybrid priority" means the combination of dynamic priority and static priority.

The idea of this ID field structure was first introduced in [12]. Then other studies [3, 14, 15] also used the similar ID field structure.

The medium access tournament is done firstly by comparing Level 2. If there are several data flows having the same *ID_dyn*, Level 1 will determine the only winner allowed to access to the medium.

Remark: With (*n-m*) *ID_sta* bits, we can determine 2^{n-m} data flows (or nodes) sharing the network. Therefore, it must be careful when we choose the number of *ID_sta* bits.

4.2.2. Specifying of the Dynamic Priority

Specifying dynamic priority part requires, firstly, to determine QoC parameter of the process control application which gives information on the transmission urgency, and secondly, to translate these urgencies into dynamic priorities (*i.e.* computation of dynamic priorities).

Two main QoC parameters using for representing the transmission urgency are steady state error *e* [3, 14] and control signal *u* [4, 15]. Some other works use the deadline [12, 13]. With these parameters, the authors proposed different functions for computation of dynamic priorities. The principle is that the higher the values of *e, u* or deadline are, the higher the dynamic priorities are.

Concerning the works using the error *e*, they used an extended ID field of 29 bits (16 bits for ID_dyn and 10 bits for ID_sta). The value of *e* is encoded directly into the ID_dyn value. The first limitation is that, we have a wide range of error value and this is not bounded (for example when the system is unstable, the error is infinite). Mapping these error values in a definite number of priority bits is not an easy task. The second limitation is that, they assume the existence of a master node knowing the current states of all controller nodes. Maintaining a global state in the whole distributed system can be problematic.

The works using the control signal u have overcome the unbounded value by a saturation value us (if *u* is higher than u_s, the dynamic priority is maximum). They used a standard ID filed of 11 bits (7 bits for ID_dyn and 4 bits for ID_sta). So, they can determine a maximum number of 16 data flows (or nodes) which is not enough to address all nodes in a NCS.

4.3. Proposal

4.3.1. Distributed Aspect

The NCS is totally distributed, *i.e.* there is not any master node.

4.3.2. ID Field

We use an extended ID field of 29 bits with 11 bits for dynamic priority part, and 11 bits for static priority part. That overcomes the limitation concerning number of *ID_sta* bits.

4.3.3. Control Parameters

Both the error *e* and the control signal *u* are used for making the dynamic priority.

4.3.4. Computation of Dynamic Priorities

The dynamic priority (noted *Prio_dyn*) is calculated by the controller using functions represented in Fig. 8 and equations (20) & (21). Here, we consider that e_s and u_s are the maximum values of *e* and *u* respectively when we consider the initial continuous control system without the network.

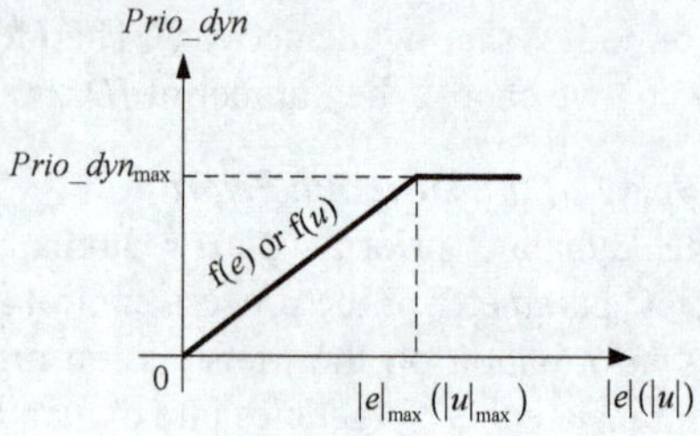

Fig 8. Functions of e, u.

$$f(e)=\begin{cases} Prio\_dyn_{max}\dfrac{|e|}{|e|_{max}}, & 0\le|e|\le|e|_{max} \\ Prio\_dyn_{max}, & |e|>|e|_{max} \end{cases} \quad (20)$$

$$f(u)=\begin{cases} Prio\_dyn_{max}\dfrac{|u|}{|u|_{max}}, & 0\le|u|\le|u|_{max} \\ Prio\_dyn_{max}, & |u|>u_{max} \end{cases} \quad (21)$$

4.3.5. Encoding Priority into ID Field

The dynamic priority part consists of 11 bits which is able to represent 2^{11} = 2048 priority levels from 0 to 2047. The minimum dynamic priority is $Prio\_dyn_{min} = 0$ corresponding to the *ID_dyn* of 11 recessive bits (bit 1).

The maximum dynamic priority is $Prio\_dyn_{max}$ = 2047 corresponding to the *ID_dyn* field of 11 dominant bits (bit 0). The relation between the *ID_dyn* and the *Prio_dyn* is as follows:

$$ID\_dyn = 2047 - Prio\_dyn \quad (12)$$

4.3.6. Implementation of the Hybrid Priority Scheme

Before we present how the hybrid priority scheme works, we should consider the implementation of a process control application on the CAN network as represented on Fig. 9.

We can see how the hybrid priority works. Firstly, concerning the static priority part, this priority of each node is specified before the system running. The subsection 2.6 shows that we have to set static priority of the f_{ca} flow (noted $Prio\_sta_{fca}$) higher than that of the f_{sc} flow (noted $Prio\_sta_{fsc}$) in order to get the best results. Here we will consider this conclusion. Secondly, concerning dynamic priority part, its implementation is as follows:

- At the instant t_k, the sensor samples the output (y_k) and gets dynamic priority ($Prio\_dyn_{k-1}$) sent from the controller in the previous period (*i.e.* period starting at t_{k-1}). After that, the sensor uses this priority ($Prio\_dyn_{k-1}$) to send its frame (containing y_k) to the controller.

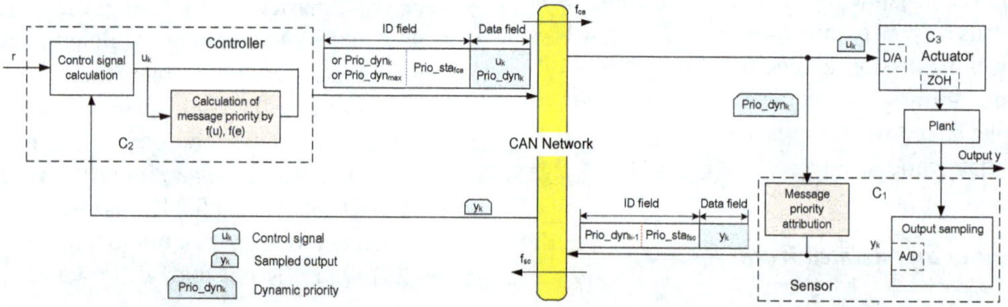

Fig 9. Implementations of process control applications on a CAN network.

Table 3. QoS ($\overline{\tau}$ ms).

	P_1	P_2	P_3	P_4	P_5	P_6	P_7	P_8	$\overline{\tau}_{max}-\overline{\tau}_{min}$
Static priority	2.4	4.8	7.2	9.6	12	14.4	16.8	19.2	16.8
Hybrid priority with *e*	5.9	7.7	9.2	10.4	13.7	12.3	14.4	10.3	8.5
Hybrid priority with *u*	9.5	9.9	11.0	10.4	12	10.2	12.6	12.9	3.4

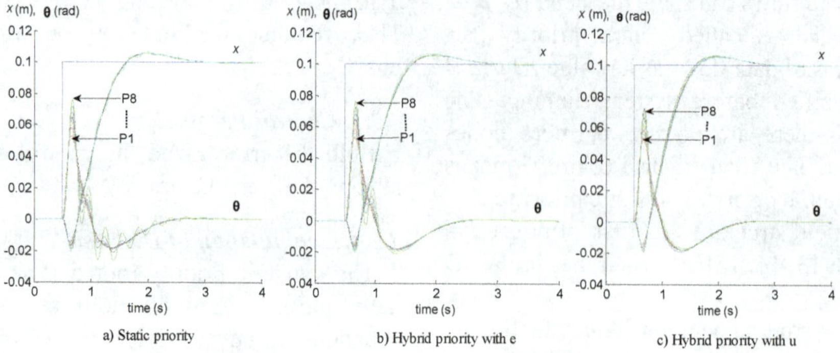

Fig 10. Time responses when non-compensation..

- After receiving the f_{sc} frame sent from the sensor, the controller computes the control signal u_k and the dynamic priority $Prio\_dyn_k$ (by equations (20) & (21)) and sends its frame on the network. Then, the actuator will get u_k and apply it to the controlled plant, while the sensor will get $Prio\_dyn_k$ to use in the next sampling period (period starting at t_{k+1}).

Concerning dynamic priority using to send the f_{ca} frame by the controller, there are two ways: the first one is that the controller use the $Prio\_dyn_k$ value which has just been computed [15]; and the second one is to use the $Prio\_dyn_{max}$ [4]. It is evident that the second way ensures that f_{ca} frame will be sent immediately after the reception of f_{sc} frame (computational time delays in the controller is negligible). Therefore, comparing to the first way, the second way performs a shorter time delay of the closed loop. In this paper, we consider the second one, *i.e.* the controller uses the $Prio\_dyn_{max}$ to send its frames.

Noting that, at the instant 0 ($t_0 = 0$), the sensor has no information about the dynamic priority from the controller. Therefore we consider that the sensor uses, at the first time, the $Prio\_dyn_{max}$.

4.4. Implementation of Process Control Application on CAN Network

4.4.1. Communication Time Delay

These time delays are network delays in the communication between the f_{sc} frame (noted τ_{sc}) and the f_{ca} frame (noted τ_{ca}). During each sampling period, time delay τ_{sc} is the time difference between sampling instant and reception instant of the f_{sc} frame by the controller; the time delay τ_{ca} is the time elapsed from the ready-to-send instant of the f_{ca} frame till the reception instant of this frame by the actuator. Therefore, communication delay in the sampling period is computed as follows:

$$\tau = \tau_{sc} + \tau_{ca} \qquad (13)$$

4.4.2. Criteria of the QoS Evaluation

In order to evaluate the QoS, we calculate first the communication time delay τ_i of the closed loop control system in each sampling period starting at t_i according to the equation (23), then we compute the average value of theses time delays during the settling time t_s by the following formula:

$$\overline{\tau} = \frac{1}{n}\sum_{i=1}^{n}\tau_i \qquad (24)$$

Where n is the number of sampling period in the settling time.

The smaller the value $\overline{\tau}$ is, the better the QoS is.

4.4.3. Criteria of the QoC Evaluation

The QoC is considered through the time responses.

4.4.4. Results

a. Quality of Service

We present on the Table 3 the QoS in term of τ of the 8 processes.

For static priority scheme, the process with higher priority has smaller time delay. We see that P_1 has the highest priority so its delay is the smallest while P_8 has the lowest priority so its delay is the biggest. It is logical.

For the hybrid priority scheme (with e and u), we obtain time delays more balanced than these with static priorities. This is the result of the predominant role of the parts "dynamic priority" compared with the parts "static priority".

The QoS balance can be observed by the difference between the maximum delay and the minimum delay in each priority scheme in the Table 3. We see that these differences are small with hybrid priorities (8.5 ms with e and 3.4 ms with u) while this value is very big with static priorities (16.8 ms).

b. Quality of Control

The QoC is represented in Fig. 10 (time responses). We see also the balances of QoC with hybrid priorities compared with static priorities. It is logical because balances of QoS induce balances of QoC. The conclusions of QoC for different priority schemes are similar to those of QoS. Precisely, for static priority scheme, the higher the priority is, the better the QoC is. And for hybrid priority schemes, the QoCs are more balanced than that of the static priority scheme.

5. Co-Design of Compensation for Communication Time Delays and the Message Scheduling

5.1. Ideas

The idea is to combine the frame scheduling scheme based on the hybrid priority and the method of compensation for communication time delay in order to have a more efficient NCSs. However, concerning the close-loop time delay compensation, in the sampling period k, we cannot consider here that the controller can use the value of the close-loop time delay of the sampling period (k-1) because now, taking into account for the dynamic priority used by the sensor task, the time delay ($\tau_{sc} + \tau_{ca}$) changes every period. Then the controller must make the delay compensation in the sampling period k by knowing the close-loop time delay of this sampling period k. We explain now this implementation.

5.2. Principle of the Implementation of the Co-Design

This principle, relatively to the sampling period starting at t_k, is represented on Fig. 11 where we indicate the content of the f_{sc} and f_{ca} frames and the computations done by the controller. The process of co-design implementation in each cycle starting at t_k as follows:

Sensor sampling state variables x and receive priority value sent from the controller in the previous period (period starting at t_{k-1}), then use the sensor priority this ($Prio\_dyn_{k-1}$)

to send a message containing x_k to the controller.

- The controller after receiving the signal from the sensor x_k, will perform the following steps:
- The Computations of dynamic priority $Prio\_dyn_k$ based on the function *f(u)* and *f(e)* (by equations (20) & (21)).
- Calculations of close-loop communication time delay τ and the parameters of the controller according to the communication time delay τ, *i.e.* implementation of the compensation for close-loop communication time delay.
- Calculations of the control signal u_k.
- Send messages including control signal u_k and dynamic priority $Prio\_dyn_k$ on the network.

Then the actuator will receive the value u_k and apply to control plant, while the sensor will get $Prio\_dyn_k$ to use for the next period (period starting at t_{k+1}).

5.3. Performance Evaluation and Summary of Obtained Results

We still consider the implementation of 8 applications (P_1, P_2, ..., P_8) studied in the previous sections.

The time responses are represented on Fig. 12. These results which are represented graphically on obviously show the balanced performances provided by the hybrid priority compared with the static priority and by the co-design compared with the static priority scheme (compensation).

From these results, we can say that, if we have a constraint of performance which cannot be exceeded (not too small), the hybrid priority allows to implement more applications than the static one and, extensionally, the bidirectional relation co-design allows to implement more applications than the static priority scheme (compensation).

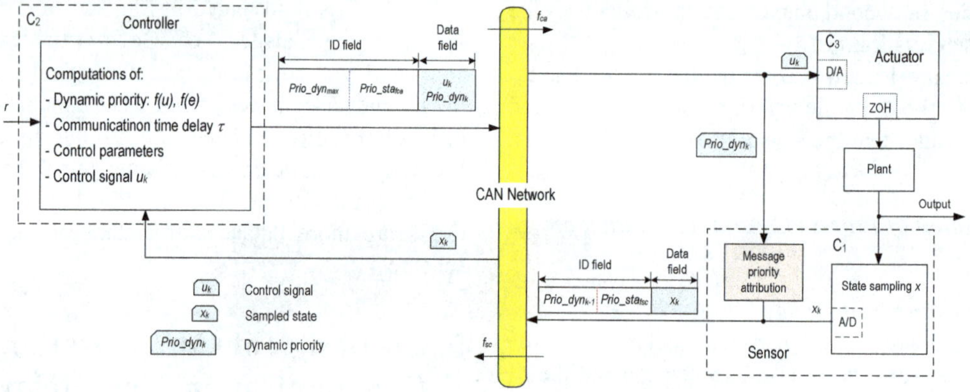

Fig 11. *Principle of the implementation of the co-design.*

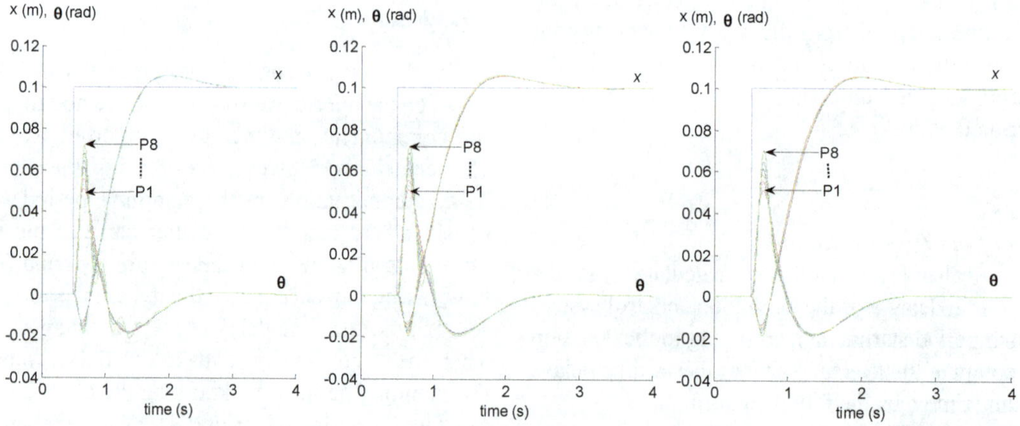

a) static priority b) hybrid priority with e c) hybrid priority with u

Fig 12. *Time responses when compensation.*

6. Conclusion

In this paper, we have presented three following points:

- The first is the implementation of the way to calculate the closed-loop communication time delay and we compensate this time delay using the pole placement design method in order to improve the QoC performances provided by static priorities only.
- The second is the implementation of the hybrid priority scheme for the message scheduling in order to improve the QoS. The hybrid priority is characterized by, on one hand, a static part representing the uniqueness of the flow, and on the other hand, a dynamic part that represents the urgency of transmission (this dynamic part is expressed from a function of the error and control signal). This hybrid priority gives, in comparison to the static priority case, balanced performances for the different process control applications.
- The final is the implementation of the bidirectional

relation QoS and QoC (*i.e.* co-design) which is the combination of the between conpensation for communication time delay and the message scheduling in order to have a more efficient NCSs design. the relation. The results, which are obtained, show the interest of the joint action of the hybrid priorities and the delay compensation (by the hybrid priorities, we introduce the balance aspect compared with the static priority; by the delay compensation, we maintain the balance aspect while improving the QoC and then we can consider the possibility to implement more applications than with the static priority scheme (compensation).

The further work should be the following points: the utilization of other compensation methods for time delays (for example, maintenance of the phase margin); the consideration of other types of controller (PID for example) and the consideration of other types of process control applications. Still furthermore, the study of this relation might be also improved by the consideration of theoretical problems (in particulars, stability conditions when the online control law parameters change from sampling period to sampling period).

References

[1] Michael S. Branicky, Vincenzo Liberatore and Stephen M. Phillips, "Networked control system co-simulation for co-design," American Control Conference, USA, Vol. 4, June 2003, pp. 3341-3346.

[2] Ye-Qiong Song, "Networked Control Systems: From independent designs of the network QoS and the control of the co-design," 8th IFAC international Conference on Fieldbuses and Networks in Industrial and Embedded Systems, Korea, May 2009, pp. 155-162.

[3] Pau Martí, José Yépez, Manel Velasco, Ricard Villà and Josep M. Fuertes, "Managing Quality-of-Control in network-based control systems by controller and message scheduling co-design," IEEE Transactions on Industrial Electronics, Vol. 51, No. 6, Dec. 2004, pp. 1159-1167.

[4] Xuan Hung Nguyen and Guy Juanole, "Design of Networked Control Systems on the basis of interplays between Quality of Control and Quality of Service," 7th IEEE International Symposium on Industrial Embedded Systems, France, June 2012, pp. 85-93.

[5] Y.B. Zhao, G.P. Liu and D. Rees, "Integrated predictive control and scheduling co-design for networked control systems," IET Control Theory & Applications, Vol. 2, Issue 1, Jan. 2008, pp. 7-15.

[6] Shi-Lu Dai, Hai Lin, and Shuzhi Sam Ge, "Scheduling and control co-design for a collection of Networked Control Systems with uncertain delays," IEEE Transactions on control systems technology, Vol. 18, No. 1, Jan. 2010, pp. 66-78.

[7] Karl J. Åström and B. Wittenmark, "Computer controlled systems: theory and design," 3th Edition, Prentice Hall, 1997.

[8] Murat Dogruel and Umit Özgüner, "Stability of a Set of Matrices-A Control heoretic Approach," 34th Conference on Decision and Control, New Orleans, USA, Vol. 2, Sep. 1995, pp. 1324-1329.

[9] Martin Ohlin, Dan Henrikssonand Anton Cervin, "TrueTime 1.5 - Reference Manual," Lund Institute of Technology, Sweden, 2007.

[10] Salem Hasnaoui, Oussema Kallel, Ridha Kbaier, Samir Ben Ahmed, "An implementation of a proposed modification of CAN protocol on CAN fieldbus controller component for supporting a dynamic priority policy," 38th Annual Meeting of the Ind. App., Vol. 1, Oct. 2003, pp. 23-31.

[11] Guy Juanole, Gerard Mouney, Christophe Calmettes, Marek Peca, "Fundamental considerations for implementing control systems on a CAN network," 6th International Conference on Fielbus Systems and their Applications, Mexico, Nov. 2005, pp. 280-285.

[12] Khawar M. Zuberi end Kang G. Shin, "Scheduling messages on Controller Area Network for real time CIM applications," IEEE Trans. Robot. Autom, Vol. 13, No. 2, Apr. 1997, pp. 310-314.

[13] Khawar M. Zuberi and Kang G. Shin, "Design and implementation of efficient message scheduling for Controller Area Network," IEEE Transactions on Computers, Vol. 49, No. 2, Feb. 2000, pp. 182-188.

[14] Manel Velasco, Pau Martí, Rosa Castané, Josep Guardia and Josep M. Fuertes, "A CAN application profile for control optimization in Networked Embedded Systems," 32nd Annual Conference onIEEE Industrial Electronics, Paris, Nov. 2006, pp. 4638-4643.

[15] Guy Juanole and Gérard Mouney, "Networked Control Systems: Definition and analysis of a hybrid priority scheme for the message scheduling," 13th IEEE conference on Embedded and Real-Time Computing Systems and Applications, Korea, Aug. 2007, pp. 267-274.

Adaptive control design for a Mimo chemical reactor

Yasabie Abatneh, Omprakash Sahu*

Department of Chemical Engineering, KIOT, Wollo University, Kombolcha, Ethiopia

Email address:
ops0121@gmail.com(O. Sahu)

Abstract: The major disadvantage of non-adaptive control systems is that these control systems cannot cope with fluctuation in the parameters of the process. One solution to this problem is to use high levels of feedback gain to decrease the sensitivity of the control system. However high gain controllers have two major problems: large signal magnitude and closed loop instability. The solution to this problem is to develop a control system that adapts to changes in the process. This paper presents the design of adaptive controller to a Multi Input Multi Output (MIMO) chemical reactor. The proposed adaptive controller is tested by using Math lab Simulink program and its performance is compared to a conventional controller for a different situation. The paper demonstrated that while the adaptive controller exhibits superior performance in the presence of noise the convergence time is typically large and there is a large overshoot. The results from the case study indicate that the use of adaptive controller can be extended to process with inverse response. For such process the adaptive controller will be superior to the conventional controller even without parameters change in the process. Although the conventional controller has the smaller response time, it is incapable of eliminating the inverse response.

Keywords: Process, Controller, Data, Model

1. Introduction

In common sense, 'to adapt' means to change a behavior to conform to new circumstances. Intuitively, an adaptive controller is thus a controller that can modify its behavior in response to the change in dynamics of the process and the character of the disturbances [1]. Adaptive control systems have been in existence for over thirty years, and a wide range of approaches have been developed [2]. The core element of all the approaches is that they have the ability to adapt the controller to accommodate changes in the process. This permits the controller to maintain a required level of performance in spite of any noise or fluctuation in the process. There are wide ranges of adaptive control methods currently in use but the objectives are the same that to provide an accurate representation of the process at all times [3]. An adaptive system has maximum application when the plant undergoes transitions or exhibits non-linear behavior and when the structure of the plant is not known. Gain scheduling is one form of adaptive control but it requires knowledge about all the process to be effective. Another alternative is to adapt the controller's parameters or when a model is available to use the model identification error to tune the controller's parameters [4]. Consequently, tuning of the controller is indirect and necessarily requires an accurate model of the process for satisfactory performance. Adaptive control systems are currently used in many operations and one of these operations is chemical process. There are two main reasons why adaptive controller is nee chemical processes. First, most chemical processes are non linear. Therefore, the linear zed models that are used to design linear controllers depend on the particular steady state (around which the process is linearized) .It is clear that as the desired steady state operation of a process changes, the 'best' values of the controller's parameters change. This implies the need for controller adaptation. Second, most of the chemical processes are non stationary (i.e. their characteristics change with time). Typical examples are the decay of the catalyst activity in a reactor and the decrease of the overall heat transfer coefficient in a heat exchanger due to fouling. These change leads again to deterioration in the performance of linear controller which was designed using some nominal values for the process parameters, thus requiring adaptation of the controller parameters. The purpose of this paper is to design and simulate a Model Reference Adaptive control (MRAC) for a multiple inputs multiple outputs chemical reactor [5]. The paper includes the following parts: section 2 provides an overview of model reference adaptive controller and design and simulation adaptive controller and comparison of performance of adaptive controller random-adaptive (conventional) controller. In the last section a case study which considered the control of Van de Vausee reactor will be demonstrated.

2. Adaptive Control Design and Simulation

This section provide three sets: the first covers the design of model reference adaptive control and its implementation in simulink. The second and the third compare the performance of an adaptive control to a conventional controller without noise and with noise respectively.

2.1. Basic Adaptive Controller

This is to provide the implementation of a basic adaptive controller using simulink. The first item that must be defined is the plant that is to be controlled. A CSTR with a general reaction A→B is the system to be controlled and the control variables are concentration of A and temperature of the reactor.

The transfer function for the two SISO systems (concentration and temperature of reaction control) are obtained as

2.1.1. Concentration Control

$$Gm(s)=\frac{1.44s^5+5.78s^4+9s^3+6.88s^2+2.53s+0.31}{s^6+4.41s^5+7.94s^4+8.42s^3+4.1s^2+1.19s+0.14} \quad (1)$$

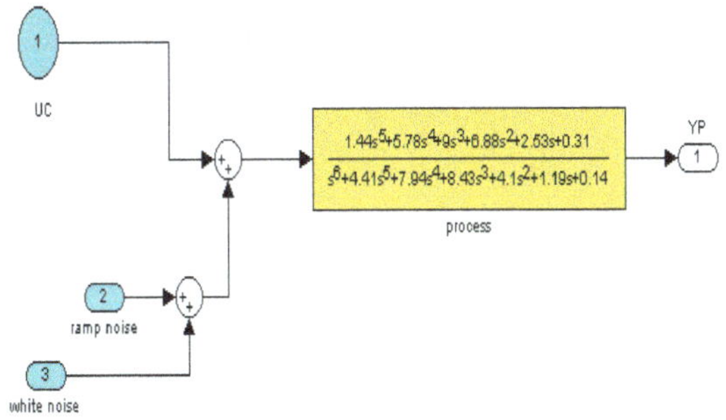

Fig 1. Simulink plant implementation

The next step is to define the model that the plant must be matched to. To determine this model we must first define the characteristics that we want the system to have. Firstly we will arbitrary select the model to be a second order model of the form:

$$Gm(s)=\frac{\omega_n^2}{s^2+2\omega_n\xi s+\omega_n^2} \quad (2)$$

We must then determine the damping ratio ξ and the natural frequency ωn to give the required performance characteristics. For the concentration control a maximum overshoot (Mp) of 5% and a settling time (Ts) of less than 2 seconds are selected. We can use the equation below to determine the required damping ratio and natural frequency of the system.

$$\xi=\frac{\ln M_p/100}{-\pi}\sqrt{\frac{1}{1+\left[\frac{\ln M_p/100}{-\pi}\right]^2}} \quad (3)$$

$$\omega_n=\frac{3}{T_s\xi} \quad (4)$$

Based upon these formulae we get ξ=0.68 and ωn=2.1986 rad/s. The transfer function for the model is therefore

$$Gm(s)=\frac{4.834}{s^2+3s+4.834} \quad (5)$$

Note that we have defined the plant we need to develop a standard controller to compare with the adaptive controller. The simulink diagram for this controller (and the plant) is shown in figure 2.

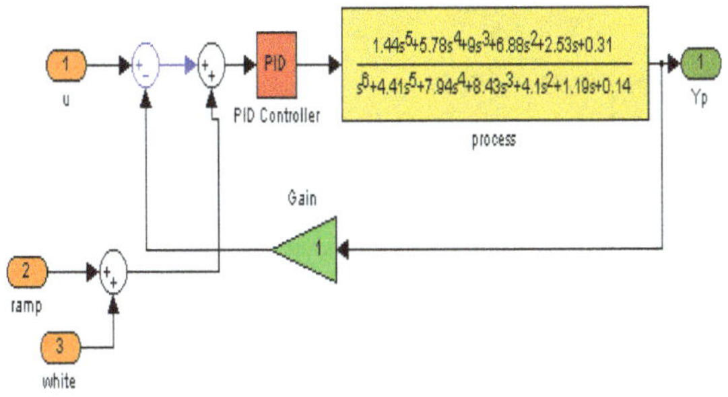

Fig 2. Simulink conventional controller

The final steps are then to implement the controller, the adaptation law and the link between the systems.

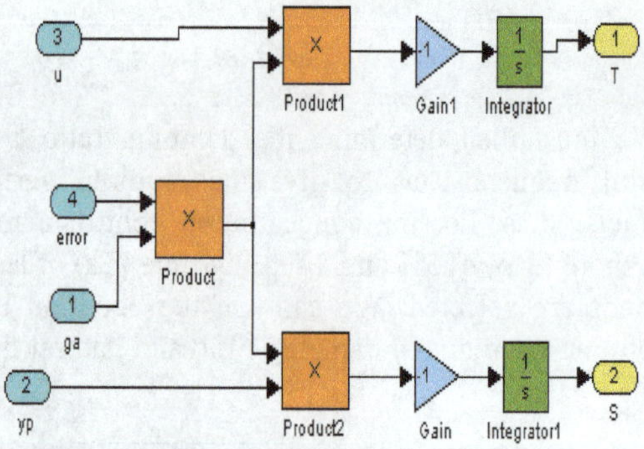

Fig 3. Simulink implementation of Lyapunov Adaptation Law

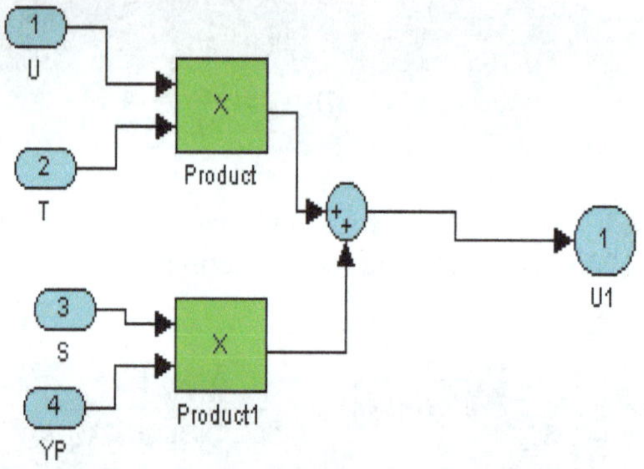

Fig 4. Simulink implementation of Controller

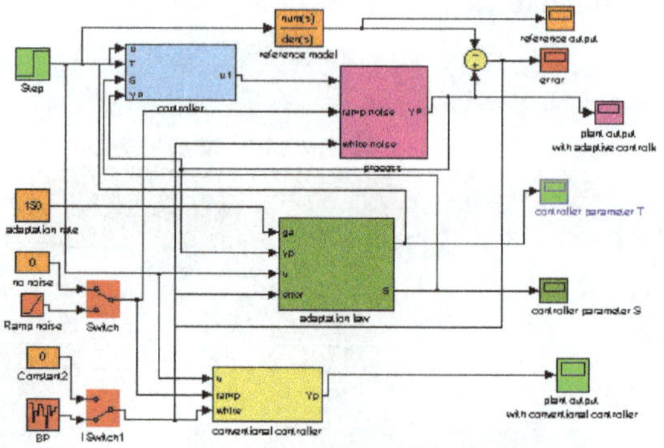

Fig 5. Simulink implementation of a Model Reference Adaptive Controller b) Temperature control

The simplified transfer function model of the process is obtained as:

$$G_p2(s)=\frac{0.3s^5+1.18s^4+1.96s^3+1.54s^2+0.62s+0.95}{s^6+4.41s^5+7.94s^4+8.42s^3+4.1s^2+1.19s+0.14} \quad (6)$$

A reference model (second order) with a maximum overshot (Mp) of 2.5% and settling time (Ts) of 1 second is chosen. And the transfer function for the model is:

$$Gm(s)=\frac{15.54}{s^2+6s+15.54} \quad (7)$$

The implementation of the complete model reference adaptive control is identical to the previous case except Gp1 and Gm1 are replaced by Gp2 and Gm2 respectively.

The completed model permits the noise to be disabled, ramp function only, white noise only or a combination of ramp and white noise. The following parameters are plotted on graph: plant output with adaptive and with conventional control, model output, error between plant and model outputs and the controller parameters.

2.2. Comparison Without Noise

Note that the model is complete; the first task we must perform is, to compare the performance of the two controllers for a step input and no noise

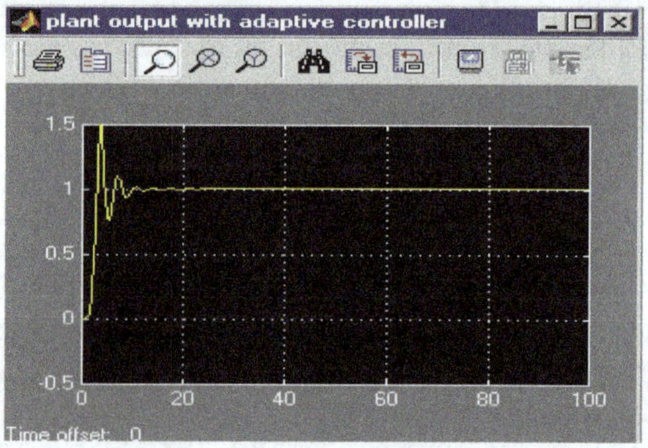

Figure 6. *Plant output (concentration) with adaptive control (step input, no noise, Gamma=0.99)*

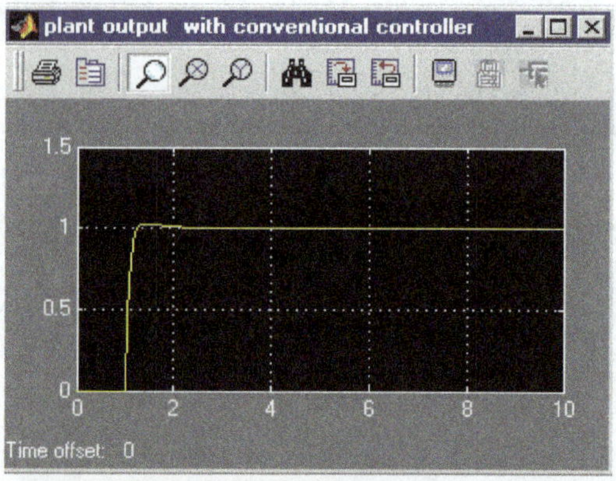

Figure 7. *Plant output with conventional control (step input, no noise, Gamma=0.99)*

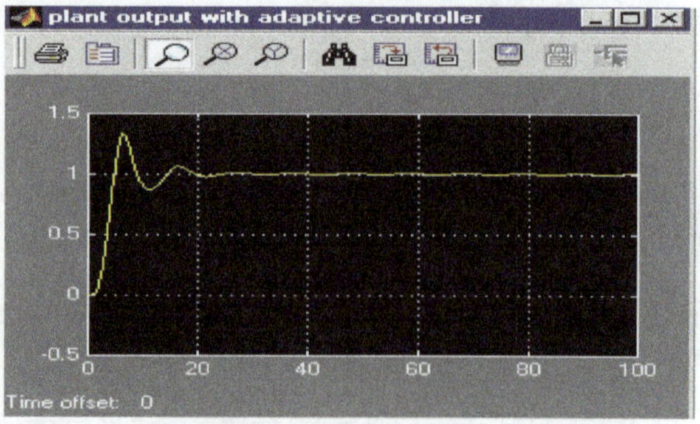

Figure 8. *Plant output (temperature) with adaptive control (step input, no noise, Gamma=0.99)*

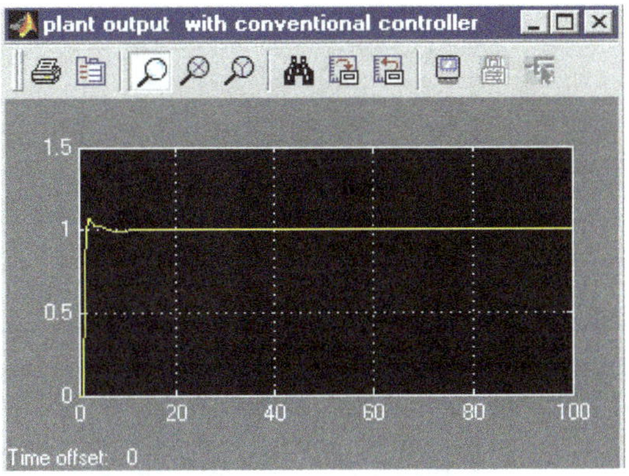

Figure 9. *Plant output temperature) with conventional control (step input, no noise, Gamma=0.99*

Looking at these graphs one of the major disadvantage of adaptive control is immediately apparent. It takes the adaptive controller nearly 20 seconds to match perfectly the output of the reference model. However the conventional controller is matched within 2 seconds. The overshoot of the adaptive controller is also excessive (of the order of 50%) while the conventional controller has an overshoot of below 3%.

One method of addressing this problem is to increase the adaptation gain (Gamma). For example, increasing the adaptation gain to 100 gives the response shown in figures 10 and 11.

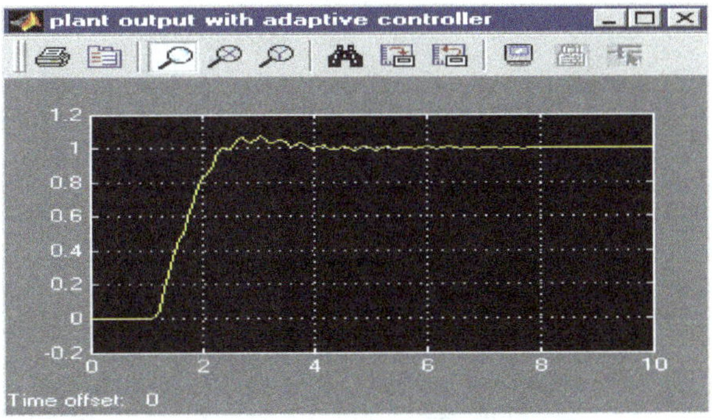

Figure 10. *Plant output with adaptive control (step input, no noise, Gamma=100)*

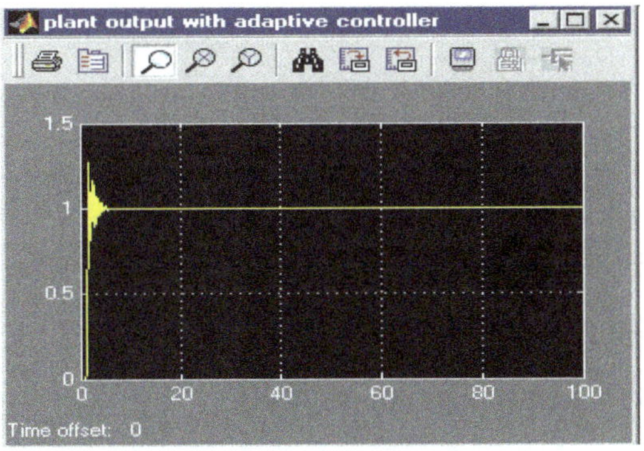

Figure 11. *Plant output with adaptive control (step input, no noise, Gamma=100)*

This has improved the overshoot to below 10% and the settling time is now less than 10 seconds. While not perfect this is a significant improvement. Further increase in the adaptation gain does not result in an improvement of the system.

2.3. Comparison with Noise (Comparison with Ramp Noise)

The next logical step is to compare the performance of the two controllers in the presence of noise in the form of ramp signal, (slope=1). The adaptation gain has been restored to 0.99.

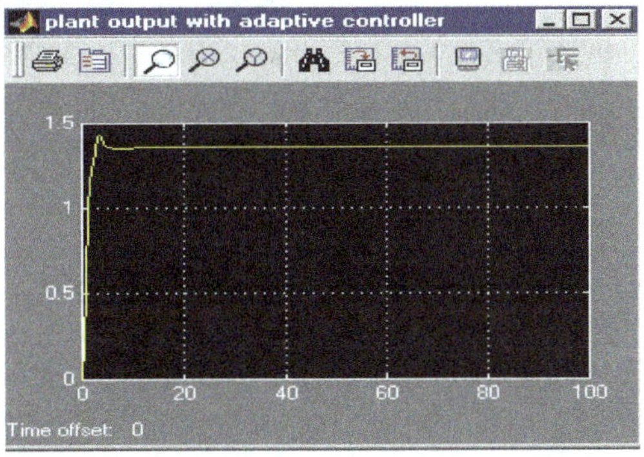

Figure 12. *Plant output (concentration) with adaptive control (step input, ramp noise, Gamma=0.99)*

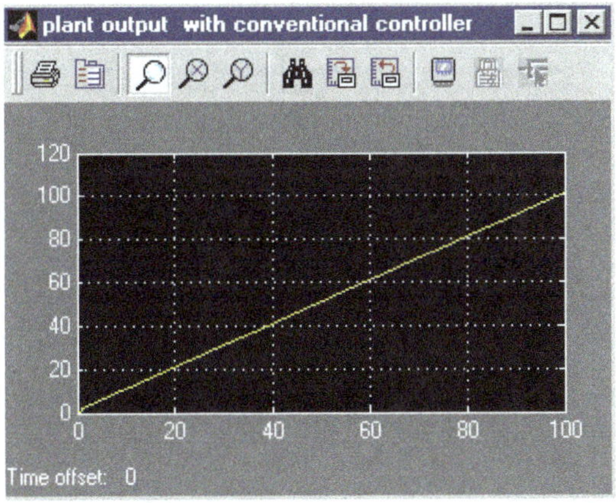

Figure 13. *Plant output (concentration) with conventional control (step input, ramp noise, Gamma=0.99)*

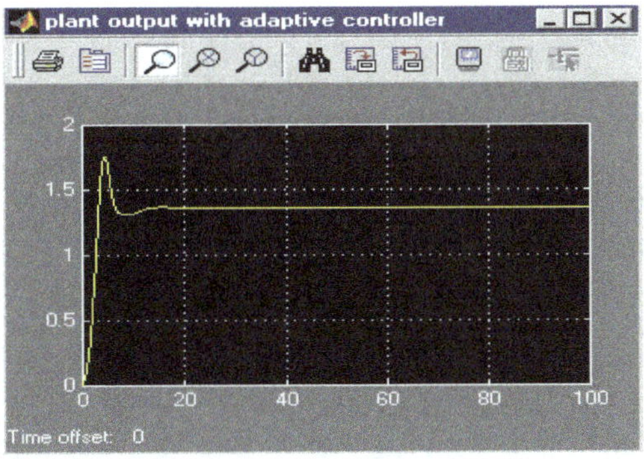

Figure 14. *Plant output (temperature) with adaptive control (step input, ramp noise, Gamma=0.99)*

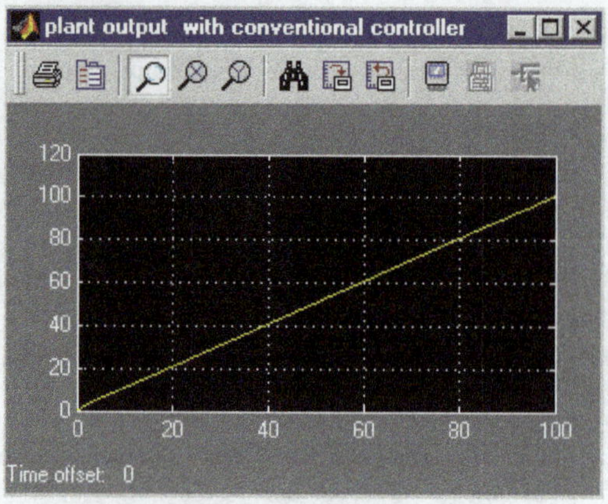

Figure 15.Plant output (temperature) with conventional control (step input, ramp noise, Gamma=0.99)

The situation begins to show the actual advantages of adaptive control. In this case the conventional controller is incapable of maintaining even a stable system. On the other hand the adaptive control manages to maintain stability. However large overshoot and offset and long settling time are present. There is also a large steady state error. As before increasing the adaptation gain to 100 reduces the overshoot to below 10%, the settling time to below 5 seconds and the steady state error to zero. This is shown below in figure.

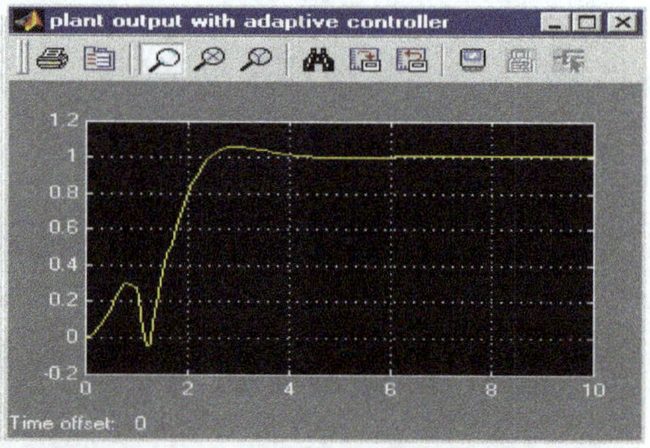

Figure 16. Plant output (concentration) with adaptive control (step input, ramp noise, Gamma=100)

2.3.1. Case Study

Up to now, we have taken a general reaction A→B. So it is essential to consider a specific reaction to demonstrate the proposed adaptive control scheme to a great existent.

This case study considers the control of the Van de Vusse reactor, the reaction scheme consisting of the following reactions:

$A \rightarrow B \rightarrow C$

$2A \rightarrow D$

A = yclopentaddiene, B= Cyclopentenol

C =Cyclopentanediol,

D= dicyclopentandiene

The actual process dynamics are described by

$$dx_1/dt = u(C_{Af} - C_{AS} - x_1) - (u_s + k_1 + 2k_3 C_{As})x_1 - k_3 x_1^2 \quad (8)$$

$$dx_2/dt = -u(x_2 + C_{BS}) + k_1 x_1 - (u_s + k_2)x_2 \quad (9)$$

Where, C_{AS} and C_{BS} denote the effluent concentration of component A and B at steady state, respectively. The state variables x_1 and x_2 are deviation variables defined by $x_1=C_A-C_{AS}$ and $x_2= C_B-C_{BS}$; u is the manipulated variable given by u=F/V – u_s, where us=Fs/V. the concentration of A in the feed stream, denoted by C_{Af} and is equal to 10mol/L. The reactor volume, V, is 7L and the rate constants are k_1= 0.8333min-s, k_2=1.6667min-s and k_3=0.1667 L.mol-1.min-1. It is known that the process is at steady state with F_s=4 L/min, C_{AS}=3mol/L and C_{BS}=1.117 mol/L initially. The control objective is to regulate the concentration of B, x_2, by manipulating the dilution rate, Here; it is assumed that the original nonlinear characteristics are unknown and only the process nominal transfer function

$$G_p(s) = \frac{0.5848(-0.3549s+1)}{0.1858s^2+0.8627s+1} \quad (10)$$

Is known and is available for design. The response for step change in the input is shown in figure 17

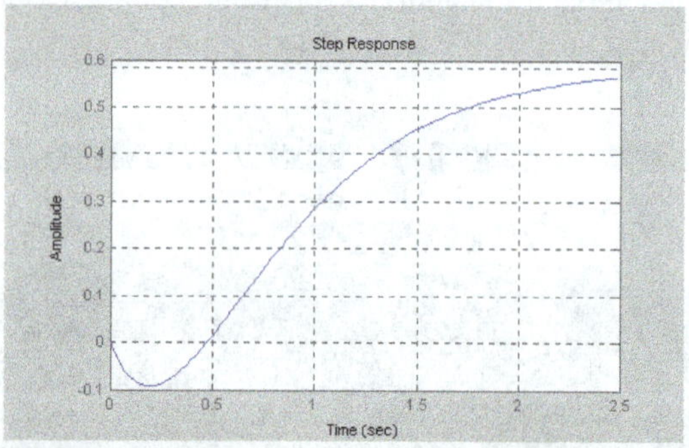

Figure 17. Response for a step change in the input

For the adaptive control system, a reference model (second order) with maximum overshoot (5%) and settling time of 2 second is chosen and the transfer function for the model is:

$$G_P(s) = \frac{4.834}{s^2+3s+4.834} \quad (11)$$

Controller setting for the non-adaptive controller is done using Ziegler-Nicholas technique and the best controller parameters are found to be Kc=3, τI=1 and τd=0.167.

Using the above information the performances of the two controllers are compared for different situation.

2.4. Comparison without Noise

First, the performances of the two controllers without noise are compared. The responses for a step change (0.1) in the input are shown in figure 18. and figure 19.

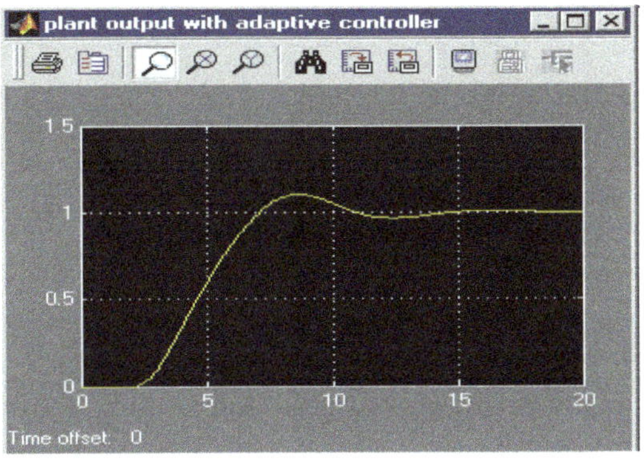

Figure 18. *Plant output with adaptive controller (step input, no noise, γ=0.5)*

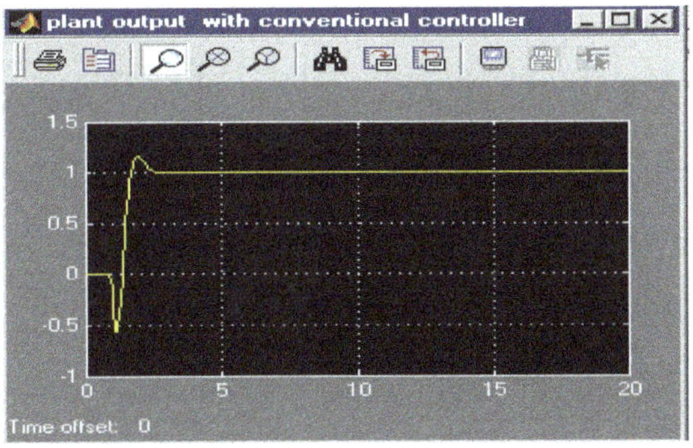

Figure 19. *Plant output with conventional controller (step input, no noise)*

Looking at these two graphs, the advantage of adaptive controller can be seen as it eliminates the inverse response. Although the conventional controller has the smaller response time (2 second) compared to the adaptive controller (7 second), it is incapable of eliminating the inverse response. The response time for the adaptive controller can be decreased at a cost of initial oscillation. Figure 20 shows the effect of increasing the adaptation rate to 0.75.

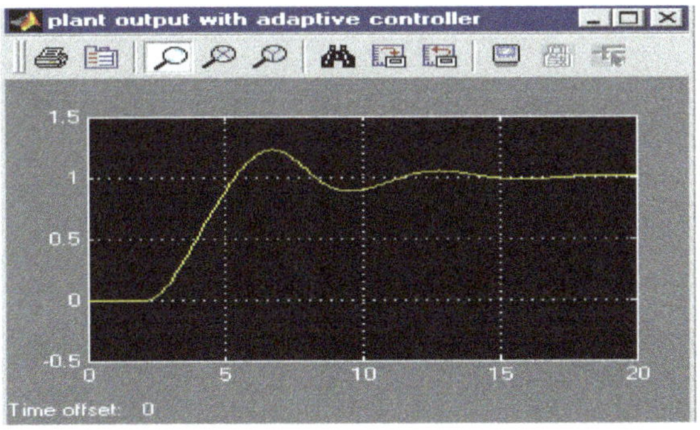

Figure 20. *Plant output with adaptive controller (step input, no noise, γ=0.75)*

2.4.1. Comparison with Ramp Noise

The next task to be done is to compare the performances when there is a disturbances or a change in the process parameters. The responses for a step change in the input with a ramp noise (slope=0.1) are shown in the next two figures.

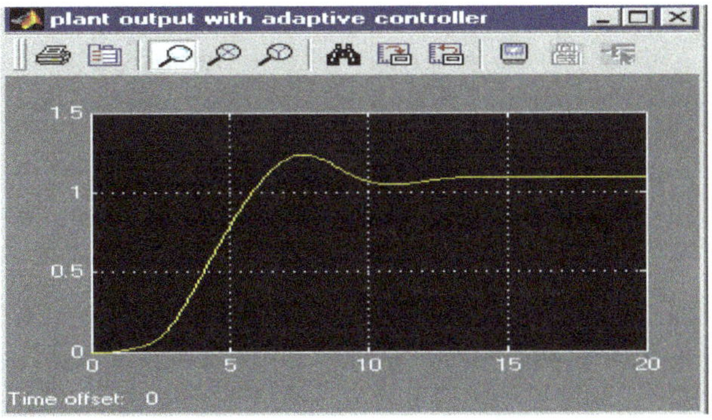

Figure 21. *Plant output with adaptive controller (step input, ramp noise, γ=0.5)*

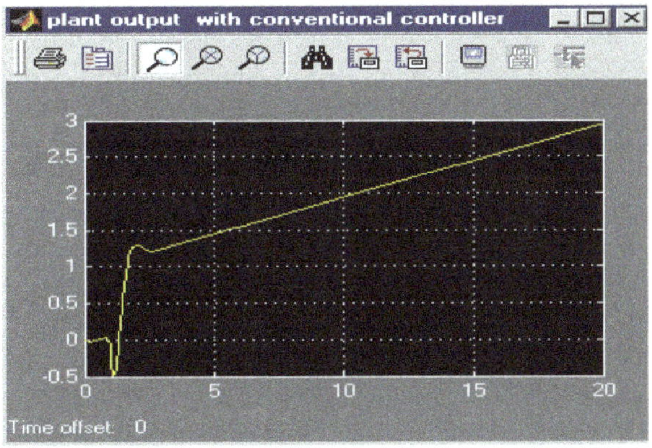

Figure 22. *Plant output with conventional controller (step input, ramp noise)*

From the above two figures it can be seen that the conventional controller is incapable of maintain even stability. Although there is a large steady state error, the adaptive controller gives a better plant response. But the steady state error can be eliminated by increasing the adaptation rate. This is shown in the next figure.

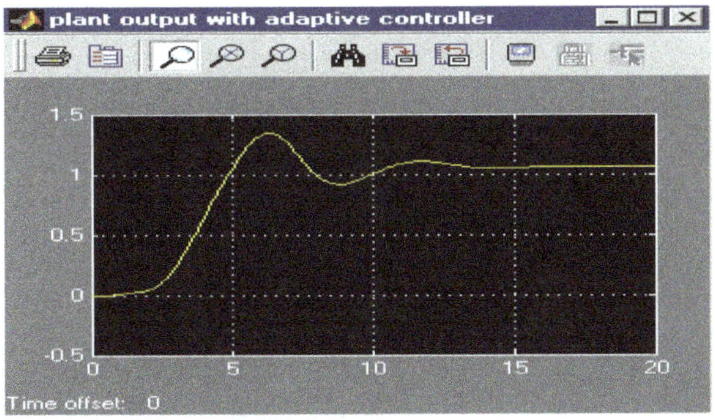

Figure 23. *Plant output with adaptive controller (step input, ramp noise, γ=0.75)*

3. Conclusion

The use of multiple model and adaptive controller are attractive especially when the process is known to transition to unknown operating states. It is also intuitive that a fixed parameter controller (conventional controller) may not provide satisfactory closed loop control. The non adaptive model can provide speed whenever its parameters are close to those of the process while the adaptive model can provide accuracy because its parameters are permitted to adapt.

The paper has aimed to provide an understanding of how to implement an adaptive controller and to compare an adaptive controller with a conventionally designed controller in various situations. The paper demonstrated that while the adaptive controller exhibits superior performance in the presence of noise the convergence time is typically large (greater than 10 seconds) and there is large overshoot. These two problems are due to the adaptive controller failing to adapt fast enough to force the plant to match the model. Increasing the adaptation rate improves the performance of the adaptive controller at the cost of increased oscillation. One interesting observation is that the presence of noise increase the time required for the adaptive controller to converge. The probable reason for this is that the presence of the noise provides the adaptive controller with more of signals to process.

The results from the case study indicate that the use of adaptive controller can be extended to process with inverse response. For such process the adaptive controller will be superior to the conventional controller even without parameters change in the process. Although the conventional controller has the smaller response time, it is incapable of eliminating the inverse response.

References

[1] B.Wayne Bequette, process dynamic modelling, analysis and simulation; Precentice Hall:NJ,1998

[2] G. Stephanopolus, Chemical process control; prentice Hall of India private limited, New Delhi, 2002.

[3] William L.Luyben Process Modeling, Simulation and Control for Chemical Engineer;MCGraw-Hill,Inc.,1990

[4] C.T. Chen; Direct Adaptive Control of Chemical Process System; Industrial Engineering Chemical Research 2001, 40, 4121-4140.

[5] R.Gundala,, K.A.Hoo, Multiple Model Adaptive control Design for a MIMOchemicalreactor,Industrial Engineering Chemical Research 38,1554-1563, 2000.

[6] D. Tyner, M. Soroush, Adaptive Temperature Control of Multi Product Jacketed, Reactor. .Industrial Engineering Chemical Research 38 4337-4344, 1999.

[7] I. Mcmanus, Adaptive Control; A QUT Avionics project, Queensland University of Technology; 2002.

[8] A. Rajagolan, G. Washegton; Simulink Tutorial, Ohio State University, 2002.

[9] M. Kozek, Mraclab-a Tool for Teaching Model Reference Adaptive Control; Vienna University of Technology, 2000.

14

Computer-aided diagnosis of pneumoconiosis X-ray images scanned with a common CCD scanner

Koji Abe[1,*], Takeshi Tahori[2], Masahide Minami[3], Munehiro Nakamura[4], Haiyan Tian[5]

[1]Kinki University, Osaka, Japan
[2]Contec EMS, Co. Ltd, Japan
[3]the University of Tokyo, Japan
[4]Kanazawa University, Kanazawa, Japan
[5]Chongqing University, Chongqing, China

Email address:
koji@info.kindai.ac.jp (K. Abe), tks_thr@outlook.com (T. Tahori), maminami@dream.com (M. Minami) , nakamura.kanazawa.u@gmail.com (M. Nakamura), haiyantian@cqu.edu.cn (H. Tian)

Abstract: This paper presents a discrimination of pneumoconiosis X-ray images obtained with a common CCD scanner. Current computer-aided diagnosis systems of pneumoconiosis have been proposed to images obtained with a special scanner such as a drum scanner or a film scanner for X-ray pictures. However, since the special scanners need a large storage space and the scanners and commitment of the imaging need high-priced costs, the systems are not practical in small clinics. In this paper, we propose features for measuring abnormalities of pneumoconiosis as variables for the discrimination. Devices in the proposed system are only a tablet PC and a CCD scanner. In images obtained with CCD scanner, abnormal levels of pneumoconiosis could depend on density distribution in rib areas. Therefore, the proposed method measures the abnormalities by extracting characteristics of the distribution in the areas. Besides, using the abnormalities, the proposed method discriminates chest X-ray images into normal or abnormal cases of pneumoconiosis. Experimental results of the discriminations for 59 right-lung images have shown that the proposed abnormalities are well extracted for the discrimination.

Keywords: Computer-Aided Diagnosis; Pneumoconiosis; Chest X-Ray Images; Medical Image Processing

1. Introduction

Pneumoconiosis is an interstitial lung disease caused by inhaling fine particles. In diagnosis of pneumoconiosis, diagnosticians judge pneumoconiosis reading small round opacities appeared on chest X-ray pictures. Pneumoconiosis is categorized into categories 0-4 according to profusion of small round opacities. Normal cases belong to category 0, where the opacities are not observed visually. And, abnormal cases belong to category 1, 2, 3, or 4, where the most serious level is category 4 and the opacities are observed most in category 4. Diagnosticians judge pneumoconiosis according to the criteria and the standard pneumoconiosis pictures prepared to every category. The criteria and the standard pictures are defined in the ILO (International Labour Organization). Like Japan, some countries have separately prepared the standard pictures and the criteria according to the criteria in ILO. However, since it is difficult for even experts on pneumoconiosis to diagnose pneumoconiosis, disagreement between diagnosticians is often happened. For the reasons, some CAD (Computer-Aided Diagnosis) systems for pneumoconiosis have been proposed.

The current CAD systems for pneumoconiosis are broadly distinguished into two ways: the one measures the abnormalities extracting features obtained by texture analysis[1], another one extracts small round opacities from lung images and measures their size and number as well as the real diagnosis by doctors[2]-[4]. All the systems were proposed to images obtained with a special scanner such as a drum scanner or a film scanner. Therefore, in order to practically use the systems in general hospitals, it is necessary to put the special scanners, or to order scanning chest X-ray pictures to a printing company in spite of high costs.

On the other hand, CAD systems for discriminating pneumoconiosis into normal and abnormal cases[5, 6] have been proposed to chest X-ray images obtained with a CCD scanner. In their evaluation experiments, abnormal lung images were obtained only from the standard pneumoconiosis pictures provided by Ministry of Health, Labour and Welfare in Japan. Since the number of the standard pneumoconiosis pictures is small for evaluations, the effectiveness of the CAD systems to actual abnormal cases is not confirmed. Moreover, the existing CAD systems[1]-[4] can not be applied to chest images obtained with CCD scanners because contrast of the shadow in the images is much unclear than the special scanners and the small round opacities are not appeared clearly. In order to practically use the CAD systems[5, 6] in general hospitals, this paper presents a discrimination of actual pneumoconiosis X-ray images obtained with a common CCD scanner. In addition, since results of the discriminations in the CAD systems[5, 6] are depended on noises such as shadows of blood vessels, ribs, and etc., this paper presents a method of reducing the affect of such noises.

2. Categorization of Pneumoconiosis X-Ray Images

The criteria for diagnosis of pneumoconiosis are defined in the Japanese Pneumoconiosis Law in accordance with criteria of pneumoconiosis in ILO. In the criteria, the severity of pneumoconiosis is indicated by profusion of small round opacities, where categories 0--4 have been established. Normal pictures belong to category 0, where the opacities are not observed visually. And, abnormal ones belong to category 1, 2, 3, or 4, where the most serious level is category 4 and the opacities are observed most in category 4. Patients diagnosed as pneumoconiosis are to be covered by the worker's compensation. In the diagnosis of pneumoconiosis, diagnosticians compare chest X-ray pictures with the standard pictures provided by ILO or Ministry of Health, Labour and Welfare in Japan. The latter one is predominately used in Japan.

Figure 1 shows the standard pictures and sketches provided by the ministry. In the sketches of category 1, 2, and 3, black specks indicate small round opacities and they increase as the category levels become higher. However, as shown in the standard pictures in Figure 1, the small round opacities do not appear clearly. In addition, since density is conspicuously high like noises at intersections of edges of other objects such as blood vessels, ribs, etc., it is very difficult for even medical doctors to read small round opacities. Although medical doctors diagnose pneumoconiosis according to the criteria and the standard pneumoconiosis pictures prepared to every category, their own experience much depends on the diagnosis because they need to observe subtle difference between a target and the standard pictures visually in the diagnosis.

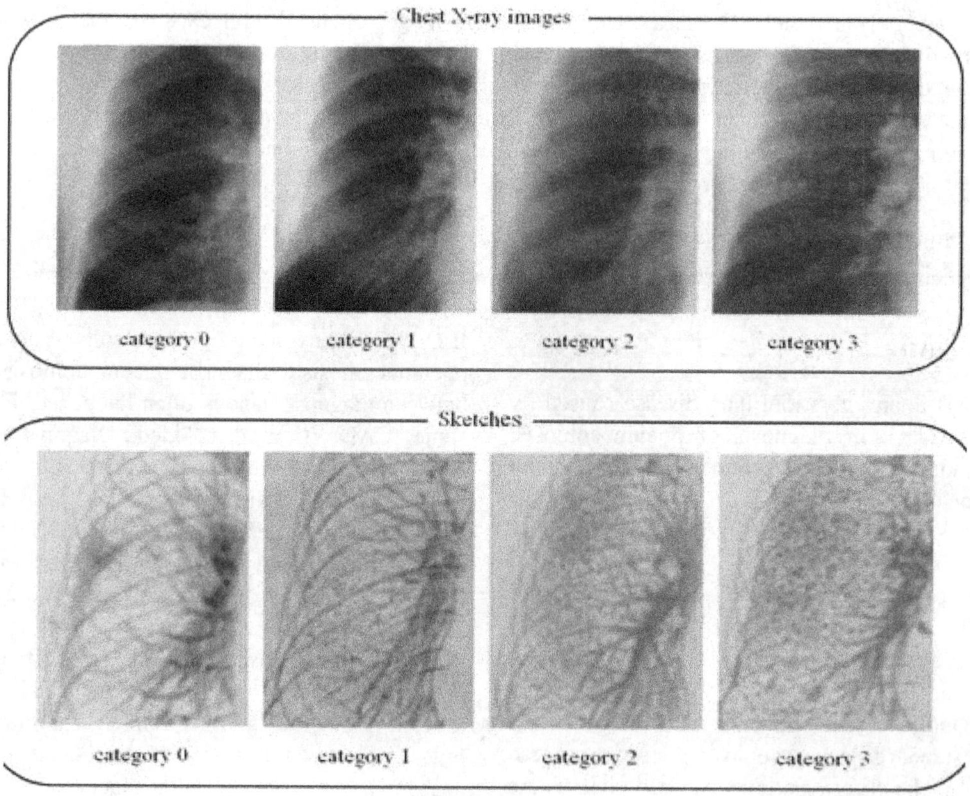

Figure 1. The standard pneumoconiosis X-ray images scanned with a drum scanner (the first line) and the handwritten sketches for each of the images drawn by a diagnostician in pneumoconiosis (the bottom line). All of them are opened from Ministry of Health, Labour, and Welfare in Japan to the public.

3. Density Distribution between Pneumoconiosis Pictures Obtained with CCD Scanner and Drum Scanner

Figure 2 shows a chest X-ray image obtained with a drum scanner and a histogram of pixel values in the part shown in the image. On the other hands, Figure 3 shows a chest image obtained with a CCD scanner and a histogram of pixel values in the part shown in the image, where the chest picture (category 2) for the image is the same one shown in Figure 2 and the location of the part is also the same as Figure 2. The chest images in Figure 2 and Figure 3 are configured with 300 dpi and 256 gray levels. Unlike the image obtained with drum scanner, we can see shadow of the ribs and small round opacities are unclear in the image obtained with CCD scanner, and the whore density values are much lower in the lung area. Table 1 shows statistics of the two histograms. From the table, we can see characteristics of the image obtained with a CCD scanner are quite different from the image obtained with a drum scanner even though their source is the same picture.

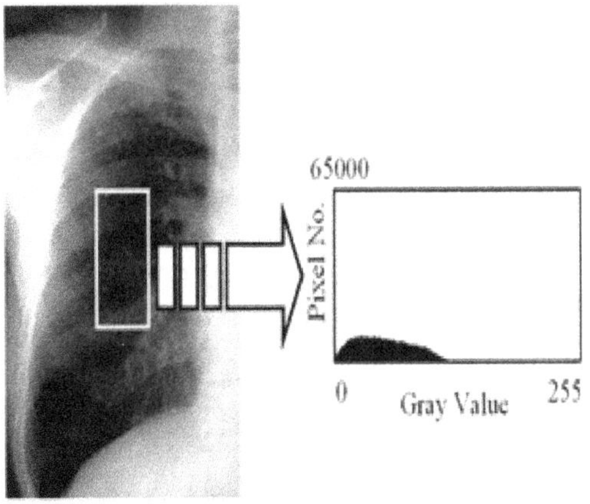

***Figure 2**. The chest image obtained with a drum scanner and the histogram obtained from the region of the white flame.*

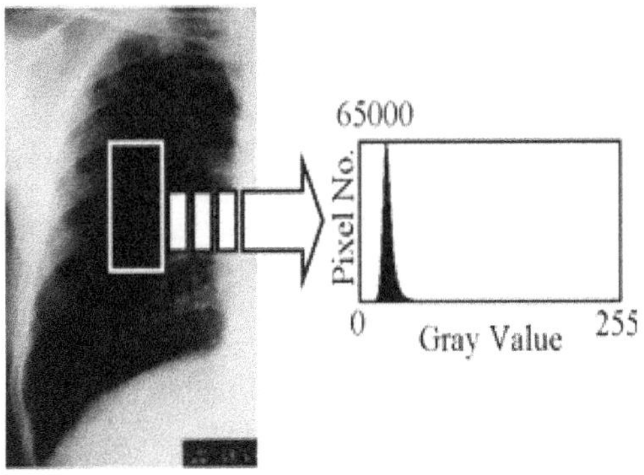

***Figure 3**. The chest image obtained with a CCD scanner and the histogram obtained from the region of the white flame.*

***Table 1**. Statistics of the histograms in Figure 2 and Figure 3.*

	Figure 2 (Drum scanner)	Figure 3 (CCD scanner)
Mean	87.4	26.5
St. Dev.	13.2	5.7
Skewness	0.8	1.9
Kurtosis	0.2	13

Thus, since CCD scanner can not accurately reproduce permeable films such as pictures, the visibility of lung in the image obtained with CCD scanner is considerably lower than drum scanner. For the reason, it is suggested that the existing system[1] can not extract small round opacities from chest X-ray images obtained with CCD scanner. Similarly, since the texture analysis is significantly affected by the image quality, texture analysis[2]-[4] could not effectively extract features from the images. Therefore, it would be necessary to propose a novel method to design a CAD for pneumoconiosis in picture images obtained with CCD scanner.

4. Proposed Method

4.1. Overview

Comparing with category 0, small high density regions assumed as small round opacities appear a lot in rib areas in images (obtained from a CCD scanner) of category 1, 2, and 3. Therefore, in the proposed method, abnormalities of pneumoconiosis are measured by extracting characteristics of the density distribution in rib areas. A method for extracting rib areas in chest X-ray images has been reported[7]. However, as shown in Figure 3, this method could not be applied for the unclear images scanned with a CCD scanner as well as the existing CAD systems of pneumoconiosis. Hence, in the proposed method, the rib areas are manually designated using a tablet PC. Since diagnosticians in pneumoconiosis judge few cases in one time, a dialogical system such as for one case in pneumoconiosis diagnosis is practical for general usage.

First, in the proposed method, the right lung area in chest X-ray pictures is digitalized by a CCD scanner. Next, the chest image is displayed on the tablet PC and the user draws curves along the edges for shadows of ribs on the image using the tablet PC. And then, the rib areas are designated using the drawn curves. Finally, abnormalities of pneumoconiosis are extracted from the rib areas. The extracted abnormalities are used as valuables for discrimination of chest X-ray images into normal or abnormal cases in pneumoconiosis.

4.2. Preprocessing

First, the right lung area in chest X-ray pictures (35 cm×35 cm) is digitalized by a CCD scanner, where all the functions on the image filtering are turned off. The digitalized chest image is configured with 300 dpi and 256 gray levels. Next, the image is resized into 1000 pixels in height without changing the aspect ratio. And then, in order to standardize the range of the density value in each image, the histogram stretching is applied to each image as below where v' is the density value after the transition and v is the density value before the transition. And, regarding $hist[i]$ as the number of density value i (i=0,1,...,255) and $hist[max]$ as the maximum value in $hist[0] \sim hist[255]$, a is a natural value from 0 to max, where $hist[a]$ has the nearest value of $hist[max]/3$ in $hist[0] \sim hist[255]$. And, b is a natural value from max to 255, where $hist[b] = 0$ and $hist[max] \sim hist[b\text{-}1]$ is greater than 0. Figure 4 shows an example of parameter $hist[max]$, a, and b.

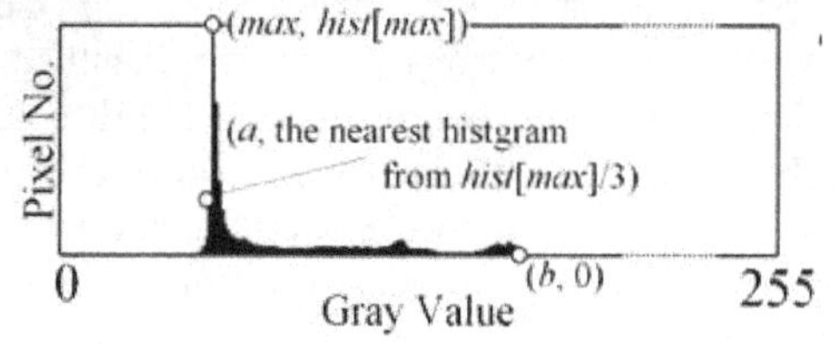

Figure 4. *The density histogram extracted from a chest X-ray image where the histogram stretching is not applied.*

4.3. The Density Histogram Extracted from a Chest X-Ray Images Where the Histogram Stretching is not Applied.

First, the chest image obtained in 4.2 is displayed on the tablet PC and the user draws 8 curves with white color along the edges for shadows of 4 ribs on the image using the tablet PC. Here, setting the thickness of the curves as 4 pixels, the drawn curves would cover the actual rib edges. Although there are 6 shadows of ribs in the image, the uppermost rib area and the bottom rib area are not designated because shadows of back bones and blade bones are often covered on the uppermost rib area and the bottom rib area is often unclear significantly. The manual for drawing the curves is described as below.

[The Manual for Drawing Curves for the Rib Edges]

(1) Open the right lung image prepared in 4.2 with paint tool.

(2) Select the round brush and configure the thickness of the circle as 4 pixels.

(3) Configure the line color as white.

(4) Start drawing the curve for an edge from the outside curve shown in Figure 5 along with shadows of the rib.

(5) Stop drawing the curve around a location where the edge can hardly be seen. Figure 5 (c) shows an example of drawn curves.

According to the manual above and the example of drawn curves in Figure 5, the user draws the curves. Figure 6 shows the procedure for designating the rib areas. First, pixel values of the drawn curves in Figure 6 are quantized as 255 (white) and the other parts are quantized between 0 and 254 in advance. Then, only the drawn curves are extracted by the labeling for pixel value of 255 as shown in Figure 6 (b). Figure 6 (c) shows the rib areas R_1-R_4 extracted from Figure 6 (b). Each of the areas is designated by the following. First, a pair of the upper and lower edge as shown in Figure 7 is regarded as neighbor edge in this paper. Then, the cutoff line in Figure 7 is drawn as it covers a neighbor edge. Next, shifting the cutoff line to the inside of the lung horizontally while the line and both of the upper and lower edge cross over, the boundary for the rib area is configured. Appling this procedure to all the pair of neighbor edges, Rm (m: $1 \sim 4$) is designated. In Figure 7, the first line is regarded as k-th scanning line (k=1) and the value k is added by 1 whenever the scanning line is shifted to the right by 1 pixel. Besides, the uppermost pixel on k-th scanning line is regarded as $j = 1$ and the value j is added by 1 whenever the pixel goes down by 1 pixel.

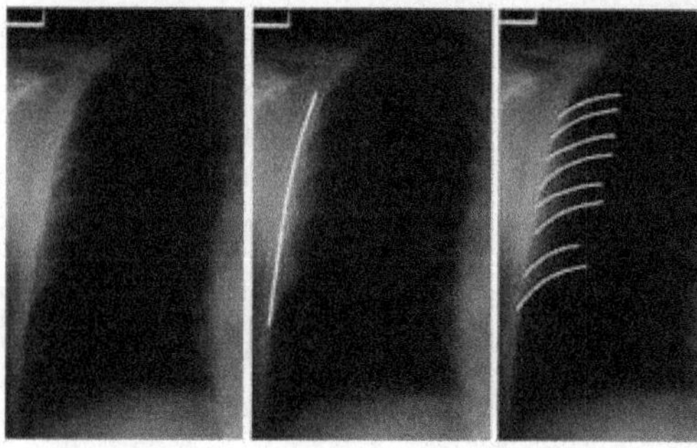

(a) original image (b) outside curve (c) drawn curves

Figure 5. *Drawn curves for the rib edges.*

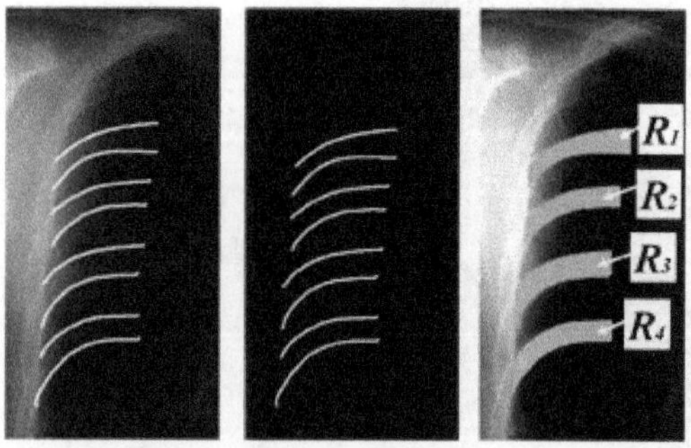

Figure 6. *Designated rib edges (R_1-R_4).*

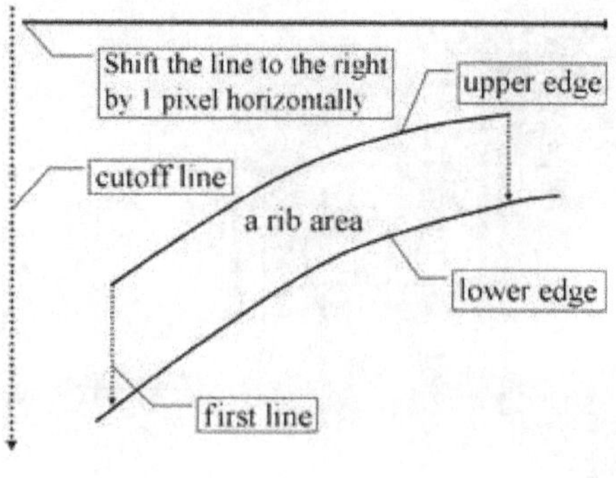

Figure 7. *Designation of a rib area.*

$$v' = \begin{cases} 0 & (0 \le v < a) \\ 255 \times \dfrac{v-a}{b-a} & (a \le v \le b) \\ 255 & (b < v \le 255) \end{cases} \tag{1}$$

4.4. Extraction of Abnormalities

Although small round opacities appear uniformly in chest X-ray pictures, they are unclear in images obtained with CCD scanner. However, it is obvious that the number of high-density small regions such as shadows of small round opacities and vessels increase according to the category level as shown in Figure 1. Hence, abnormalities of pneumoconiosis are extracted based on difference between pixel values in the regions and others. However, in the case when shadows of ribs overlap, abnormalities become high due to difference of the pixel value between two pixels even when high-density small regions are not covered on k-th scanning line. Figure 8 shows two pixel values on 8th scanning line in a rib area of category 1. As Figure 8 shows, although there are no small round opacities on the line, the difference of the pixel values between two pixels is up to 60. Figure 9 (a) shows density values of j-th scanning spot $v[j]$ on the scanning line. And, Figure 9 (b) shows the dif ference of pixel values between $v[j]$ and $v[j+1]$ in Figure 9 (a). From Figure 9 (b), we can see the density value becomes lower as j increases until $j = 24$. However, from $j \geqq 25$, both of the positive and negative value are appeared 7 times alternatively and we can see the density distribution from $j = 25$ is different from that of $j = 1 \sim 24$. Considering the characteristic, the proposed method divides the set of all the pixels on k-th scanning line into two, as its boundary is configured on the pixel where the density distribution changes. Then, the variance of pixel values in each of the sets is calculated. The boundary is configured on each of k-th scanning lines. And then, in order to enhance the difference of density distributions in two sets, the histogram stretching is applied to k-th scanning line in Rm. Figure 9 (c) shows density values on the scanning line where the histogram stretching is applied. Then, regarding $tm[k]$-th pixel counted from the uppermost edge as the boundary, the discriminant analysis is conducted to divide the set of $j = 1 \sim tm[k]-1$ and $j = tm[k]+1 \sim heiRm[k]$.

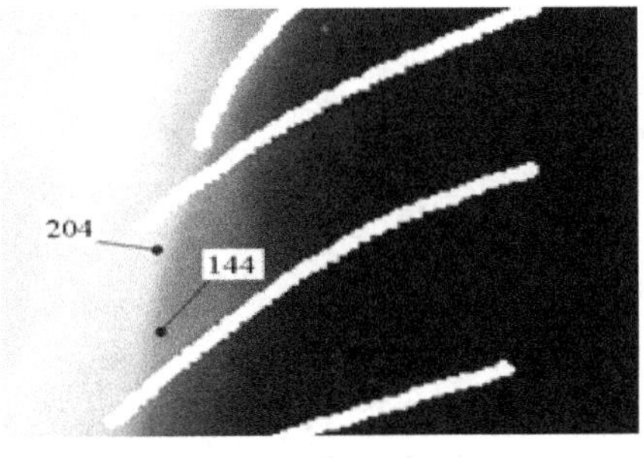

***Figure 8**. Pixels values at some locations in a chest X-ray image (category 0) obtained with a CCD scanner.*

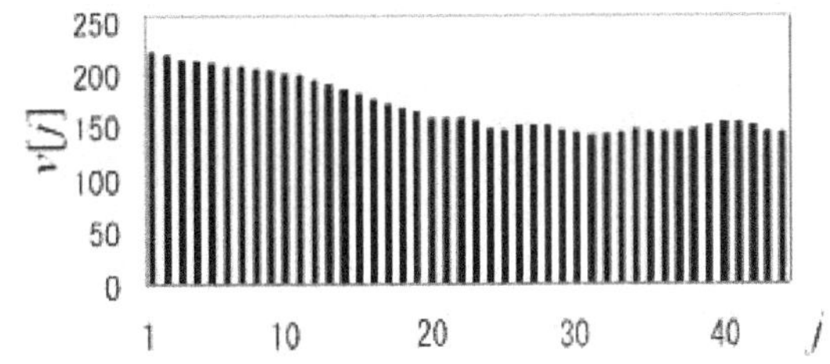

(a) Density values on the 8-th scanning line in a rib area.

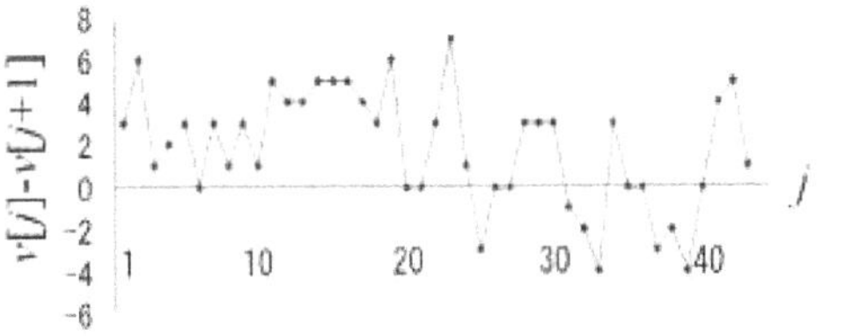

(b) Difference of density values between the neighbor pixels in (a).

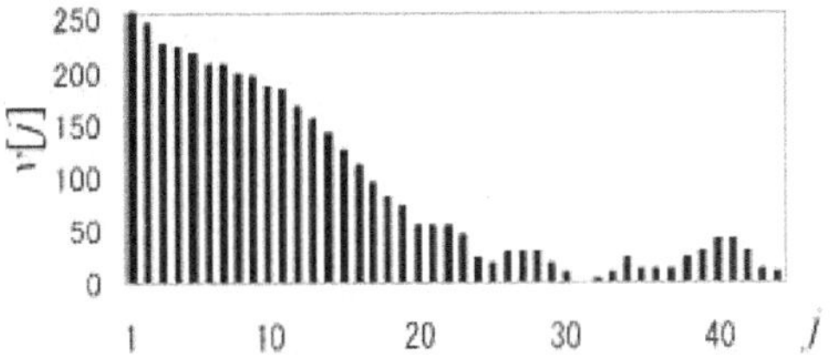

(c) Density values after the histogram stretching applied to (a).

***Figure 9**. Pixels values at some locations in a chest X-ray image (category 0) obtained with a CCD scanner.*

Figure 10 shows parameters for extracting the abnormality of k-th scanning line in Rm, where $heiRm[k]$ is the width of the k-th scanning line. As the scanning is conducted from the upper edge to the lower edge along with the scanning line, the abnormality of k-th scanning line in Rm is defined as

$$Abn(Rm[k]) = \left[\frac{1}{heiRm[k]-1} \left(\sum_{j=1}^{heiRm[k]} |v[j] - avg[k,j]|^2 \right) \right]^{\frac{1}{2}} \tag{2}$$

Where

$$avg[k,j] = \begin{cases} \text{if } 1 \le j < tm[k], \\ \quad \left(\sum_{i=1}^{tm[k]-1} v[i] \right) / (tm[k]-1) \\ \text{if } j = tm[k], \\ \quad v[j] \\ \text{if } tm[k] < j \le heiRm[k], \\ \quad \left(\sum_{i=tm[k]+1}^{heiRm[k]} v[i] \right) / (heiRm[k]-tm[k]) \end{cases} \tag{3}$$

Next, as the first scanning line in *Rm* shifts pixel by pixel to the most inside of the scanning line horizontally, the abnormality in *Rm* is defined as

$$Abn(Rm) = \frac{1}{widRm}\left(\sum_{k=1}^{widRm} Abn(Rm[k])\right) \quad (4)$$

where *widRm* is the number of the horizontal shifts. Then, the abnormality in the whole rib areas is defined as

$$AbnR = \frac{1}{L_R}\left(\sum_{m=1}^{4}\sum_{k=1}^{widRm} Abn(Rm[k])\right) \quad (5)$$

where

$$L_R = \sum_{m=1}^{4} widRm \quad (6)$$

In addition, the maximum value among $Abn(R_1)$ - $Abn(R_4)$ is represented as *AbnRMAX*. Thus, *AbnR* represents overall abnormalities of pneumoconiosis in rib areas and *AbnRMAX* represents local abnormalities of pneumoconiosis in all the rib areas. The proposed abnormalities are based on doctor's experience that they consider comprehensive appearance of small round opacities in lung as well as appearance of each opacity. Figure 11 shows pixel values at some locations in a chest picture of category 0. From the figure, we can see the density value becomes lower towards the inside of the lung. Hence, if we extract abnormalities of pneumoconiosis not using the vertical scanning line, *Abn*(*Rm*[*k*]) would rather depend on density distribution in lung.

Seeing the concept of the proposed method, the proposed method could be regarded as a texture analysis method focusing on difference of contrast between small areas. It means the proposed method could be applied in problems to conduct texture analysis. Especially, differing from general texture analysis methods such as fractal dimension, run length matrix, etc., the proposed method decreases dependence of the horizontal gradation in an area to the abnormalities since the proposed method measures the variance of pixel values in each vertical scanning line. In addition, the proposed method can be applied even if the gradation in an area is occurred in the vertical direction and the scanning lines are in horizontal direction.

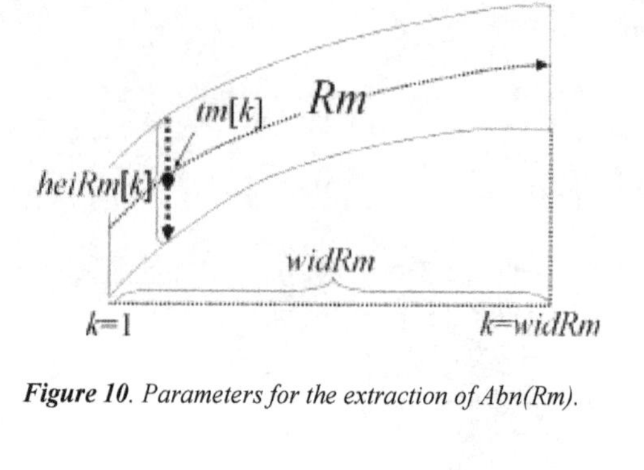

***Figure 10**. Parameters for the extraction of Abn(Rm).*

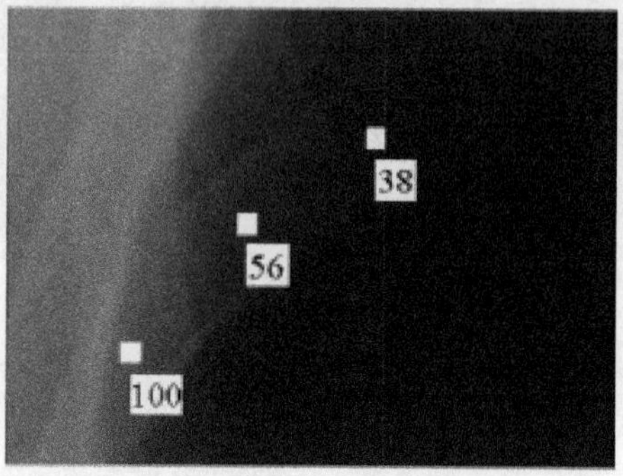

***Figure 11**. Pixel values at some locations in a chest picture of category 0 obtained with a CCD scanner.*

5. Experimental Results

5.1. Results of the Abnormalities

The proposed abnormalities were extracted from 59 chest X-ray pictures. The X-ray pictures were scanned with a CCD scanner (Canon:CanoScan 8800F) and only the right lung in every of the pictures was obtained (resolution: 300 dpi, gray level: 8 bits, format: bitmap image). The number of X-ray images which belong in category 0 is 47, the number in category 1 is 5, the number in category 2 is 4 and the number in category 3 is 3. To obtain rib areas, each of an expert diagnostician and three non-experts (represented as userA, userB, and userC) in pneumoconiosis diagnosis drew the rib edges on all the chest images with a tablet PC (ThinkPad 200Tablet 4184-F5J). The total time used for drawing all the images was 48 minutes in the case of userA, 56 minutes in the case of userB, and 40 minutes in the case of userC. The total time for the expert could not be measured because he/she drew them in spare moments from his regular work. Figure 12 shows the abnormalities *AbnR* and *AbnRMAX* extracted from all the images. From the figure, we can see both of *AbnR* and *AbnRMAX* have been increased gradually according to increase of the category level.

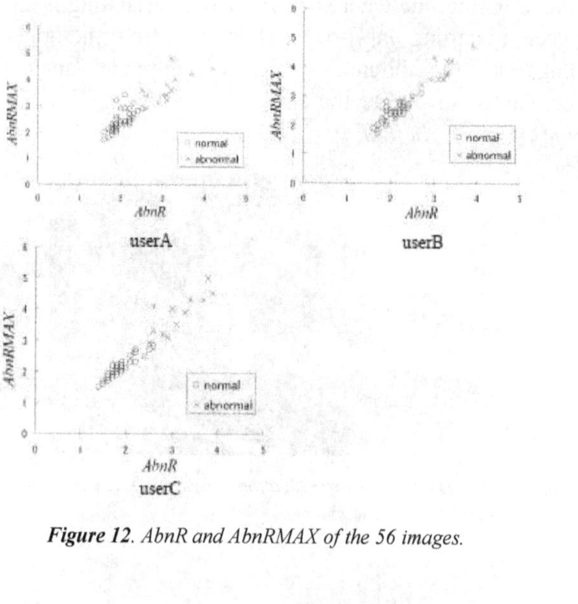

***Figure 12**. AbnR and AbnRMAX of the 56 images.*

5.2. Discrimination of Chest X-Ray Images into Normal or Abnormal Cases

Since patients diagnosed as pneumoconiosis are to be covered by the worker's compensation, the discrimination into normal and abnormal cases in pneumoconiosis diagnosis is especially important for medical doctors. In the checkup of pneumoconiosis, if the diagnostician judges a patients picture as abnormal case, the picture comes on a committee of experts in pneumoconiosis. Then, when it is difficult to discriminate a target picture, disagreement between the diagnostician and the committee is often happened. For the reason, since it is difficult for even experts on pneumoconiosis to diagnose pneumoconiosis, CAD systems for pneumoconiosis have been required. Therefore, we examine performances of the proposed method by the discrimination of the X-ray images into normal or abnormal (i.e., pneumoconiosis) cases by regarding the abnormalities as variants, respectively. As the way of the discrimination, Linear Discriminant Random Trees (RT), a Neural Network (NN), a Support Vector Machine (SVM) were applied to the proposed abnormalities. The number of the trees in RT was set as 10. And, NN was designed as a three-layered perceptron (input layer: 2, hidden layer: 3, output layer: 1), where sigmoid function and the random search method by Matyas were used. Comparing with normal case, since the number of abnormal cases is small, training set and test set were chosen by the cross validation[8] in all the discriminations. The method of cross validation is often applied to increase the number of trials when the number of samples is small for evaluations. In fact, every discrimination was conducted as the following procedure:

(1) Choose N images from all the images as test data, and use the other images as training set.

(2) Discriminate the test data between category 0 and the other categories.

(3) Repeat the procedure from (1) to (2) to every combination changing the test data.

First, Table 2 shows the average recall (*Recall*) in the discriminations when $N = 1$. And, Table 3 shows the average precision (*Precision*) in the discriminations when $N = 1$. The brace notation in Table 3 shows the number of images used for calculation of *Recall* and *Precision*. *Recall* and *Precision* are defined as where, regarding H as the target category, $N_{category}$ is the number of images in H, $N_{correct}$ is the number of images discriminated correctly in H to all the test data (59 images), and N_{output} is the number of images discriminated as H to all the test data.

Next, in order to increase the number of trials for the discrimination, we examined performance of the proposed method when $N = 3$. Hence, ${}_{59}C_3 = 32509$ evaluation functions were prepared in the discrimination. Here, each image is included in test data for ${}_{58}C_2 = 1653$ times and the total number of trials is $59 \times 1653 = 97257$. The average recall (R_{avg}) and precision (P_{avg}) in the discrimination are shown in Table 4 and Table 5, respectively. R_{avg} and P_{avg} are defined

$$R_{avg} = \frac{|A_{output} \cap A_{category}|}{|A_{category}|} \times 100 \tag{9}$$

$$P_{avg} = \frac{|A_{output} \cap A_{category}|}{|A_{output}|} \times 100 \tag{10}$$

where A_{output} is the set of numbers in all the trials discriminated as H, where each of all the trials (97257) is numbered to avoid overlap. Hence, $|A_{category}|$ is the total number of images correctly discriminated in all the trials.

Table 2. *Recall and Precision for the existing method (N=1).*

		Normal		Abnormal	
User	Method	Recall	Precision	Recall	Precision
A	RT	93.62%	81.48%	16.67%	40.00%
	NN	97.87%	80.70%	83.33%	50.00%
	SVM	78.72%	80.43%	25.00%	23.07%
B	RT	87.23%	85.41%	41.67%	45.45%
	NN	70.21%	80.46%	33.33%	22.22%
	SVM	93.62%	88.00%	50.00%	66.70%
C	RT	93.61%	86.27%	41.67%	62.50%
	NN	93.62%	88.00%	50.00%	66.70%
	SVM	91.49%	89.58%	58.33%	63.63%

Table 3. *Recall and Precision for the proposed method (N=1).*

		Normal		Abnormal	
User	Method	Recall	Precision	Recall	Precision
A	RT	95.74%	95.74%	83.33%	83.33%
	NN	95.74%	95.74%	83.33%	83.33%
	SVM	95.74% (45/47)	95.74% (45/47)	83.33% (10/12)	83.33% (10/12)
B	RT	95.74% (45/47)	95.74% (45/47)	83.33% (10/12)	83.33% (10/12)
	NN	97.87%	97.87%	91.67%	91.67%
	SVM	97.87% (46/47)	97.87% (46/47)	91.67% (11/12)	91.67% (11/12)
C	RT	97.87% (46/47)	97.87% (46/47)	91.67% (11/12)	91.67% (11/12)
	NN	100.0%	97.92%	91.67%	100.0%
	SVM	97.87% (46/47)	100.0% (46/46)	100.0% (12/12)	92.31% (12/13)

5.3. Discussion

First, Table 6 shows the mean and standard deviation (SD) of *AbnR* and *AbnRMAX* extracted from each of the normal and abnormal images drawn by each user. From Table 6, we can see both of two abnormalities in abnormal cases are higher than those of normal cases, and the proposed abnormalities are appropriately extracted for the discrimination of pneumoconiosis.

Next, we consider results of the discrimination shown in 5.2. First, the result of the discriminations in Table 3 and 5 is similar to each other. Besides, from experimental results obtained by the existing method shown in Table 2 and 4, we can see both of *Recall* and *Precision* are increased by the proposed method as shown in Table 3 and 5. Although this is not rigorous comparison, since *Recall* of the discrimination in the existing systems[1]-[4] is Approximately from 60% to 80%, we can say the proposed method has high performance in the discrimination of pneumoconiosis. On the other hand, the accuracy of border line in diagnosis of pneumoconiosis is unclear because screening examina- tion is not conducted on pneumoconiosis. As an example, in screening examination of the Central Committee on Quality Control of Mammographic Screening, diagnosticians need to find breast cancer at 80% accuracy and judge the others as not cancer at 80% accuracy. Considering the case above, we can say the proposed method has high performance even from clinical point of view.

Table 4. *Recall and Precision for the existing method (N=3).*

		Normal		Abnormal	
User	Method	Recall	Precision	Recall	Precision
	RT	91.26%	81.45%	18.59%	35.19%
A	NN	98.30%	80.91%	9.2%	57.85%
	SVM	80.36%	62.20%	31.87%	29.29%
	RT	88.64%	84.02%	33.99%	43.32%
B	NN	99.03%	79.80%	1.8%	32.50%
	SVM	71.27%	82.24%	39.72%	26.09%
	RT	96.22%	87.36%	45.51%	75.43%
C	NN	94.50%	88.84%	83.50%	71.28%
	SVM	91.34%	89.57%	58.33%	63.22%

Table 5. *Recall and Precision for the proposed method (N=3).*

		Normal		Abnormal	
User	Method	Recall	Precision	Recall	Precision
	RT	95.81%	95.75%	83.35%	83.56%
A	NN	98.12%	95.55%	82.11%	91.79%
	SVM	95.97%	95.89%	83.90%	84.17%
	RT	96.36%	96.02%	84.37%	85.53%
B	NN	98.58%	97.33%	89.39%	94.14%
	SVM	97.87%	97.87%	91.67%	91.67%
	RT	97.57%	97.81%	91.44%	90.58%
C	NN	99.06%	97.99%	92.06%	96.17%
	SVM	97.80%	99.73%	98.97%	92.02%

Table 6. *Statistics of AbnR and AbnRMAX.*

Category	User	AbnR		AbnRMAX	
		Mean	SD	Mean	SD
	A	2.0	0.2	2.4	0.4
Normal	B	2.1	0.2	2.4	0.3
	C	1.8	0.3	2.0	0.3
	A	3.2	0.4	3.7	0.5
Abnormal	B	3.1	0.3	3.8	0.5
	C	3.1	0.5	3.8	0.6

We consider images discriminated incorrectly in experiments shown in 5.2. There are 2 images discriminated incorrectly as normal case when both of $N = 1$ and $N = 3$. Each of the images is represented as abnormal ① and abnormal② respectively in this paper. Figure 13~15 show the abnormalities *AbnR* and *AbnRMAX* extracted from all the images drawn by each user. abnormal① was discriminated incorrectly when $N = 1$ expect in the discrimination by SVM in the case of userC. Table 7 shows the total number of cases that abnormal ① was discriminated correctly in all the 1653 trials. From Figure 13~15, abnormal① has the lowest abnormality in all the images regardless of user. Figure 16 shows the standard deviation $SD[k]$ obtained from density values in all the pixels on k-th scanning line in R_4. And, Figure 17 shows $Abn(Rm[k])$ extracted from the density values in all the pixels on k-th scanning line in R_4. Although shadows of ribs overlap each other on scanning lines from $k = 1$ to 24, the average of $SD[1]$~$SD[24]$ is 10.2 in Figure 16 and the average of $Abn(Rm[1])$~$Abn(Rm[24])$ is 4.6. Since both of $SD[k]$ and $Abn(Rm[k])$ are based on average of density values in a set of pixels on a scanning line, we can see noises appeared from $k = 1$ to 24 in Figure 16 are reduced as shown in Figure 17. However, sometimes $Abn(Rm[k])$ does not become higher in the case when small round opacities on k-th scanning line appear only on the set of pixels, namely $j = 1$~$tm[k]$-1 or $j = tm[k]+1$~$heiRm[k]$ because the opacities are regarded as noise. For example, although the noise does not exist on 39 th scanning line in R_4, $SD[39]$ is 8.2 in Figure 16 and $Abn(Rm[39])$ is 5.0 in Figure 17. There is a gap of the average density value between the set of pixels $j = 1$~$tm[39]$-1 and $j = tm[39] + 1$ ~$heiRm[39]$ on the 39th scanning line, the former one is 87.7 and the latter one is 71.5.

Table 7. *Number of the cases that abnormal ① was discriminated correctly.*

User	Method	Correct results / Trials	
	RT	4/1653	0.24%
A	NN	184/1653	11.13%
	SVM	3/1653	0.01%
	RT	0/1653	0%
B	NN	3/1653	0.01%
	SVM	0/1653	0%
	RT	12/1653	0.7%
C	NN	365/1653	22.08%
	SVM	1448/1653	87.60%

Table 8. *Number of the cases that abnormal ② was discriminated correctly.*

User	Method	Correct results / Trials	
	RT	2/1653	0.12%
A	NN	147/1653	8.89%
	SVM	111/1653	6.71%
	RT	1653/1653	100%
B	NN	1621/1653	98.06%
	SVM	1653/1653	100%
	RT	1630/1653	98.61%
C	NN	1641/1653	99.27%
	SVM	1653/1653	100%

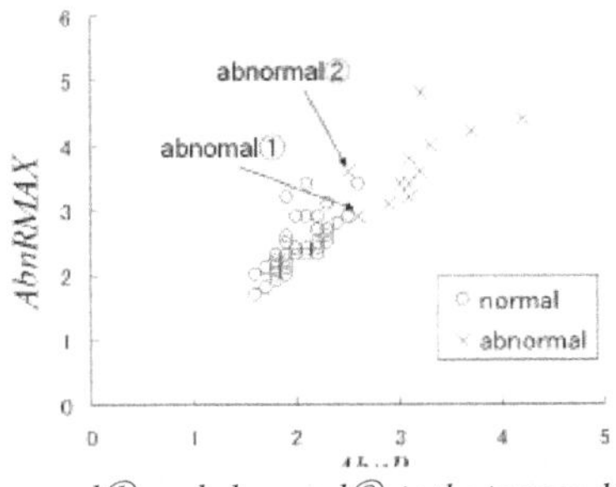

***Figure 13**. abnormal① and abnormal② in the images drawn by userA.*

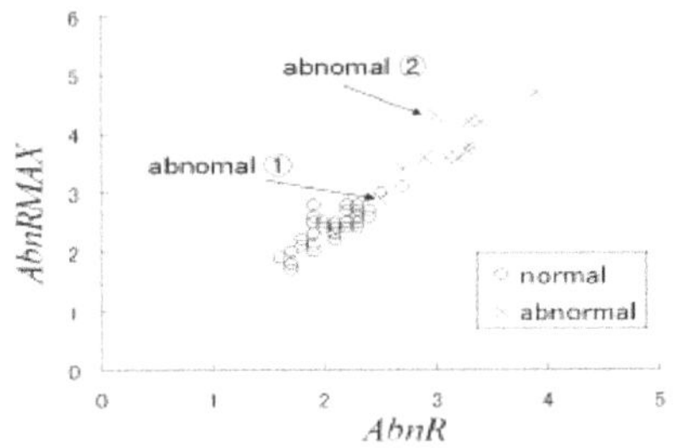

***Figure 14**. abnormal① and abnormal② in the images drawn by userB.*

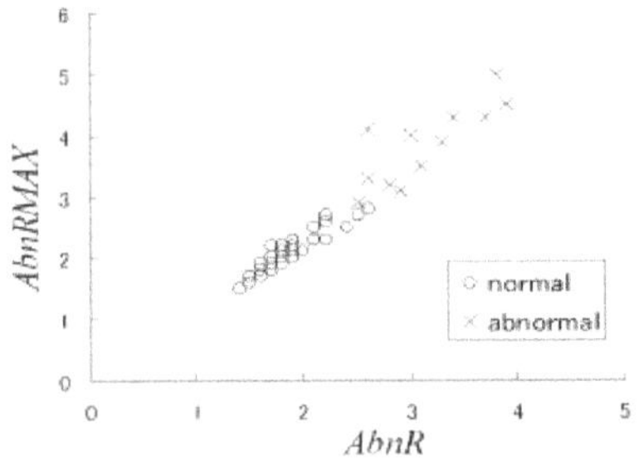

***Figure 15**. abnormal① and abnormal② in the images drawn by userC.*

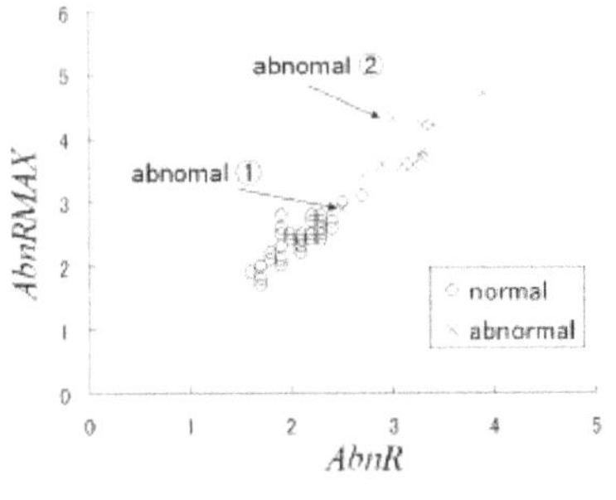

***Figure 16**. SD[k] extracted from a rib in abnormal①.*

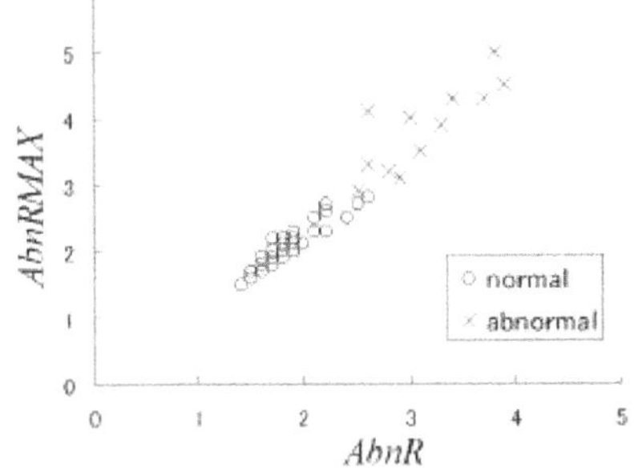

***Figure 17**. Abn(Rm[k]) extracted from a rib in abnormal①.*

The gap indicates a small round opacity is appeared mostly on the set of pixels $j = 1 \sim tm[39]\text{-}1$, and variance of the density values in the set would be low. In this case, by detecting overlap of shadows of ribs on the scanning line, we would like to select $SD[k]$ or $Abn(Rm[k])$ as the abnormality on the scanning line. abnormal ② was discriminated incorrectly by all the tools when $N = 1$. The total number of trials to abnormal② is 1653 when $N = 3$, and Table 8 shows the number of correctly discriminated cases by each tool in each user. From Figure 13～15, we can see the normal cases and abnormal cases can be broadly grouped, and abnormal② for userA in Figure 13 is placed around boundary between the two groups of normal case and abnormal case. From Table 8, since average correct ratio of abnormal ② is significantly different between userA and the others, abnormal② for userA could be discriminated incorrectly according to the accuracy of the drawn curve. The above example is considered to be an example of the case that the proposed abnor-malities are depended on the accuracy of drawn curves. Hence, as a future subjective, to reduce the significant difference of drawn curves between users, we need to consider the way of displaying the manual and samples for drawing curves for the rib edges.

6. Conclusions

To enhance cost-performance and mobility in CAD systems for pneumoconiosis, this paper has presented a method for extracting abnormalities of pneumoconiosis from chest X-ray images obtained with a CCD scanner. The abnormalities have been extracted from rib areas based on density distribution in the areas. Then, regarding the abnormalities as variants in the discrimination of pneumoconiosis, this paper has examined performance of the proposed abnormalities with the common discriminations. Experimental results of the discriminations for 59 right-lung images have shown that both of the recall and precision has become more than 90% to the normal lungs and 80% to the abnormal lungs.

As future subjects, it is necessary to design an interface for drawing rib edges or propose a method for extracting rib areas automatically. And then, considering cases of discrimination failure, it is necessary to detect overlap of shadows of ribs on scanning lines, and to consider the manual and samples for drawing rib edges as the proposed method less depends on user.

References

[1] R. P. Kruger, W. B. Thompson, and A. F. Turner, "Computer diagnosis of pneumoconiosis," Trans. on Systems, Man. And Cybernetics, vol. SMC-4, no. 1, pp. 40–49, Jan. 1974.

[2] A. M. Savol, C. C. Li, and R. J. Hoy, "Computer aided recognition of small rounded pneumoconiosis opacities in chest X-rays," IEEE Trans Pattern Anal. March. Intell., vol. 2, no. 5, pp. 479-482, Sep. 1980.

[3] J. Wei and H. Kobatake, "Detection of rounded opacities on chest radiographs using convergence index filter," Proc. ICIAP, pp. 757-761, Sep. 1999.

[4] H. Kondo and T. Kouda, "Computer-aided diagnosis for pneumoconiosis using neural network," Proc. IEEE Symposium on Computer-Based Medical Systems, pp. 467-472, Jul. 2001

[5] M. Nakamura, K. Abe, and M. Minami, "Quantitative evaluation of pneumoconiosis in chest radiographs obtained with a CCD scanner," Proc. of ICADIWT 2009, pp. 673-678, London, UK, Ang. 2009.

[6] M. Nakamura, K. Abe, and M. Minami, "Extraction of features for diagnosing pneumoconiosis from chest radiographs obtained with a CCD scanner," Journal of Digital Information Management, vol. 8, no. 3, pp. 147-152, Jun. 2010.

[7] M. Loog and B. Ginneken, "Segmentation of the posterior ribs in chest radiographs using iterated contextual pixel classification," IEEE Trans. on Medical Imaging, vol. 25, no. 5, pp. 602-611, May 2006.

[8] F. Mosteller, "A k-sample slippage test for an extreme population," The Annals of Mathematical Statistics, vol. 19, no. 1, pp. 58-65, Mar. 1948.

15

On the warning system of obstacle avoidance of embedded electronic guide dogs

YUE Xiangyu

School of Management and Engineering, Nanjing University; Nanjing Jiangsu; 210093; PR China

Email address:
yuexiangyu168@gmail.com

Abstract: With the rapid development of China's transportation, the frequency of traffic accidents is also high. This not only restricted the development of China's transportation greatly, but also threatened to people's safety seriously. In particular, the accidents caused by the blind due to there are more frequent reproduction, so their traffic safety has become a big issue to solve urgently. In the program, a new type of "warning system of obstacle avoidance of embedded electronic guide dog" has been developed on the basis of careful analysis of all kinds of present anti-collision warning systems, which has a core micro-controller, 32-bit ARM7 microprocessor, and takes the embedded operating system uCLinux as its platform. Such warning system of obstacle avoidance of embedded electronic guide dog can effectively eliminate the impact of the traffic environment and the subjective factors of the blind, warning in advance for the travelling blind in time, effectively avoiding obstacles such as vehicles, to reduce traffic accidents caused by the their blindness. This humane technology innovation is the specific embodiment of environmental science and technology aesthetic theory in the field of scientific and technological innovation. It has a positive and promoting role to the development of transportation and blind-man welfare in China.

Keywords: Embedded System, Electronic Guide Dog, Avoiding Obstacles, Warning, ARM, USB Communication, Uclinux

1. Introduction

During last summer, I went blind orphanage as a volunteer, and personally experienced the inconvenience and hardship of the blind when they travel. Especially when I heard the news that a blind man had been hit and killed when crossing the street just several days before because of his blindness, my compassion was once again inspired, and I was very eager to do my best to help the blind. I think that if electronic guide dogs guide them to avoid obstacles such as vehicles, the blind people are able to avoid such accidents, aren't they? So I had an idea that I use the scientific knowledge of embedded system grasped by me to invent a kind of "electronic guide dog". After that I began to design the structure of the entire system, which made the overall implementation of "electronic guide dog" take a welcome step.

In fact, this "warning system of embedded electronic guide dog avoiding obstacles" is similar to a car anti-collision avoidance system. Currently, some of the international anti-collision avoidance system development has made some achievements, and the more successful countries are Japan, Germany, and the United States. In China, the research and development in this area have relatively large gap with developed countries, and not until recent years did a number of research institutes and universities begin to study it. The research on "warning system of embedded electronic guide dog avoiding obstacles" in this paper is an important part of intelligent electronic guide dog system, the study of which is mainly based on the embedded system, anti-collision warning system, and the USB technology is applied to warning system of embedded electronic guide dog avoiding obstacles.

2. The Design of Embedded Electronic Guide Dog of Obstacle-Avoiding Early Warning System

The design of obstacle-avoiding early warning system of embedded electronic guide dog is an inheritance and innovation, based on the design of traffic information collecting system, which is generally used in vehicle anti-collision at present. Nowadays, the main types of the design of traffic information collecting system, generally used in vehicle anti-collision, are shown as follows, that is, using single chip microcomputer for data analysis, PC-centric (industrial control computer), using DSP for data analysis and so on. [1] In the design process of the obstacle-avoiding early warning system of embedded electronic guide dog, we firstly has carried on a comprehensive evaluation and analysis for each scheme, and then put forward a new idea that it introduces the open source embedded real-time operating system uCLinux, with a high-powered ARM core as the core processor.[2] We take this as the foundation, expanding a variety of the functions required, for instance, the function of USB communication interface, and human-machine interface. In this way, the design of obstacle-avoiding early warning system of embedded electronic guide dog has the following advantages. It has a simple design, and small volume, so that it is convenient to carry the "dog". Also, it is easy to extend the functions, which is convenient for field operations, greatly reducing the cost of the traffic information collecting system in the obstacle-avoiding early warning system of electronic guide dog.

What's more, the design of obstacle-avoiding early warning system of embedded electronic guide dog has utilized the core processor, based on the realization of the system function, and development conditions. Its operating system is thought from the perspectives of practical application, development tool, instantaneity, technical services, and price, etc. Its main control unit hardware uses the ARM microprocessor as the core, processing and handling the signal of each sensor. [3] The alarm system of guide dog adopts the combination of light and sound. Sound will produce auditory stimulus for the blind. Meanwhile, light and sound may make an auditory and visual stimulation for other pedestrians, arousing the attention of the blind and pedestrians to let them take appropriate actions at the same time to avoid accidents. Moreover, the obstacle-avoiding early warning system of embedded electronic guide dog also selects USB protocol to transfer data, saving the collected data in the hardware after being managed.

The overall design plan of obstacle-avoiding early warning system of embedded electronic guide dog mainly is the velocity and distance measurement module, the power circuit module, the LED screen show module, USBcommunication circuit module, and clock circuit module. The hardware of the obstacle-avoiding early warning system of embedded electronic guide dog can reference the Figure 1:

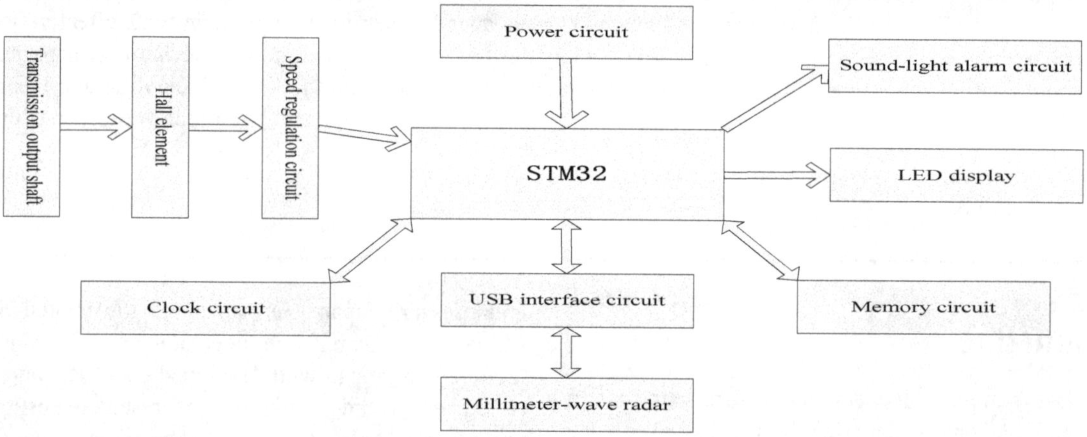

Figure 1: *Schematic diagram of hardware system structure [4]*

3. The Application of Embedded Electronic Guide Dog of Obstacle-Avoiding Early Warning System

In the anti-collision early warning system, referenced in the design of obstacle-avoiding early warning system of embedded electronic guide dog, there mainly exits four steps, that is, information collection, information processing, information judgment, and warning information. The techniques of information collection are ultrasonic wave, laser, infrared ray, machine vision and interactive method. Information processing is mainly to analyze the collected information, usually using ARM, MCU and DSP microprocessor, etc.[5] And information judgment is based on the information, like the model of the guide dog, opportunity, weather, the distance between the guide dog and the obstacle, relative acceleration, relative speed and so on. Through the technique of information fusion, this step will make a real-time dynamic measurement and identification of its risk or safety state.

Presently, the types of the generally used anti-collision warning systems mainly are radar anti-collision warning system, ultrasonic anti-collision warning system, laser anti-collision warning system, infrared anti-collision warning system, machine vision anti-collision warning system, and interactive intelligent anti-collision warning system. The selection of anti-collision warning system for the design of obstacle-avoiding early warning system of embedded electronic guide dog should start from the characteristics of the highway network and the street network construction in our country, combing with the characteristics of the obstacle-avoiding early warning system of electronic guide dog, as well as the construction of our country's highway and street traffic integrated management system. We should optimize the choice of the obstacle-avoiding early warning system technique of embedded electronic guide dog, based on China's national situation. Therefore, in the design process of the obstacle-avoiding early warning system of embedded electronic guide dog, after the comprehensive and comparative analysis of the anti-collision warning techniques, we have compared the optimization principles of the anti-collision warning techniques, and finally adopt the approach of millimeter wave radar sensor + ARM microprocessor.

The design of obstacle-avoiding early warning system of embedded electronic guide dog also adopts the generally used embedded processor. Presently, the generally used embedded processors mainly are ARM, Power PC, MIPS, Motorola 68000 series, etc. The full name of ARM is Advanced RISC Machines. The ARM architecture follows the principle of reduced instruction set computer (RISC). It has a variety of merits, like small size, high performance, low power consumption, and cheap cost. It extensively uses the registers with a fast speed of instruction execution. And its way to addressing is much easier and more flexible, and its operating efficiency is high. Also, it could complete most of the data manipulation in the register. Due to the operating system is an aggregation of the system program, people don't have to consider the differences between different hardware, and could achieve the target of writing application software through the system interface provided uniformly. The embedded system will further compile all of its procedure codes, including the operating system code, driver's code and application code, into a whole paragraph of executable code and in inserted them into the hardware. In this way, the embedded operating system in embedded system is more like a set of function library. And the embedded operating system should satisfy basic requirements of the reliability, real-time performance, and tailing capability. At present, the embedded RTOS (Real Time Operate System) mainly includes two categories--commercial and free. Thereinto, the commercial RTOS includes WINCE and VxWotks; while the free RTOS has Linux (uCLinux and RT Linux included) and uC/OS - II. UCLinux aimed at the micro-control field, designing the Linux system, which is specially designed for the CPU without MMU (such as the S3C44B0X adopted in this project), and it has done a lot miniaturization work for the Linux kernel. These are the most remarkable features that uCLinux owned. Besides, its other main features are universal Linux API, core file less than 512KB, core+file system less than 900KB, and complete TCP/IP. Also, it supports a large number of other network protocols and various file systems, including NFS, EXT2, ROMFS, JFFS, ms-dos, FAT16/32, etc.

In addition, the design of obstacle-avoiding early warning system of embedded electronic guide dog also makes a full use of USB. USB is a new kind of computer peripheral communication interface standard, which abandons the defects of the traditional computer series/parallel, with advantages of the reliability of data transmission, hot plug (plug and play), convenient extension, and low cost. The reference model is shown in Figure 2.

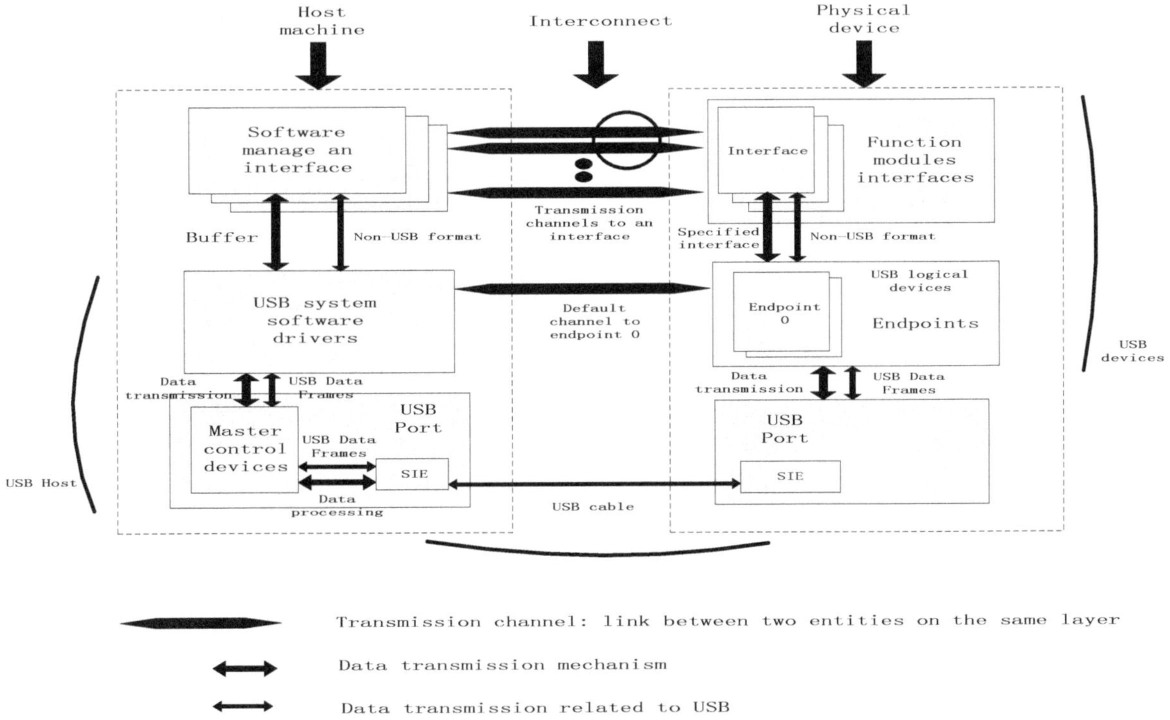

Figure 2: *the structure diagram of USB reference model*[4]

4. Conclusion

Today, with the rapid development of transport and vehicles, the frequency of traffic accidents has been being high, which not only greatly restricted the development of transport, but also made people travel in great danger. The traffic safety problem of the blind is an urgent issue to solve. The warning system of embedded electronic guide dog avoiding obstacles studied in the project can effectively eliminate the influence of traffic environment and blind subjective factors, send the pre-trip alarm for the blind in time, effectively avoid obstacles such vehicles, to reduce traffic accidents caused due to the blindness. This humanized technology innovation is the embodiment of environmental science and technology aesthetics theory in the field of science and technology innovation practice. This has not only a positive role in promoting the development of Chinese transport and the blind welfare, but also important practical significance and practical value for us to build a harmonious society and a beautiful China.

Brief Introduction to the Author

Yue Xiangyu (1992-), who is a male, Han nationality, was born in Weifang, Shandong, with BA, Nanjing University, mainly engaged in electrical information and automation, aesthetics and other aspects of learning and research, specializing in modern industrial embedded control systems, networked control systems, intelligent control. He presided over one national college student science and technology innovation project, taking part in the research on one "Eleventh Five-Year Plan" education and science key project of the Education Ministry, and one soft science research project of Shandong Province. He has 12 science and technology papers published in Chinese and oversea academic journals. He has won the 1st scholarship of China, the 1st scholarship of people, the 1st and the 2nd prizes of China Education Robotics Competition, the 1st prize of China Mathematical Modeling Contest, the top award of scientific research achievement of Nanjing University, two China science and technology patents and the honorary title, "three-good-student" of Jiangsu Province. In addition, as a representative of Nanjing University, he went to the National University of Singapore to participate in an academic exchange program. Address: School of Management and Engineering, Gulou Campus of Nanjing University, 22, Hankou Avenue, Gulou District, Nanjing, Jiangsu Province, PR China; Zip: 210093.

References

[1] R. Weil, Jwooton, A. Garcia-Ortiz. Traffic Incident Detection: Sensors and Algorithms [A]. Math Comput Modeling [J], 1998, (27): 257 - P291.

[2] Zhu Weiwei, Yang Jianming. uClinux: an embedded Linux system[A]. Ship Electronic Engineering [J], 2003,12 (3): 3-5.

[3] GL Duckworth. A comparative study of traffic monitoring sensors [A]. Proceedings of the IVHS AMERICA annual meetingmAtlanta [C], 1994: 283-293.

[4] Ji Xiaojiang, Zhao Jing. Vehicle Rear Pre-Alarm System[A]. Enterprise technology development[J]. 2011, (12): 8-9.

[5] Guo Min. Vehicle detection technology and comparative analysis of vehicle detector[A]. Chinese Transportation Information Industry[J], 2003, (10): 55-56.

16

Development of a model for simultaneous cost-risk reduction in JIT systems using multi-external and local backup suppliers

Faraj El Dabee, Romeo Marian, Yousef Amer

School of Engineering, University of South Australia, South Australia, Australia

Email address:
eldfy001@mymail.unisa.edu.au(F. E. Dabee), romeo.marian@unisa.edu.au(R. Marian), Yousef.amer@unisa.edu.au(Y. Amer)

Abstract: In many organisations, Just-In-Time (JIT) implementation plays a significant role in minimizing their excessive costs, and increasing their efficiency. However, the risks accompanying JIT strategies are often overlooked and affect system processes disrupting the entire chain of supply. This paper proposes an inventory model that can simultaneously reduce costs and risks in JIT systems. This model is developed in order to ascertain an optimal ordering strategy for procuring raw materials by using multi-external suppliers and local backup supplier to reduce the total cost of the products, and at the same time to reduce the risks associated with JIT supply within production systems. The effectiveness of the developed model is tested using an example problem with inbuilt disruption. A comparison between the cost of using the JIT system and using the inventory system shows the superiority of the use of the inventory policy.

Keywords: Lean Manufacturing, Just-In-Time (JIT), Production System, Cost-Risk Reduction, Inventory Model, External Supplier, Local Backup Supplier

---

1. Introduction

In today's competitive global markets, customers seek to obtain their supplies whilst simultaneously obtaining the cheapest prices irrespective of where they are produced. This leads organisations to implement new techniques, in order to reduce the costs of their products and to insure their position in the marketplace [1]. Lean manufacturing is a philosophy, which may be used to assist production systems to reduce their waste, and to increase the activities that add value to the end product to increase its attraction within its market segment [2]. This approach is conceptually simple which has earned it wide popularity. By understanding these foundational concepts and principles, lean manufacturing may be more easily applied [3].

The main task of the lean manufacturing system is to locate the major sources of waste which are then to be eliminated by the application of a large number of tools such as JIT and production smoothing [2]. JIT is considered as one of the significant lean manufacturing tools that can be used within organisations leading to improvement on a continuous basis including the flow of materials and information, management of human resources, improved throughputs, costs reduction, and elimination of wastes and non-value added activities [4].

Most international organisations have implemented lean manufacturing tools such as Just-in-Time (JIT) in their processes to reduce their costs and to improve their efficiencies [1]. However, they tend to ignore the risks arising from these goals. These risks will impact on their processes disrupting the entire supply chain.

The main objective of this paper is to develop an inventory model for simultaneously reducing costs and their effects in JIT systems. The goal is to determine an optimal ordering strategy for obtaining supplies within the production systems using both external and local backup suppliers. This strategy is crucial in case of the occurrence of unforeseen disruptions such as natural and man-made disasters, and economic crises to achieve a high product quality and total financial and operational actions within the supply chain. The flexibility to access various suppliers can significantly and positively affect lead time and performance [5].

The rest of this paper is organised as follows: Section 2 reviews some of the literature on JIT, and cost and risk modelling. The problem illustrating JIT implementation is described in section 3. Section 4 presents the proposed model formulation to reduce costs and their risks in JIT systems. In section 5, a simplified problem is provided to illustrate the application of the developed model. Section 6 discusses the findings from implementing the developed model in a simplified problem. Finally, section 7 summarises and concludes this paper.

2. Literature Review

Just-in-Time (JIT) is a Lean manufacturing tool that can be utilised to improve organisations' efficiency. It is a manufacturing pull system, which can be used for planning and controlling operations, in order to produce, and supply the required products at the correct place, when they are required, and at the right ordered amounts [6], [7]. The main principles of JIT include: high quality, small lot sizes, and regular deliveries in short lead times, close contact with suppliers [8]. The appropriate use of JIT in manufacturing can reduce waste and increase productivity, efficiency, profit, and customer satisfaction [9], [10]. According to Tourki [10], some critical principles such as people involvement, training and education, supplier relations, waste elimination, Kanban or pull system, uninterrupted work flow, and total quality control are used for successful implementation of JIT system. In addition, JIT is highly beneficial for many companies, as the literature indicates that the efficiency gained from the implementation of JIT in production processes translates in accelerated productivity. Inventory levels for manufacturing dropped from 50 days to 40 days during 1999 and 2000 in United States. This indicates the importance of JIT implementation for manufacturing production by achieving operational efficiency [11]. Furthermore, it is a critical tool that can also be utilized for the purpose of managing the external activities associated with an organisation including that of purchasing, as well as distribution. Three elements included in case of JIT are: JIT production, JIT distribution and JIT purchasing [2].

Recently, researchers have searched for an economic quantity model for production systems following a JIT approach for ordering raw materials and the shipping processes. Different models can be utilised for the purpose of ensuring reduction in the level of cost and risk in case of JIT systems. For instance, one such model type that can be utilised for achieving cost efficiency is the lot size reduction model. This model emphasizes that by ensuring reduction in the lot size, it can become possible to achieve a reduction with respect to the level of the cost required in performing the delivery of finished products to final consumers [12]. Fahimnia et al. [13] developed a mixed integer formulation for optimising a two-echelon supply network. They concluded that by implementing the developed model in a case study, it is clear that by considering all production costs prove the effectiveness of this model in the real applications. A higher lot size unnecessarily increases cost and some components of risk, while reducing others. As a result, the lot size risk reduction model can be utilised in order to ensure an optimum lot size and thereby, efficient management of risk from the lot size can ultimately become possible to achieve cost efficiencies. An operation model may also be used for the purpose of JIT scheduling which explains each and every process included in the JIT system. Thus, by way of identifying the stages of JIT systems, necessary actions can be taken for the purpose of achieving cost efficiency in the operation [14].

Sarker and Khan [15] developed a general cost model for the two-stage batch environment taking into consideration a limited rate of production. This model can be utilised to ascertain the product batch-sizes and order-sizes of raw materials, so reducing the total cost that meet the same batches of products, at fixed intervals, to the buyers. Yang and Pan [12] investigate a JIT purchasing model where a single vendor supplies a single purchaser with a product. Their work presents an integrated inventory model, which minimizes the sum of the ordering cost, holding cost, quality improvement and crashing cost by optimizing the order quantity, lead time, process quality and the number of deliveries to provide a lower total cost, higher quality, smaller lot size and shorter lead time. Therefore, applying JIT methods such as small lot size, lead time reduction and quality improvement play a significant role in achieving JIT purchasing goals. In their article, a stochastic model, which includes two stages, was developed by Carneiro et al. [16] to optimise investment portfolios within an oil supply chain in Brazil. Three sources of uncertainty are considered by adopting the conditional value-at-risk (CVaR) as a risk measure within six oil refineries, in order to minimise the expected net present value (ENPV) in the supply chain. Additionally, Julka et al. [17] propose a unified, flexible, and scalable framework for modelling, monitoring and management of refinery supply chains. This framework has two basic elements: object modelling of supply chain flows and agent modelling of supply chain entities. Three classes of agents, emulation, query, and project agents are used for methodologies required for decision-support systems. It is essential to define the optimum production lot size and the order quantities of associated raw materials simultaneously. This could be done by treating the production and purchasing as modules of a single system, minimizing the total cost of the system [15].

As systems become increasingly integrated, any disruption cannot be arrested in the functional area of origin and propagated through the production and distribution system. The reduction of waste (muda), as inventory or extra production capacity, exposes adjacent activities and may affect the whole supply chain. In his article, Tomlin [18] investigates some features of the organisation, its supplier(s), and its products such as supplier reliability, and supplier failure correlation and their impacts on the organization's preference. He also mentions that common dual sourcing can protect organisation from any disruption impacts due to receiving deliveries from both in case of one supplier is disrupted. Simchi-Levi et al. [19] point out the risks associated with a JIT system in cases of unforeseen disasters disrupting supply chain such as what eventuated with some auto manufacturers following Sept. 11, 2001. They emphasise that sharing risks throughout supply chain parties has a significant impact on them.

Dimakos and Aas [20] presented a new method to model the required total economic capital required, in order to strengthen a financial organization against possible losses. The system was implemented in the Norwegian financial group DnB's system for risk management. It is concluded that the total economic capital was reduced by 20% of the actual rate for a one year. Also, Gaivoronski et al. [21] presented an approach for considering a cost–risk balanced process to manage the scarce water resources in conditions of uncertainty. A new technique was modelled relating to a re-optimization phase that allows users to organise emergency strategies by adopting the barycentric value as a new target, which resulted in drastic risk reduction in resource delivery. In addition, Jose [22] clarifies how risk management sources in a project´s innovation can be better managed through a modelling process. Although the innovation management relevance is uncertain, several methods of risk management have been proposed. This article focuses on the formation and management of uncertainties in a context and the deployment of risk management techniques. By using a general model of innovation to manage the parameters of risk creation, the risk management process is applied to a specific case. El Dabee et al. [23] developed a mathematical model to reduce the total cost of the products, and at the same time to reduce the risks arising from this cost reduction within production systems by using external suppliers for supplying raw materials to the production systems. They concluded that comparing the use of a JIT system with the use of a specific amount of inventory during a limited period of time had a significant impact on the production system.

According to the literature review related to JIT, all developed models were used to reduce either cost or risk independently. It is clear that risks have an adverse impact in organisations' performance, which causes an increase in their total costs and at the same time reduces their efficiency. Therefore, risks should be assessed by identifying, evaluating, and measuring them, in order to reduce the undesired effects they cause within these organisations.

3. Problem Description

In this paper, it is assumed that a distribution network consists of multiple external suppliers. This is due to pricing variances for the same product in different markets. The materials are transported from different manufacturers to the production system, which in turn produces the final product for sale to wholesale or retail outlets. Also, the raw materials are replenished instantaneously to the production system to meet JIT requirements. To avoid any risks that arise from possible disruptions occurring to the external supply chain, it is assumed that the production system is capable of obtaining the raw materials required for full production up to the finished product from local backup suppliers at a higher cost but in a shorter lead time.

4. Model Development

All notations and assumptions, decision variables, parameters, and mathematical formulations will be described as follows:

4.1. Assumptions

The model formulation is based on the following assumptions:

- The ordering cost of raw materials is a at fixed rate for each order regardless of the order size;
- The utilities cost of the final product is a percentage of total cost of the product that can be changed by the inventory batch size;
- The final product price is at a fixed rate regardless of the inventory batch size;
- The raw materials are supplied by the regular external supplier if there is no disruption occurs;
- The raw materials can be purchased from the local backup supplier when one or more of the regular external suppliers are disrupted;
- The cost of raw materials from the local backup supplier SLB can be considered as a percentage of their cost when they are purchased from the regular external suppliers depending on its reliability (RS);
- The worker cost required for producing the final product per time unit is a fixed rate per time unit;
- The risk cost arising from the likelihood of risk occurrence is a percentage rate depending on its impact on the production system;
- The duties cost is incurred if raw materials can be supplied by an external supplier; and
- The transfer price required to procure raw material from the regular external supplier can be considered as a percentage of its total cost CM.

4.2. Notations

The following notations are used in the proposed model:

C_T: Total cost required to produce one product in monetary unit (MU);

C_M: Raw material cost required for producing one product (MU);

C_O: Ordering cost of raw materials (MU);

C_H: Holding cost of raw materials within the production

system warehouses (MU);

C_{UM}: Unit cost of the raw material at the beginning of that cycle (MU);

C_R: Risk cost arising from disruption occurrence (MU);

C_{Li}: Labor cost rate per labor time in operation *i* (MU/hr);

C_{tr}: Transportation cost for delivering raw materials to the production system (MU);

C_P: The purchasing cost of the raw materials that are required to produce the product (MU);

C_U: Utilities cost of the final product (MU);

C_D: Duties cost arising from procuring raw material from an external supplier (MU);

C_{UH}: The cost that is carried per unit during each cycle (MU);

TP_i: Transfer price required for procuring raw material i from an external supplier *i* (MU);

D_i: Duty rate (%) per price of raw material *i* supplied by an external supplier (MU);

tp: The percentage rate of raw material cost (MU);

$T_{S,\ n,\ m}$: Tensor for transportation cost per critical measurement (MU);

S: Origin of ordered raw materials;

V: Destination of required raw materials;

m_i: Transportation mode for transporting raw material *i* to its customer;

tm: Critical transportation measurement of raw materials shipped using transportation mode *m*;

SE_i: Raw material external supplier *i*;

SLB_i: Raw material local backup supplier *i*;

IF: Indicator function for duty with a value 1 or 0. 1 if the supplier and the production facility are in the same country and 0 otherwise;

M_i: raw material types required in producing one unit of product *i*;

LH: Likelihood of occurrence for risk in the supply chain;

I: Impact of risk occurrence in the supply chain; and

%TRS: Total risk score percentage value.

4.3. Parameters

d_P: Customer demand for the final product in a period (unit);

N_O: Number of operations required for producing one product (unit);

N_W: Number of workers required to produce one product (unit);

N_h: Number of working hours for producing the final product (unit);

N_P: Number of parts required to produce one product (unit);

N_S: Number of external suppliers required to supply raw materials to the production system (unit);

C_W: Worker cost required for producing the final product per time unit (MU);

R_S: The reliability of supplier reflects the availability for supplying raw materials at the planned time (0- 1);

h_i: Operation time required to produce a product *i* (hr); and

P_i: Final price per unit of final product *i* sold to the customer (MU).

4.4. Decision Variables

Q_M: The quantity of raw materials ordered in each patch (unit); and

LT: Lead-time in time unit taken between placing and receiving the placed order (day).

4.5. Model Formulation

A general cost model is developed considering supplier of raw material point of view. This model is utilised to ascertain an optimal ordering strategy for obtaining raw materials batch size using both external and local backup suppliers to minimize the total cost of the final products and its risk effect in JIT systems. It is built to determine the total cost of producing the final product within production systems. The total cost of this product can be found by:

$$C_T = C_{RM} + C_W + C_U + C_R \tag{1}$$

Also, for the regular external and local backup supplier, C_{RM} includes the sum of costs C_O, C_H, C_P, C_{tr}, C_D and *TP*. Therefore, it can be calculated as:

$$C_{RM} = C_O + C_H + C_P + C_{tr} + C_D + TP \tag{2}$$

Where, C_O as the cost of ordering and receiving an amount of raw materials each order that can be calculated as:

$$C_O = \sum_{i=1}^{N_P} C_{O_i} \tag{3}$$

Also, the rate of C_H equals:

$$C_H = \sum_{i=1}^{N_P} C_{UH_i} \tag{4}$$

C_P is the unit cost of the raw material at the beginning of that cycle C_{UR} that equals:

$$C_p = \sum_{i=1}^{N_P} C_{UM_i} \tag{5}$$

C_{tr} as a component of C_M can be calculated as:

$$C_{tr} = t_m \times \sum_{i=1}^{N_S} T_{S_i, V, m} \tag{6}$$

C_D is the duty cost arises from supplying raw materials by a regular external supplier S_{Ej} to the production system. It means that for local backup supplier S_{LB}, there are no duties arising from supplying raw materials to the production system. It can be calculated as:

$$C_D = \sum_{i=1}^{N_P}\sum_{i=1}^{N_S} C_{M_i}(1 - IF_j) \times D_j \quad (7)$$

TP as a transfer price for procuring raw material from a regular external supplier S_{Ei} can be calculated as:

$$TP = \sum_{i=1}^{N_P}\sum_{i=1}^{N_S} tp_j \times C_{M_i} \quad (8)$$

Therefore, C_{RM} can be calculated as follows:

$$C_{RM} = \sum_{i=1}^{N_P} C_{O_i} + \sum_{i=1}^{N_P} C_{UH_i} + \sum_{i=1}^{N_P} C_{UM_i} + t_m \times \sum_{j=1}^{N_S} T_{S_j,V,ml} + \sum_{i=1}^{N_P}\sum_{j=1}^{N_S} C_{M_i}(1 - IF_j) \times D_j + \sum_{j=1}^{N_S}\sum_{i=1}^{N_P} t_{P_j} \times C_{M_i} \quad (9)$$

However, C_P, as the unit cost of raw material i (C_{UMi}) procured from local backup supplier S_{LBj} at the beginning of that cycle can be calculated as:

$$C_p = \sum_{i=1}^{N_P} C_{UM_i} \times R_{SLB} \quad (10)$$

Also, the worker cost C_W can be found as:

$$C_W = \sum_{i=1}^{N_O} C_{W_i} = \sum_{i=1}^{N_O} C_{L_i} \times h_i \quad (11)$$

In addition, C_U is the utilities cost that can be considered as a raw material cost percentage of the final product. It equals:

$$C_U = \sum_{i=1}^{N_P} \%C_{RM_i} \quad (12)$$

C_{pt} is the cost of the part of raw material that equals

$$C_{Pt} = C_O + C_H + C_P + C_{tr} + C_D + TP + C_W + C_U \quad (13)$$

Furthermore, C_R as a risk cost can be calculated by the following equation:

$$C_R = \sum_{i=1}^{N_P} \%TRS_i \times C_{M_i} = \sum_{i=1}^{N_P} (LH \times I / Max(LH \times I)) \times C_{M_i} \quad (14)$$

Finally,

$$C_T = C_{Pt} + C_R \quad (15)$$

C_T can be calculated in case of using the regular external supplier for procuring raw materials as follows:

$$C_T = \sum_{i=1}^{N_P} C_{O_i} + \sum_{i=1}^{N_P} C_{UH_i} + \sum_{i=1}^{N_P} C_{UM_i} + t_m \times \sum_{j=1}^{N_S} T_{S_j,V,ml} + \sum_{i=1}^{N_P}\sum_{j=1}^{N_S} C_{M_i}(1 - IF_j) \times D_j + \sum_{j=1}^{N_S}\sum_{i=1}^{N_P} (1 + t_{P_j}) \times C_{M_i} + \sum_{i=1}^{N_O} C_{L_i} \times h_i + \sum_{i=1}^{N_P} \%C_{M_i} + \sum_{i=1}^{N_P} (LH \times I / Max(LH \times I)) \times C_{M_i} \quad (16)$$

Also, when raw materials are supplied by the local backup supplier, C_T can be found as:

$$C_T = \sum_{i=1}^{N_P} C_{O_i} + \sum_{i=1}^{N_P} C_{UH_i} + \sum_{i=1}^{N_P} C_{UM_i} + t_m \times \sum_{j=1}^{N_S} T_{S_j,V,m} + \sum_{i=1}^{N_O} C_{L_i} \times h_i + \sum_{i=1}^{N_P} \%C_{M_i} + \sum_{i=1}^{N_P} (LH \times I / Max(LH \times I)) \times C_{M_i} \quad (17)$$

According to [23], the proposed model was tested in the simple process for assembling a brushless DC electric motor (BLDC). It was used to ascertain the decision variables effect on other studied parameters within the production system.

5. Example Problem

The mathematical model proposed in section 4 has been tested with a simple assembly process for a landline phone. It uses multiple, identical operations to assemble five individual parts M_i into the finished product (N_P= 5) namely, transmitter, receiver, push-button, ringer or audible indicator, and a small assembly of electrical parts (circuit-board). It is assumed that a production system purchases raw materials in a fixed size from three different regular external suppliers (N_S= 3). These raw materials are delivered at a fixed interval of time when they are needed (JIT system). Parts 1 and 2 are supplied by the supplier S_{E1}, which need three weeks (LT) to arrive, parts 3 and 4 are supplied by the supplier S_{E2}, which require five weeks to arrive, and Part 5 is supplied by the supplier S_{E3}, which take four weeks to arrive. The production system includes four operations conducted by four workers (W_1, W_2, W_3, and W_4 respectively). The number of working hours N_h is 8 hours a day during 5 days per week, each worker has a fixed wage C_{Wi} valued 12 monetary unit (MU)/ hour. Operation 1 assembles parts 1 and 2 and transfers them to operation 2, which assembles parts 3 and 4, and then transfers them to operation 3. Operation 3 assembles parts 5, and finally transfers the product to operation 4, which tests and places the final product in packaging before sending it to the sales department. The production facility produces 70 units per day, and it purchases raw materials from the three different regular external suppliers S_{E1}, S_{E2}, and S_{E3} (if no disruption occurs) and two local backup suppliers S_{LB1} and S_{LB2} (in the event of one or more of the regular suppliers experiencing disruption). Each order is 1050 units from Parts 1 and 2, 1750 units from Parts 3 and 4, and 1400 units from Part 5 respectively. These order quantities can meet customer needs during a fixed time-period under normal supply conditions. It is assumed that the utilities cost C_U is equal to 10% of the raw material cost of the final product. It is also assumed that Parts 1, 2, 3, and 4 can be supplied to the production system by S_{LB1} with a rate of 150% of their cost when they are purchased from the regular suppliers S_{E1} and S_{E2} and Part 5 can be procured from S_{LB2} with a rate of 160% of its cost if they are purchased from the regular suppliers S_{E3}. Finally, the end customer purchases the final product by 82 MU. Figure 1 shows the supply chain for this production system.

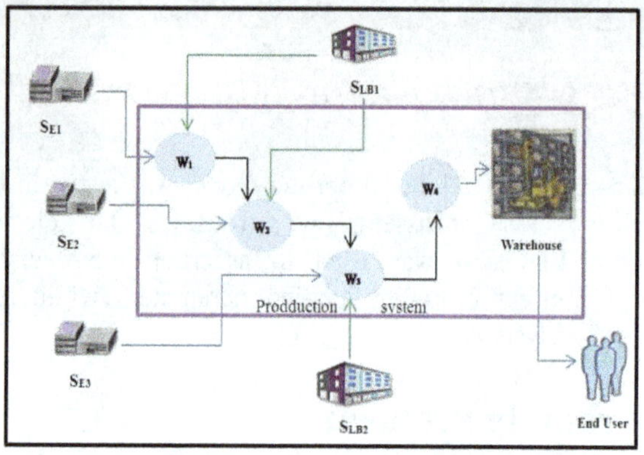

Figure 1. *The supply chain for the production system*

However, many risks result from delays in the delivery-time of these materials to the production system. These may arise from risks caused by physical, social, legal, operational, economic and political factors. These factors can affect and disrupt the production system and all the supply chain parties. Therefore, this paper studies the effects of these factors on the case study of a production facility.

The next step is to identify supply chain risks facing the production facility. Table 1 includes the main supply chain risks potentially facing the production/ marketing of landline phones and their impact within the production system. Risk identification was prepared based on what is perceived as the effect of a disruption or change in demand on the production facility. It may also be approached by investigating all possible root causes of supply chain issues. According to [24], risk can be assessed by two common approaches; the likelihood of the occurrence of an (undesirable) event, and the negative ramifications of the event. Therefore, the total risk score can be calculated by multiplying those scores together.

The risks H_1, H_2, and H_3 may result from increasing the lead time of raw materials of external suppliers S_{E1}, S_{E2}, and S_{E3} respectively to arrive at the manufacturing plant at the planned time. The likelihood of the occurrence of such risks might arise as a result of some factors such as natural and man-made disasters, and economic crises (currency evaluation/ strikes). All of these mentioned risks will disrupt the production system, and at the same time will affect the other parties in the supply chain. However, impacts of these factors can be avoided by keeping a sufficient inventory within the production facility. An inventory is an important supply chain driver because changing inventory policies can dramatically improve the supply chain's efficiency and responsiveness that makes it able to maintain its permanent production during the disruption time.

Table 1. *Risk assessment of the landline phone in production system*

Risk Symbol	Risk	Product Effect	Likelihood (1 - 5)	Impact (1 - 5)	% Total Risk Score
H1	External supplier 1 cannot supply raw materials on the scheduled time.	All product	2	2	4/25= 16%
H2	External supplier 2 cannot supply raw materials on the scheduled time.	All product	2	4	8/25= 32%
H3	External supplier 3 cannot supply raw materials on the scheduled time.	All product	2	3	6/25= 24%

The main cost drivers in a landline phone are: transmitter, receiver, push-button, ringer or signaler, and populated circuit board. They are shown in Table 2 as a percentage rate of the total cost of the phone. This table also illustrates the cost percentage rate, incurred duties, and transfer price for each supplier.

Table 2. *Cost drivers in Landline phone*

Supplier	Raw material type	Cost percentage (%)	% Supplier rate	% Duties rate	% Transfer price (TP)
S1	Transmitter	20	40%	5%	4%
	Receiver	20			
S2	Push-button	19	36%	4%	3%
	Ringer/ alerter	17			
S3	Electric board	24	24%	3%	2.5%

6. Results and Discussion

In this section, the proposed model will be used to ascertain the effect of decision variables on other parameters examined within the production system. The findings of this paper are organised in three cases as follows:

6.1. Case I

The impact of lead time on cost types of the final product will be investigated for a scenario of having disruption from an external supplier. This prompts sufficient stock keeping from the external supplier to prevent any likelihood of stock running out.

The findings illustrated in Figure 2 show that if the supplier 1 has disruption for any reason, keeping different amount of raw materials in warehouses (1-35 days) have direct impact on the total cost arising from the risk cost associated with the supplier. Keeping raw materials in the warehouses have high impact on the earned profit. From this figure, it is clear that the production system is able to procure sufficient raw materials to produce the final product for 29 days with an appropriate profit. That is because the external suppliers offer sales discount to their customers for purchasing any extra amount of raw materials. Therefore, it is clear that the net profit increases in the beginning of each week and then gradually decreases until the week ends.

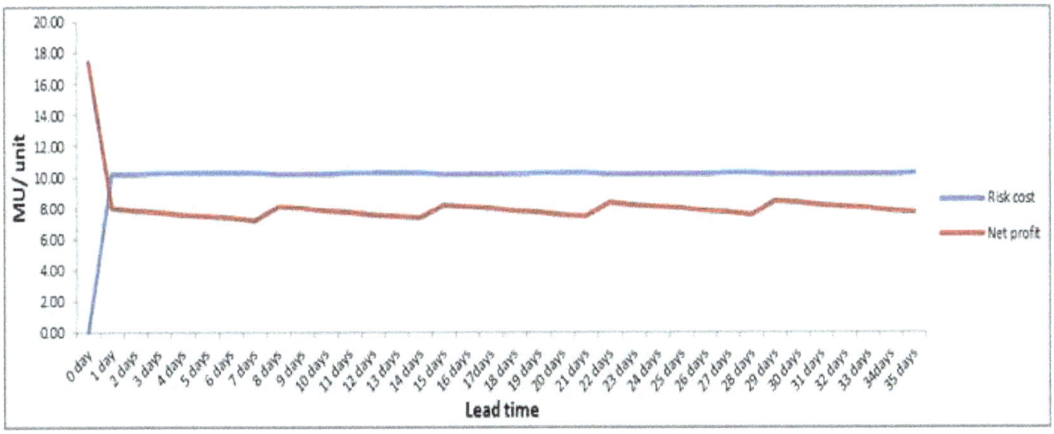

Figure 2. *Lead time and its impact on net profit and risk cost arising from external supplier 1 disruption*

Table 3 demonstrates the results of some cost types calculated using the developed model equations. The maximum duration used for keeping a limited amount of raw materials is 5 weeks based on lead time for external suppliers. It shows that if any disruption affects supplier 2 who supplies some an amount of raw material types used for production, then keeping safety stock of these raw materials in warehouses (1-5 weeks) at different periods of lead time have a direct impact on risk cost.

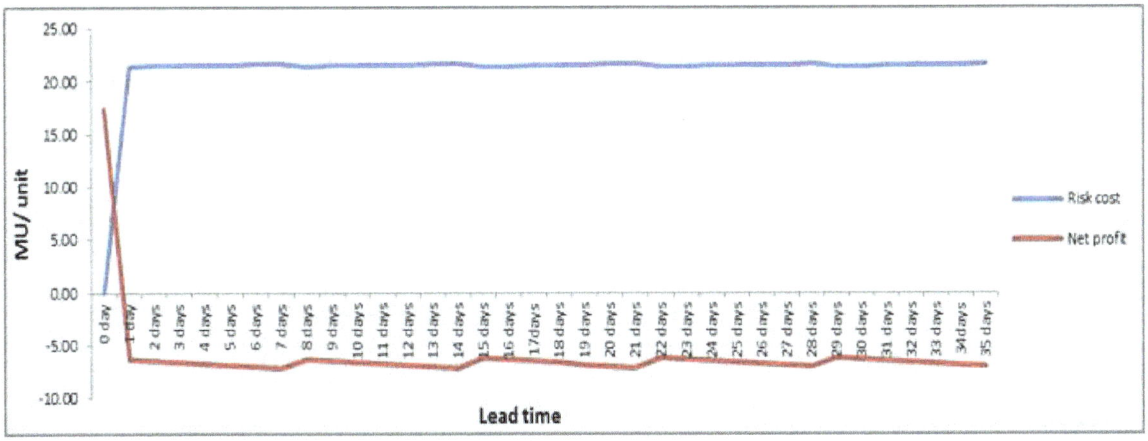

Figure 3. *Lead time and its impact on net profit and risk cost arising from supplier 2 disruption*

Table 3 shows increase in the utilities and risk costs, whereas the purchasing cost decreases. However, the ordering, transportation, duties, transfer price, and worker costs are fixed. Surprisingly, Figure 3 shows that the safety stock amount for 1-35 days give a negative profit rate.

From Figure 4, it is also clear that there is a striking impact on the risk cost, when supplier 3 is disrupted. This is because of the impact of the supplier risk cost arising from this disruption.

Table 3. *Effects of lead time on cost types arising from disrupting external supplier 2*

Cost type	MU/ unit					
	0 week	1 week	2 weeks	3 weeks	4 weeks	5 weeks
Ordering cost	0.9	0.9	0.9	0.9	0.9	0.9
Holding cost	0	0.7	1.4	2.1	2.8	3.5
Purchasing cost	40	39.28	38.56	37.84	34.4	36.4
Transportation cost	10	10	10	10	10	10
Duties	1.66	1.43	1.43	1.43	1.43	1.43
Transfer price	1.31	0.89	0.89	0.89	0.89	0.89
Utilities cost	5.37	5.3	5. 3	5.3	5.3	5.3
Worker cost	5.49	5.49	5.49	5.49	5.49	5.49
Risk cost	0	20.43	20.42	20.42	20.41	20.4
Total cost	64.58	84.27	84.24	84.22	84.19	84.16
Net profit	17.42	-2.27	-2.24	-2.22	-2.19	-2.16
Sales price	82	82	82	82	82	82

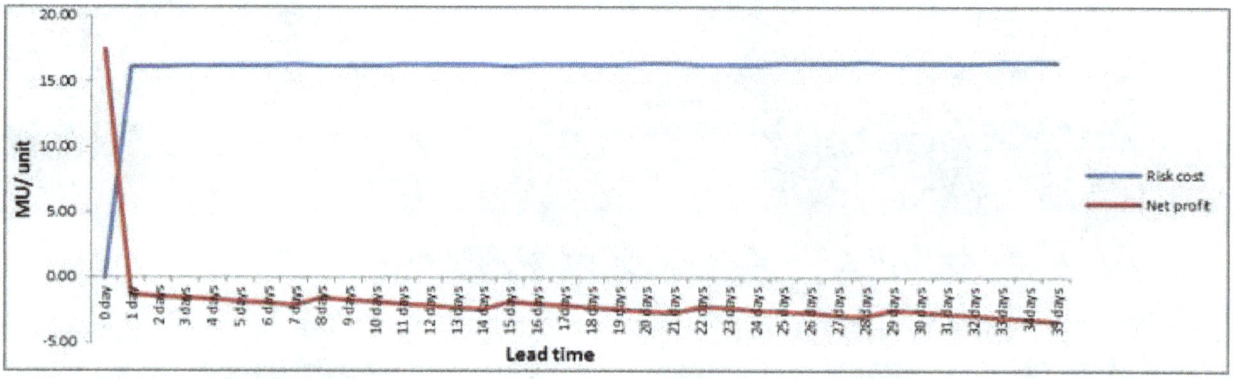

Figure 4. *Lead time and its impact on net profit and risk cost arising from supplier 3 disruption*

6.2. Case II

Keeping the same base case in-point as the first, the impact of lead time on cost types of final product will be investigated where stock is procured from local backup supplier. This case assumes that the external supplier is not able to meet supplier demand due to the disruption.

By using local backup suppliers for supplying the required raw material in the event of any disruption occurring from the three external suppliers, stoppage of production caused by a lack of raw materials can be easily avoided. However, this will increase the purchasing and risk cost that depends on the reliability of these suppliers. Figure 5 shows the effects of lead time on the net profit and the risk cost arising from the disruption caused by external supplier 1. This prompts the use of local backup supplier 1 to supply the required amounts of raw materials in different periods of lead time.

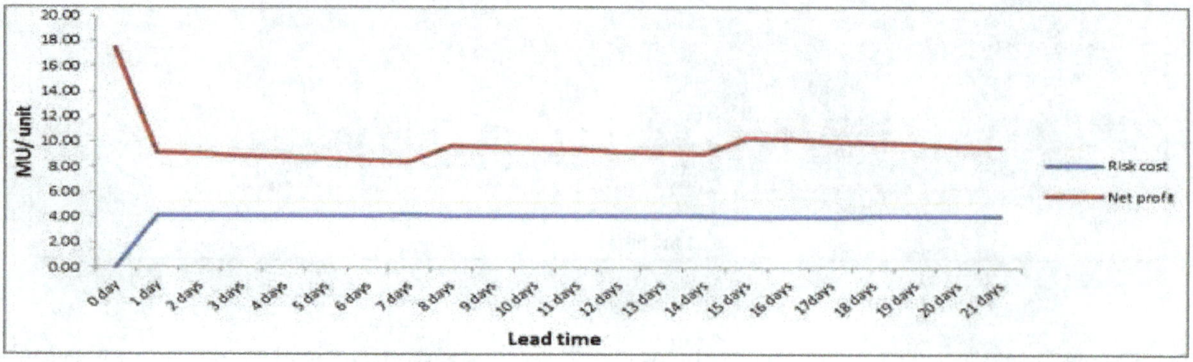

Figure 5. *Lead time and its impact on net profit and risk cost arising from supplier 1 disruption using local backup supplier 1*

Figure 6 also illustrates the lead-time impact on the net profit and the risk cost arising from the disruption caused by external supplier 2. This prompts the use of local backup supplier 1 to supply the required amounts of raw materials in different periods of lead time.

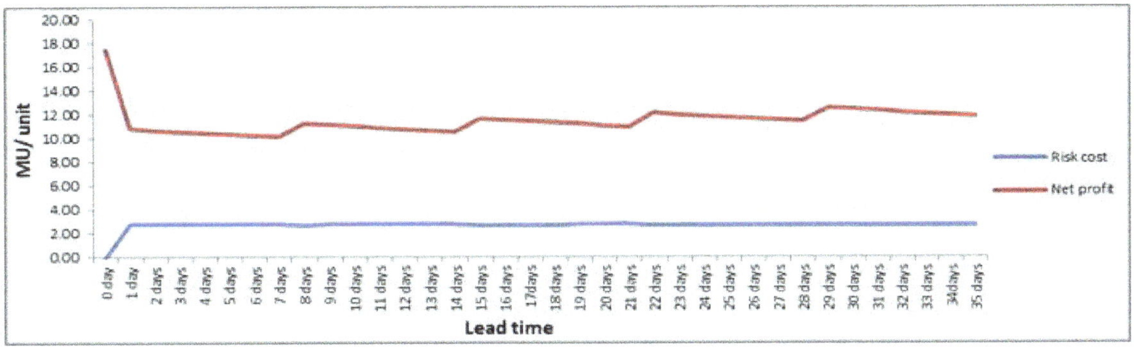

Figure 6. Lead time and its impact on cost types arising from external supplier 2 disruption using local backup supplier 1

In Figure 7, it is clear that the lead time has marked impact on the total cost arising from the disruption occurring from supplier 3 if the local backup supplier 2 is used to supply the required amount of raw materials in different periods.

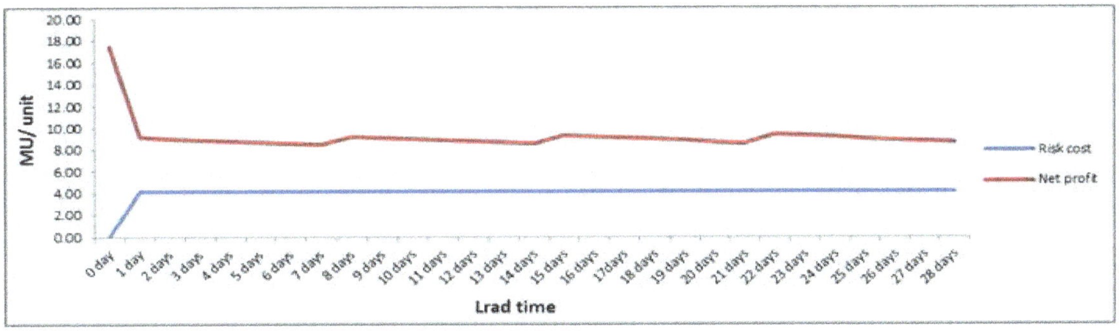

Figure 7. Lead time and its impact on cost types arising from supplier 3 disruption using local backup supplier

6.3. Case III

In this case, the two cases are compared to find the optimum quantity of required raw materials that give an appropriate profit during the disruption period.

Figure 8 illustrates the comparison between the net profit and risk cost arising from producing final product if disruptions occur from external supplier 1. This compares the case of solely relying on an external supplier 1 or using local backup supplier 1. It can be observed that if supplier 1 has disruption, the cost arising from keeping inventory during this time using local supplier is less than the cost using the same supplier. Therefore, it can be observed that working with a 3 weeks inventory from a local backup supplier during the disrupted time gives a reasonable profit for the production system.

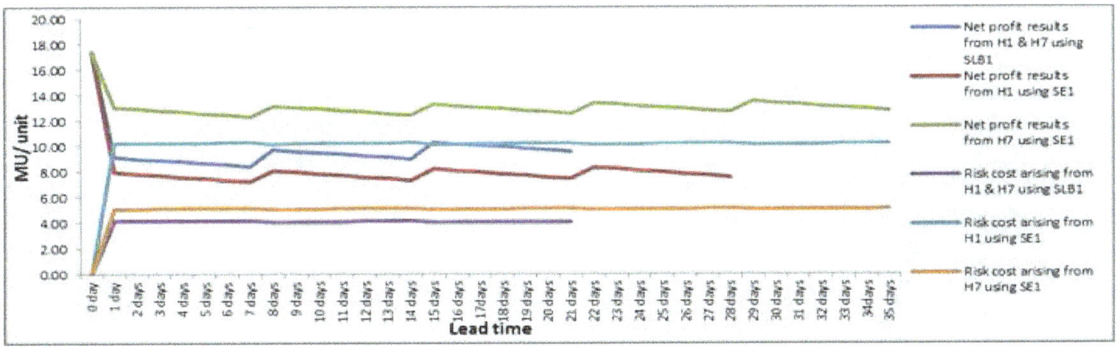

Figure 8. Comparison between the net profit and risk cost arising from supplier 1 disruption using the disrupted supplier and local backup supplier 1

Figure 9 illustrates a similar comparison for the case of supplier 2 and a local backup supplier. It is clear that if supplier 2 is disrupted, the cost arising from keeping inventory during this time using local supplier is also less than the cost using the same supplier.

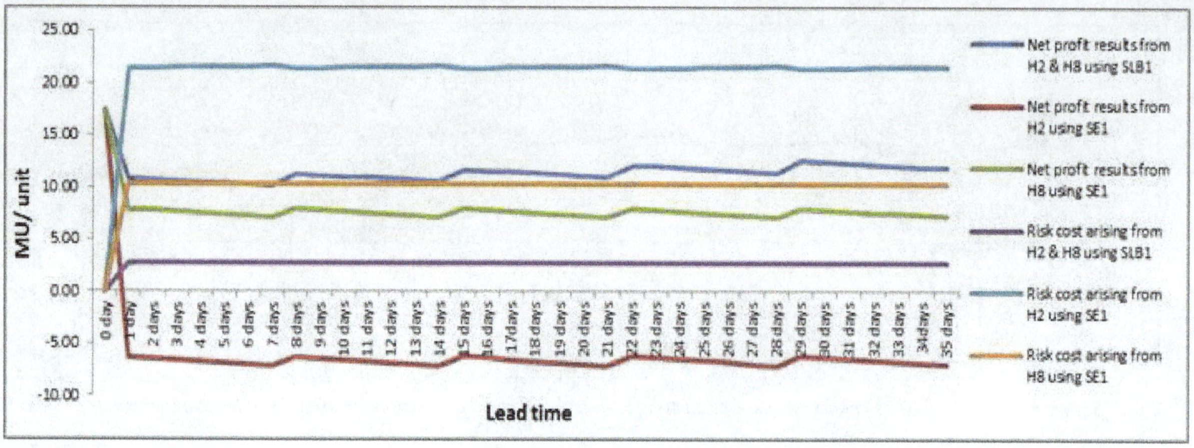

Figure 9. Comparison between the net profit and risk cost arising from supplier 2 disruption using the disrupted supplier and local backup supplier 1

The same result has been found in case of supplier 3 is disrupted from supplying raw materials to the production system. Figure 10 shows that by comparing the total cost arising from keeping safety stock amount within the production facility using the regular external supplier 3 and local backup supplier 2, the risk cost arising from keeping inventory during this time using local supplier 2 is also less than the risk cost using the same supplier.

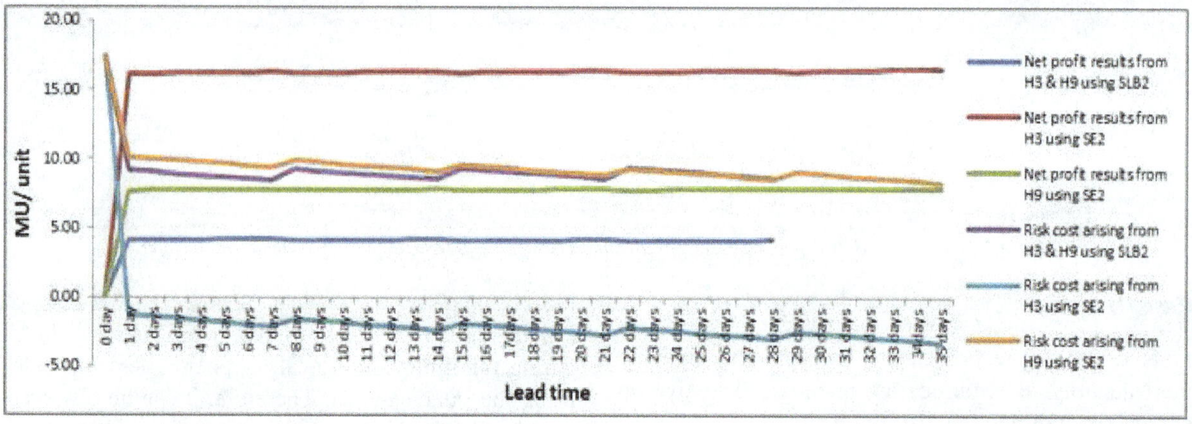

Figure 10. Comparison between the net profit and risk cost arising from supplier 3 disruption using the disrupted supplier and local backup supplier 2

7. Conclusion and Further Research

This paper presented a mathematical model for a simultaneous cost-risk reduction in JIT systems. It was developed to determine an optimal strategy for supplying raw materials to the production systems by using regular multi-external and local backup suppliers in case of the occurrence of likely disruption such as natural and man-made disasters, and economic crises. By implementing the model in a simplified example, it is concluded that comparing the use of a JIT system with the use of a specific amount of inventory during a limited duration had a significant impact on the production facility especially, by using the local backup supplier during the disruption time. This means that by using strictly JIT, the production system will be stopped completely during supply disruption. However, by keeping a sufficient inventory, the production system can produce its final products but with a limited profit. Thereby JIT principles can be effectively applied for satisfying customer requirements at a minimum inventory cost with a minimum level of risk.

Due to the stochastic character of supply chain operations, it seems that the developed mathematical model needs to be supplemented with a simulation model as a validation tool to describe the dynamic nature of supply chain management. Hence, the authors plan to consider this point of view in future research where a simulation modelling will be deployed to find the outputs of some components of supply chain management system. This will enhance the level of model accuracy for real application systems.

References

[1] C. Canel and B. M. Khumawala, "International facilities location: a heuristic procedure for the dynamic uncapacitated problem," International Journal of Production Research, vol. 39, pp. 3975-4000, 2001.

[2] F. Abdullah, "Lean Manufacturing Tools and Techniques in the Process Industry with a Focus on Steel," PhD thesis, University of Pittsburgh, 2003.

[3] A. Badurdeen, "What is Lean Manufacturing?," 2006, [Online], http://www.learnleanblog.com/2006/12/what-is-lean-manufacturing.html [Accessed 07.11.2011].

[4] B. Fahimnia, R. Marian and B Motevallian, "Analysing the hindrances to the reduction of manufacturing lead-time and their associated environmental pollution, International Journal of Environmental Technology and Management (IJETM), ISSN 1466-2132, vol. 10, no. 1, pp. 16-25, 2009.

[5] Y. Monden, Japanese Cost Management. World Scientific, 2000.

[6] T. C. E. Cheng and S. Podolsky, "Just-in-time manufacturing: an introduction," New York, Chapman & Hall, 1996.

[7] R. A. Hokoma, M. K. Khan and K. Hussain, "The Present Status of Quality and Manufacturing Management Techniques and Philosophies within the Libyan Iron and Steel Industry," TQM Journal, vol. 22, no. 2, pp. 209-221, 2010.

[8] Z. X. Chen and B. R. Sarker, "Multi-vendor integrated procurement-production system under shared transportation and just-in-time delivery system," The Journal of the Operational Research Society, vol. 61, pp. 1654-1666, 2010.

[9] Y. Li, K. F. Man, K. S. Tang, S. Kwong and W. H. Ip, "Genetic algorithm to production planning and scheduling problems for manufacturing systems," Production Planning & Control, vol. 11, no. 5, pp. 443-458 , 2000.

[10] T. Tourki, "Implementation of Lean within the Cement Industry," PhD thesis, De Montfort University, 2010.

[11] L. Eldenburg, "Cost Management: Measuring Monitoring and Motivating Performance," John Wiley & Sons, 2007.

[12] J. S. Yang and J. C. H. Pan, "Just-in-Time Purchasing: an Integrated Inventory Model Involving Deterministic Variable Lead Time and Quality Improvement Investment," International Journal of Production Research, vol. 42, no. 5, pp. 853-863, 2004.

[13] B. Fahimnia, L. Luong and R. Marian, "An integrated model for the optimisation of a two-echelon supply network," Journal of Achievements in Materials and Manufacturing Engineering, vol. 31, pp. 477-484, 2008.

[14] A. Chae and H. Fromm, Supply Chain Management on Demand. Springer, 2005.

[15] R. A. Sarker and L. R. Khan, "An Optimal Batch Size for a Production System Operating under Periodic Delivery Policy," Computers & Industrial Engineering, vol. 37, pp. 711-730, 1999.

[16] M. C. Carneiro, G. P. Ribas and S. Hamacher, "Risk Management in the Oil Supply Chain: A CVaR Approach," Industrial & Engineering Chemistry Research, vol. 49, no. 7, pp. 3286-3294, 2010.

[17] N. Julka, R. Srinivasan and I. Karimi, "Agent-Based Supply Chain Management—1: Framework," Computers and Chemical Engineering, vol. 26, pp. 1755-1769, 2002.

[18] B. Tomlin, "Disruption-Management Strategies for Short Life-Cycle Products," NAVAL RESEARCH LOGISTICS, vol. 56, pp. 318-347, 2009.

[19] D. Simchi-Levi, L. Snyder and M. Watson, "Strategies for Uncertain Times. Supply chain Management Review," 2002, [Online]. http://www.lehigh.edu/~lvs2/Papers/SCMR.pdf [Accessed 06 November 2012].

[20] X. K. Dimakos and K. Aas, "Integrated Risk Modelling," Statistical Modelling, vol. 4, pp. 265-277, 2004.

[21] A. A. Gaivoronski, G. M. Sechi and P. Zuddas, "Balancing Cost-Risk in Management Optimization of Water Resource Systems under Uncertainty," Physics and Chemistry of the Earth, vol. 42-44, pp. 98-107, 2012.

[22] G. V.-H. Jose, "Modelling Risk and Innovation Management," Advances in Competitiveness Research, vol. 19, no. 3&4, pp. 45-57, 2011.

[23] F. El Dabee, R. Marian and Y. Amer, "An Optimisation Model for a Simultaneous Cost-Risk Reduction in Just-in-Time Systems," in the 11th Global Congress on Manufacturing and Management, GCMM2012, Auckland, New Zealand, 2012, pp. 190-201.

[24] D. Bogataj and M. Bogataj, "Measuring the Supply Chain Risk and Vulnerability in Frequency Space," International Journal of Production Economics, vol. 108, no. 1, pp. 291-301, 2007.

Dynamic traffic signal phase sequencing for an isolated intersection using ANFIS

Kingsley Monday Udofia[1], Joy Omoavowere Emagbetere[2], Frederick Obataimen Edeko[2]

[1]Department of Elect/Elect/Computer Engineering, University of Uyo, Uyo, Nigeria
[2]Department of Electrical/Electronic Engineering, University of Benin, Benin City, Nigeria

Email address:
kingsleyudofia@uniuyo.edu.ng (K. M. Udofia), miracle5ng@yahoo.com (J. O. Emagbetere), fFrededeko@yaho.co.uk (F. O. Edeko)

Abstract: This paper presents a traffic signal phase sequencing using adaptive neuro-fuzzy inference system (ANFIS) technique. The system is designed to emulate traffic expert on the selection of the appropriate phase to be given right-of-way at an isolated intersection based on the prevailing traffic situation. Inputs (queuelength and waiting time of vehicles) from traffic detectors are used to determine the selection of the next green phase. We evaluated the developed model for five different common traffic scenarios using MATLAB. The results obtained indicates that the developed model adaptively and effectively selects a phase to be given next green signal after considering the traffic situation and the nature of the intersection in question.

Keywords: Adaptive, Neuro-Fuzzy Inference System, Phase Sequencing, Vehicle Traffic Control, Isolated Intersection

1. Introduction

The use of traffic signals is considered to be one of the most effective ways to control traffic at intersections [1]. Traffic signals are used to assign the right-of-way to intersecting traffic streams for the purpose of ensuring that all streams are served safely and without excessive delay.

The working of the traffic signal control currently deployed in many intersections is based on predetermined and fixed signal phase sequencing [2]. The streams constituting each phase and the order in which the corresponding phases come on are fixed. The right of way are always given to the phases in a fixed sequence irrespective of whether there is traffic in a phase or not, thus leading to unnecessary delays of traffic at the intersection [2]. This has instigated various ideas and scenarios by traffic engineers to solve the traffic problem [3 - 7]. To design an intelligent and efficient traffic control system, a number of parameters that represent the status of the traffic conditions must be identified and taken into consideration [8].

In this paper, we adopted adaptive neuro-fuzzy inference system (ANFIS) to model a real time, adaptive traffic signal phase control for an isolated intersection. ANFIS uses a hybrid training method to automatically generate fuzzy rules according to a given input-output datasets. The proposed case study is a real four-way intersection, located in the urban area of Uyo (Nigeria), with severe traffic congestion.

2. ANFIS Architecture

ANFIS works by applying neural learning rules to identify and tune the parameters and structure of a Fuzzy Inference System (FIS). A typical architecture of an ANFIS, in which a circle indicates a fixed node, whereas a square indicates an adaptive node, is shown in figure 1. For simplicity, we assume that the examined FIS has two inputs and one output. For a first-order Sugeno fuzzy model, a typical rule set with two fuzzy "if then" rules can be expressed as follows:

$$\textit{Rule } 1\text{: If } x \text{ is } \mathrm{A}_1 \text{ and } y \text{ is } \mathrm{B}_1, \text{then } z_1 = f_1(x) = p_1x_1 + q_1x_2 + r_1$$

$$\textit{Rule } 2\text{: If } x \text{ is } \mathrm{A}_2 \text{ and } y \text{ is } \mathrm{B}_2, \text{then } z_2 = f_2(x) = p_2x_1 + q_2x_2 + r_2$$

Where x and y are the two crisp inputs, and Ai and Bi are the linguistic labels associated with the node function.

As indicated in figure 1, the system has a total of five

layers. The functioning of each layer is described as follows [9].

Input node (Layer 1): Nodes in this layer contains membership functions. Parameters in this layer are referred to as premise parameters. Every node i in this layer is a square and adaptive node with a node function:

$$O_i^1 = \mu_{A_i}(x) \ for \ i = 1,2 \tag{1}$$

Where x is the input to node i, and Ai is the linguistic label (short, long, etc.) associated with this node function. In other words, O_i^1 is the membership function of Ai and it specifies the degree to which the given x satisfies the quantifier Ai.

Rule nodes (Layer 2): Every node in this layer is a circle node labeled ∏, whose output represents a firing strength of a rule. This layer chooses the minimum value of two input weights. In this layer, the AND/OR operator is applied to get one output that represents the results of the antecedent for a fuzzy rule, that is, firing strength. It means the degrees by which the antecedent part of the rule is satisfied and it indicates the shape of the output function for that rule. The node generates the output (firing strength) by cross multiplying all the incoming signals:

$$O_i^2 = \mu_{A_i}(x) \text{ x } \mu_{B_i}(x) \ for \ i = 1,2 \tag{2}$$

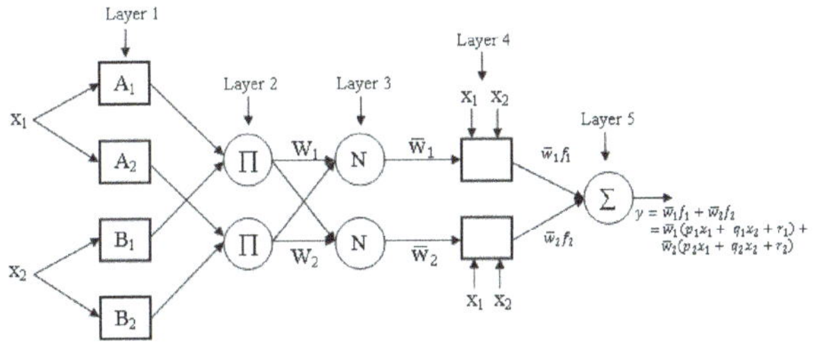

Figure 1. *ANFIS architecture.*

Average nodes (Layer 3): Every node in this layer is a circle node labeled N. The ith node calculates the ratio between the ith rule's firing strength to the sum of all rules' firing strengths. Every node of these layers calculates the weight, which is normalized. For convenience, outputs of this layer are called normalized firing strengths.

$$\bar{w}_i = \frac{w_i}{w_1 + w_2}, \quad for \ i = 1,2 \tag{3}$$

Consequent nodes (Layer 4): This layer includes linear functions, which are functions of the input signals. This means that the contribution of ith rule towards the total output or the model output and/or the function defined is calculated. Every node i in this layer is a square node with a node function:

$$O_i^4 = \bar{w}_i f_i = \bar{w}_i(p_i x_1 + q_i x_2 + r_i) \tag{4}$$

Where wi is the output of layer 3, and {pi, qi, ri} is the parameter set of this node. These parameters are referred to as consequent parameters.

Output node (Layer 5): The single node in this layer is a fixed node labeled Σ, which computes the overall output by summing all incoming signals:

$$O_i^5 = overalloutput = \sum_i \bar{w}_i f_i = \frac{\sum_i w_i f_i}{\sum_i w_i} \tag{5}$$

The tuning or training procedure for ANFIS is achieved based on the batch learning technique using input–output training dataset. During training ANFIS optimizes the adjustable parameters by comparing ANFIS output with trained data. Each period of training is divided into two phases. In the first phase, the consequent parameters are adjusted with Least-squares method and in the second phase, the premise parameters are adjusted with gradient descent (back propagation) method.

3. Methodology

In this paper, we adopted adaptive neuro-fuzzy inference system (ANFIS) to model a real time, adaptive traffic signal phase control for an isolated intersection. ANFIS uses a hybrid training method to automatically generate fuzzy rules according to a given input-output datasets.

In the design, the following requirements were taken into consideration:

- to reduce the delay time of waiting vehicles;
- to avoid traffic congestion (queue lengths); and
- to avoid conflicts at the intersection.

There are two major ways of controlling phase sequence in order to avoid conflicts at traffic intersections, namely, fixed and variable phase sequences. In fixed phase sequence, the sequencing of the phases are pre-determined and fixed despite the prevailing traffic conditions at the intersection. Whereas, the sequencing in a variable phase sequence is determined by the existing traffic situations. The variable phase sequence has the flexibility to fully adapt to the traffic flow fluctuations.

In this work, a variable phase sequence control method was adopted. This method uses queuelengths and waiting time as input parameters. The decision output is to select a phase with worst traffic condition in an intersection.

In order to effectively control traffic at an intersection adaptively, the ANFIS model used two sets of input parameters, namely, the queuelengths (q), and waiting time of each phase (wt) in an intersection. These inputs parameters are obtained for each phase using vehicle detectors. Figure 2 shows the architecture of the proposed ANFIS-based traffic signal phase controller.

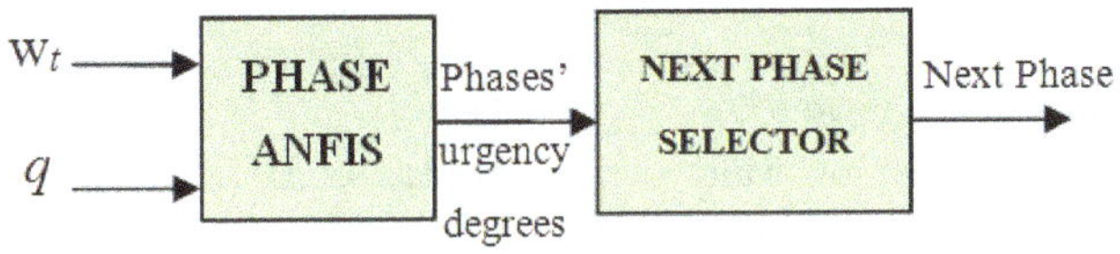

Figure 2. *Block diagram of the proposed ANFIS-based traffic signal phase controller.*

The queuelength (q) indicates the number of vehicles in a phase during a red light phase, and is given as the sum of the residue vehicles since the last green signal and the arrival during the current red signal as given by equation (6).

$$q = q_{gr} + \sum V_r^i \tag{6}$$

where q_{gr} is the number of vehicles that did not exit the intersection in the green phase and $\sum V_r^i$ is the sum of arrived vehicles in the ith second for the red phase. In general queuelength depends on the number of vehicles arriving and leaving an approach at a given time interval as given by equation (7) [15].

$$q_n(t+1) = q_n(t) - D_n(t) + A_n(t) \tag{7}$$

where $q_n(t+1)$ is the number of queued vehicles at time t, $q_n(t)$ is the residual queue from previous periods, and $A_n(t)$ and $D_n(t)$ are the vehicles arrivals and departures within time interval [t, (t+1)].

The waiting time (Wt) indicates the duration of time, vehicles have waited in the queue since the elapse of the last green phase signal and is computed using equation (8). This input parameter is considered so as to avoid the vehicles waiting too long for the green signal.

$$W_t = \begin{cases} t_{v1} \ if \ q_0 = 0 \\ t_R \ if \ q_0 > 0 \end{cases} \tag{8}$$

where t_{v1} is the time between the arrival of the first vehicle in a queue during the current red phase and the next green phase signal, t_R is the current red phase time measured from the end of the last green phase signal, and q_0 is the number of vehicles in the queue at the end of the last green phase signal.

The phase ANFIS was designed to use waiting time and queuelength to determine the urgency degree (Urg_{Ph}) separately for each phase. To adaptively and effectively determine the phase in an intersection to be given the right of way at a particular time taking cognizance the prevailing traffic situation, certain considerations were made. It is expected for a phase with long queuelength to have high probability of having the next green signal. In cases where there are vehicles, though few, which have stayed in another phase for a period of time longer than necessary, the decision to chose the phase for the next green signal becomes complex. To determine the optimum solution to this situation the urgency degree of each phase (Urg_{Ph}) which combines the individual worsening effect of these two factors must computed as given in equation (9).

$$Urg_{Ph} = W_e + Q_e \tag{9}$$

where W_e is the effect of waiting time, and Q_e is the effect of phase queuelength.

The effect of waiting time W_e was considered in the design such that the same actual waiting time w_t but different maximum waiting time $w_{t(max)}$, cannot have the same worsening effect; the one with lower $w_{t(max)}$ should have higher effect as expressed as in equation (10).

$$W_e = \frac{w_t}{w_{t(max)}} \tag{10}$$

The worsening effect of waiting time W_e is additive. Hence, equation (10) can be rewritten as equation (11).

$$W_e = +\frac{w_t}{w_{t(max)}} \tag{11}$$

The effect of queuelength was measured by equation (12), where Q_e is the worsening effect of queuelength, q is the actual queuelength and $q_{(max)}$ is the phase's maximum queuelength capacity. Two phases with the same q but different $q_{(max)}$ cannot have the same worsening effect; the one with lower $q_{(max)}$ have higher effect.

$$Q_e = \frac{q}{q_{(max)}} \tag{12}$$

Thus, the worsening effect of queuelength Q_e is additive. Hence, equation (12) was rewritten as;

$$Q_e = +\frac{q}{q_{(max)}} \tag{13}$$

Substituting equations (11), (13) into equation (9), we have an expression for the overall effects of traffic factors affecting a phase at a time, called urgency degree (Urg_{Ph}), as given in equation (14).

$$Urg_{Ph} = \frac{w_t}{w_{t(max)}} + \frac{q}{q_{(max)}} \tag{14}$$

Since a phase with no vehicle in queue does not need green signal, for equation (14) to be successfully adapted in traffic control, the phase urgency degree Urg_{Ph}, must be zero (0) whenever the phase's queuelength q, is zero (0). Thus, equation (14) is rewritten as in equation (15).

$$Urg_{Ph} = \begin{cases} \left[\frac{w_t}{w_{t(max)}} + \frac{q}{q_{(max)}}\right], for \ q > 0 \\ 0, \quad for \ q = 0 \end{cases} \tag{15}$$

$w_{t(max)}$ and $q_{(max)}$ can be determine based on the prevailing factors of the intersection's traffic such as traffic density and lane length.

Equation (15) is used to construct the input/output datasets for the training the ANFIS models as described in section 4.

The Next Phase Selector module uses urgency degrees of all phases in the intersection to determine the phase to be given the next green signal. It selects the phase with the maximum urgency degree the intersection (16).

$$Urg_{NxtPh} = \max_{1 \le i \le n}\{Urg_{Ph}(i)\} \tag{16}$$

where Urg_{NxtPh} is the urgency degree of the Next phase, $Urg_{Ph}(i)$ is the urgency degree of individual phases constituting the intersection, and n is the number of the phases considered.

4. ANFIS Models Designing

It is our target to develop a traffic control model that will automatically and adaptively take traffic decisions just like expert/professional traffic wardens. Although ANFIS model could achieve this, there is need to develop and train the ANFIS with datasets that reflect expert warden's judgments at different traffic situations.

The input-output training dataset for each phase ANFIS model was developed from parameters obtained from a case study intersection located in Uyo, Nigeria using equation (15). The intersection schematic is shown in figure 3 with four phases labeled A, B, C and D. Phases A and C are double lanes while phases B and D are single lane. To develop the dataset used in training each phase ANFIS:

- the range of waiting time, w_t, was 0 – 500 seconds;
- the range of queuelength, q, was 0 – 300 vehicles;
- the threshold waiting time, $w_{t(max)}$, was two minutes (120 seconds); and
- the threshold queuelength, $q_{(max)}$, was 34 vehicles for double-lane phases (A and C) and 17 vehicles for single-lane phase (B and D).

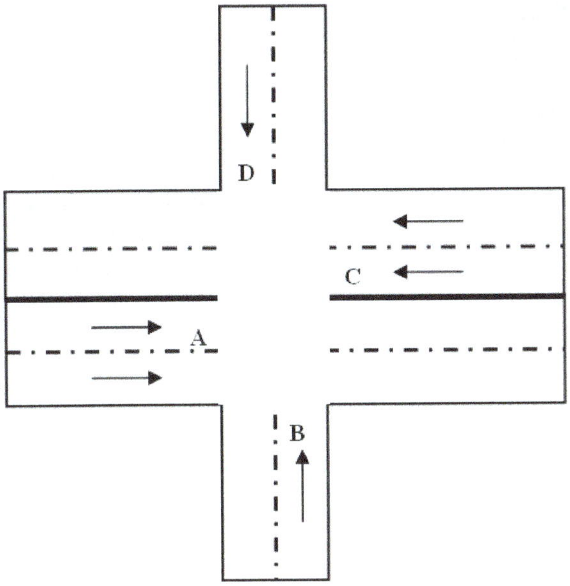

Figure 3. The schematic diagram of the case study intersection.

In designing the ANFIS we used MATLAB Anfis editor. Each phase ANFIS model passed through the following four steps:

- Load data;
- Generate fuzzy inference system (FIS);
- Train FIS; and
- Test FIS model.

The ANFIS model used Grid Partition method for FIS generation. Grid partition divides the data space into rectangular subspaces using axis-paralleled partition based on pre-defined number of membership functions and their types in each dimension. The generated FIS includes two inputs and one output. The input variables are: waiting time (w_t) and queuelength (q). The type of membership function used is the generalized bell because of its smoothness and non-linearization ability coupled with the degree of freedom to adjust the steepness at the crossover points. The number of membership functions for waiting time (w_t) and queuelength (q) is six each. The output field is the urgency degree. Membership function type of output variable is linear. Figures 4 and 5 show tuned membership functions of each input in the ANFIS model. The structure of the tuned FIS is shown in figure 6 and contains 36 rules.

The ANFIS model used hybrid optimization method (least-squares error and back propagation gradient descent methods) to train the membership function parameters to emulate the training data.

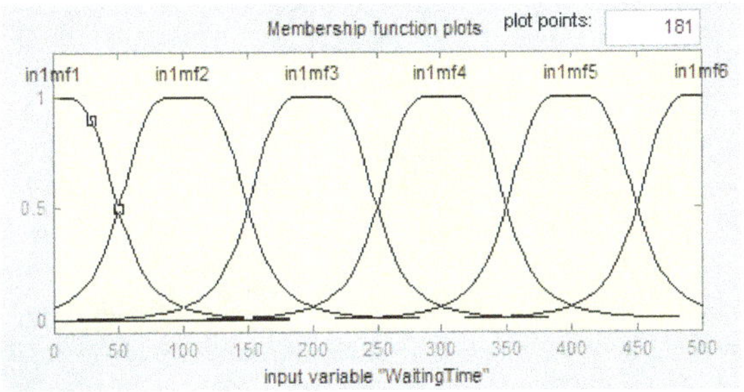

Figure 4. Tuned membership functions for waiting time.

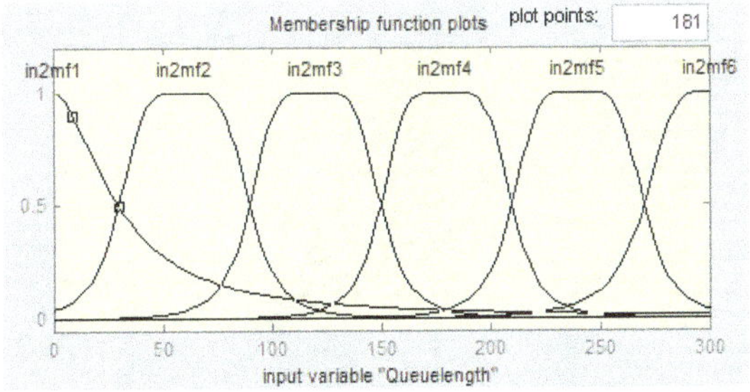

Figure 5. Tuned membership functions for queuelength .

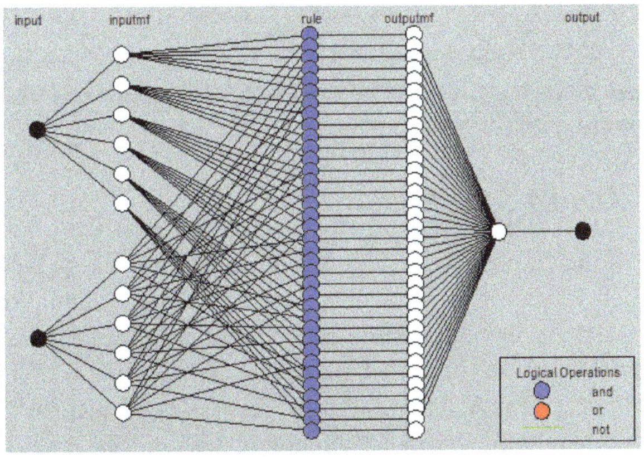

Figure 6. Structure of the Phase ANFIS model.

A block diagram of the trained phase ANFIS model is shown in figure 7. It can be seen that the model has two input variables (waiting time and queuelength) and an output variable (urgency degree). The input-output views of the ANFIS model in figures 8 and 9 illustrate how the Urgency degree will respond to varying values of waiting time and queuelength.

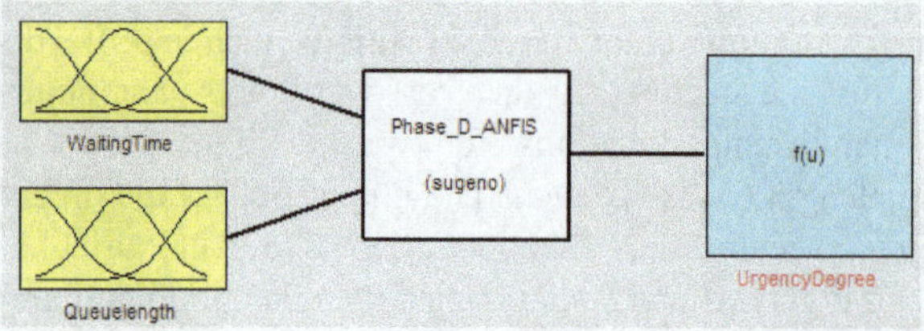

Figure 7. *Structure of the trained Phase ANFIS model.*

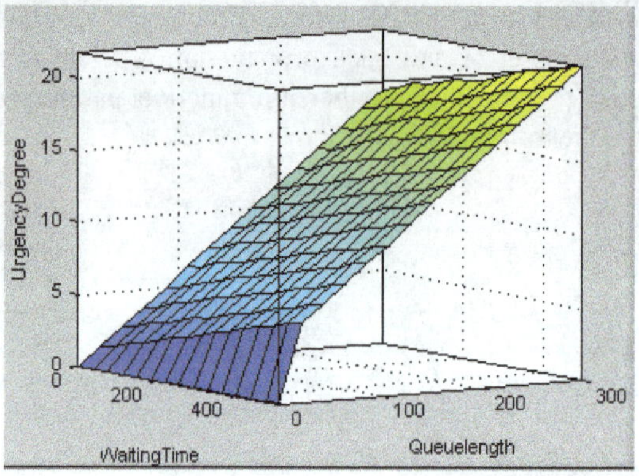

Figure 8. *3D plot of waiting time and queuelength (inputs variables) Vs Urgency degree (output variable).*

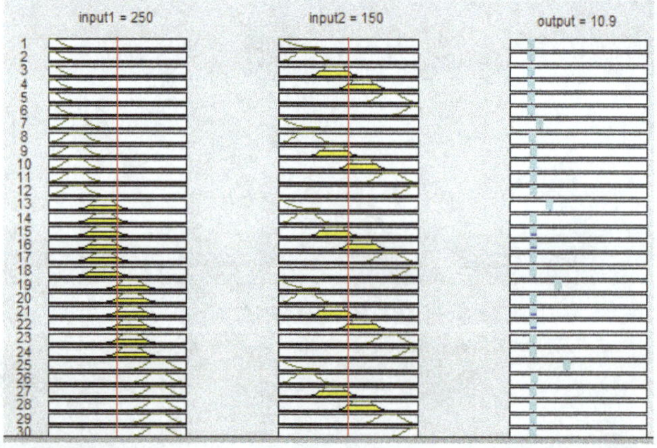

Figure 9. *MATLAB rule viewer showing fuzzy rules and membership functions.*

5. Result and Discussion

We perform five different evaluations of the ANFIS models based on real traffic scenarios. In each evaluation, we vary the inputs (waiting time and/or queuelength) of a phase(s) and we observed the effect in the choice of the next phase. The MATLAB Anfis rule viewer of each of the phases was used to extract the urgency degree for each set of input variables.

CASE 1: A phase with a high queuelength compared to other phases

In this evaluation, the effect of having a high queue in a phase compared to the other phases of the intersection, in the selection of the next green phase is considered. Assuming that phase C has queuelength of 100 vehicles which is high compared to other phases as can be seen in table 2.

Table 2. *Effect of Congested phase on next phase selection.*

ANFIS Model	Waiting Time (s)	Queuelength (veh)	Urgency Degree	Selected Phase
Phase A	100	30	1.79	} Phase C
Phase B	100	20	2.01	
Phase C	100	100	3.77	
Phase D	100	25	2.38	

As can be observed from table 2, phase C which has the highest urgency degree of 3.77 is selected as the next phase to be given green signal. This signifies that the phase that is most congested among all phases in an intersection will be preferred for the next green phase.

CASE 2: A single-lane phase and a double-lane phase with high queuelength of same magnitude

In this evaluation, a situation where there are high queue of the same volume of vehicles in a single-lane phase and a double-lane phase is considered. Assuming that phase A (double lanes) and phase B (single lane) have a queuelength of 100 vehicles each as in table 3.

Table 3. *Effect of number of lanes in a route on next phase selection.*

ANFIS Model	Waiting Time (s)	Queuelength (veh)	Urgency Degree	Selected Phase
Phase A	100	100	3.77	} *Phase B*
Phase B	100	100	6.72	
Phase C	100	40	2.01	
Phase D	100	25	2.30	

As can be observed from table 3, despite the fact that both phases A and B have the same waiting times and queuelengths, phase B which is a single-lane route has a higher urgency degree of 6.72 and thus, is selected as the next phase to be given green signal. This is because the same number of vehicles in a single-lane route will occupy double the distance of that of a double-lane route; thus a single-lane route will be said to be experiencing more congestion.

CASE 3: No vehicle in a phase

In this evaluation, a case where there is no vehicle waiting in a particular phase is considered. Assuming that there is no vehicle in phase D as given in table 4.

Table 4. *Effect of having no vehicle in a phase on next phase selection.*

ANFIS Model	Waiting Time (s)	Queuelength (veh)	Urgency Degree	Selected Phase
Phase A	100	30	1.79	} *Phase B*
Phase B	100	21	2.09	
Phase C	100	40	2.01	
Phase D	100	0	0	

As can be observed from table 4, phase D with no vehicle has an urgency degree of 0 and hence, can never be selected as the next phase to be given green signal. Whenever there is no vehicle waiting in a route, it will be useless assigning green signal to that phase.

CASE 4: A Phase with low queuelength and very high waiting time

In this evaluation, the effect of a phase with low queuelength and high waiting time in the selection of the next green phase is considered. Assuming that phase C has 5 vehicles with a waiting time of 400 seconds as given in table 5.

Table 5. *Effect of a phase with low volume of vehicles and very high waiting time on next phase selection.*

ANFIS Model	Waiting Time (s)	Queuelength (veh)	Urgency Degree	Selected Phase
Phase A	100	60	2.60	
Phase B	100	20	2.01	} *Phase C*
Phase C	400	5	3.41	
Phase D	100	30	2.65	

As can be observed from table 5, though phase C has a very low queuelength, its urgency degree is the highest because of the very high waiting time of vehicles and hence, is selected as the next phase to be given green signal. Although a phase has few vehicles compared to other phases, it should not be denied a right of way most especially having waited for a long time.

CASE 5: The current green signal phase has a very high queuelength compared to other phases

In this evaluation, a case where the phase currently being served green signal has a very high queuelength compared to other phases in an intersection is considered. Assuming that phase A is the current green phase with queuelength of 200 vehicles and zero (0) waiting time.

Table 6. *Effect of Congested phase on next phase selection.*

ANFIS Model	Waiting Time (s)	Queuelength (veh)	Urgency Degree	Selected Phase
Phase A	0	200	5.88	
Phase B	100	21	2.09	} *Phase A*
Phase C	100	40	2.01	
Phase D	100	25	2.30	

As can be observed from table 6, though phase A is the current green phase, it has the highest urgency degree because of the congestion level compared to other phases and hence, its green signal is extended. If the phase currently being served green signal is congested compared to the other phase, it is wisdom to extend the green signal.

6. Conclusion

The conventional traffic control system uses a fixed phase sequence where the order of corresponding phases is fixed irrespective of the prevailing traffic conditions at an intersection. Thus, the tendency of assigning green signal to a phase with no vehicle or an over-congested phase, irrespective of the fact that other phases have low traffic. These are issues that do not promote economic growth in any society as man-power hours are lost, hence productivity is greatly affected, aside the personal discomforts experienced by the motorists.

In this paper, we developed a model that adaptively and effectively control phase sequencing at traffic isolated intersection using ANFIS. A mathematical model for the design of input-output datasets used in the training and tuning of each phase ANFIS model was also developed. We evaluated the developed model for five different common traffic scenarios using MATLAB. The results obtained indicated that the developed model effectively selects a phase to be given next green signal after considering the traffic situation and the nature of the intersection in question.

References

[1] E.A. Mueller. Aspects of the history of traffic signals. IEEE Transactions on Vehicular Technology, 19(1):6 –17, February 1970.

[2] G.E.M.D.C. Bandara et at.: Application of fuzzy logic in intelligent traffic control systems, National University of Singapore, CIRAS, 2003

[3] M. R. A. Purnomo et al.: Development of a low cost smart traffic controller system, European Journal of Scientific Research, 2009, 32(4): 490 – 499

[4] Niittymaki J. et al.: Fuzzy Traffic Signal Control and a New Interface Method - Maximal Fuzzy Similarity. In: Proc., The 13th Mini-EURO Conf. (Handling Uncertainty in the Analysis of Traffic and Transportation Systems) and the 9th Mtg. EURO Working Group on Transportation Intermodality, Sustainability and Intelligent Transportation Systems, Bari, Italy, 2002, 716–728.

[5] Zhang L, Li H, Prevedouros P D. Signal control for oversaturated intersections using fuzzy logic. In: Proc. of 84th Transp. Res. Bd. Ann. Mtg., Washington, D.C., 2005

[6] Nakatsuyama M, Nagahashi H, Nishizuka N. Fuzzy logic phase controller for traffic junctions in the one-way arterial road. In: Proc., IFAC 9th Triennial World Cong., Budapest, Hungary, 1984, 2865–2870

[7] Tan, K.K., Khalid, M., Yusof, R.: Intelligent Traffic Lights Control by Fuzzy Logic. Malaysian Journal of Computer Science 9(2), 29–35 (1996)

[8] Pappis C. and Mamdani E.: A fuzzy logic controller for a traffic junction, IEEE Trans. Systems, Man, and Cybernetics SMC-7, 1977, 7(10): 707–717.

[9] J.S.R. Jang: ANFIS: Adaptive-network-based Fuzzy Inference Systems. IEEE Trans, Syst, Man Cybern., 23(3), pp. 665–685, 1993.

[10] Wenteng M.: A Real-time Performance Measurement System for Arterial Traffic Signals. A Ph.D thesis, Graduate School, University of Minnesota, 2008.

Classification credit dataset using particle swarm optimization and probabilistic neural network models based on the dynamic decay learning algorithm

Reza Narimani[1,*]**, Ahmad Narimani**[2]

[1]Department of Financial Engineering, University of Economic Sciences, Tehran, Iran
[2]Department of Economics, University of AllamehTabatabae'i, Tehran, Iran

Email addresses:
reza_narimani@yahoo.com (R. Narimani), ahmad_narimani67@yahoo.com (A. Narimani)

Abstract: This paper describes a credit risk evaluation system that uses supervised probabilistic neural network (PNN) models based on the Dynamic Decay learning algorithm (DDA). The PNN-DDA has two parameters called positive and negative threshold. This learning algorithm trains very quickly. Thus it makes sense that we use a meta-heuristic algorithm such as particle swarm optimization to optimize these parameters. When using the meta-heuristic algorithm such PSO, the tuning process of parameters is implemented wisely. Thus in this paper we also obtained optimum threshold. Two credit datasets in UCI database are selected as the experimental data to demonstrate the accuracy of the proposed model. The result shows that this new hybrid algorithm outperforms the most common used algorithm such as multi-layer neural network.

Keywords: Probabilistic Neural Network Particle Swarm Optimization, Dynamic Decay Algorithm, Classification

---

1. Introduction and Related Work

One another of first studies that became well-known in credit risk measurement was Z-score that is obtained from multi variable scoring model (Altman, 1968). This model is multiple discriminant analysis (MDA) that by using important financial ratios tries to distinguish between bankrupt and non-bankrupt firms. With attention to this fact that most unpaid loan are related to the firms that will have financial distress in future, so credit risk prediction is possible by this model. In this way, Saunders and Allen, 2002, used this model for credit risk of the firms that borrowed loan from banks. Their investigation showed that MDA has high performance of prediction capability.

In the context of credit risk measurement, we can mention another important study was conducted by the work of Elmer and Borowski, 1988. They predict loan repayment ability by multi-layer perceptron neural network model. Their input variables ware the same as the variables used in the model Z-Altman model. They compared the results from neural network model and the z-Altman model and found that the capability of credit scoring prediction of neural network is higher than the z-Altman model. Among the studies that were conducted in the area of credit risk measurement model, Morgan, 1998 work on designing a credit risk model can be mentioned.

Traditional statistical methods such as linear discriminant models (Reichert, Cho, & Wagner, 1983), logit and probit models (Beaver, 1966 and Ohlson, 1980 & Henley, 1995) in the past two decades were the most applicable models but the use of models based on neural networks increased the accuracy of these predictions significantly.

In any of these traditional methods, what causing the problem is that these models are bound to skewness that exist in the sample selected. Thus, the final model is constructed based on the data selected and reduce the accuracy of the model. For this reason, the most recent non-parametric methods such as the nearest model k Neighbor (Henley & Hand, 1996), neural networks (Desai, Crook, & Overstreet, 1996; Malhotra & Malhotra, 2002; West, 2000; Charalambous and Charitous & Kaourou, 2000; Lee and

Han & Kwon, 1996; Boritz and Kennedy, 1995), and classification and regression trees (Davis, Edelman, & Gammerman, 1992), and Support Vector Machines (SVMs) (Min and Lee, 2005; Shin and Lee & Kim, 2005), and genetic algorithm (Shin and Lee, 2002; Varetto, 1998) are used.

Artificial neural networks (ANN) are one of the strongest data mining tasks which can be used for finding the nonlinear nature and pattern of data. They are inspired by the biological network of neurons in the human brain (Mira & Sanchez-Andres, 1999).

Hsieh, 2005, present a hybrid approach based on clustering and neural network techniques. They used clustering techniques to preprocess the input samples with the objective of indicating unrepresentative samples into isolated and inconsistent clusters, and used neural networks to construct the credit scoring model.

Khashman, 2010, described a credit risk evaluation system that uses supervised neural network models based on the back propagation learning algorithm. In this paper, the dataset used for evaluation was German credit. So the input layer size was 24, but they use three neural networks with 18, 23 and 27 hidden layer size respectively.

Hybrid model have more flexibility that can mimic the nonlinearity of behavior of data relationships. These models don't have the limits of traditional classification models.

The researchers use PNN model for pattern recognition context such as: marketing (Kazemi, 2013), signal processing (Übeyli, 2008), medical/biochemical field applications (Mantzaris, 2011; Hajmeer, 2002), civil (Tam, 2004).

The most interesting algorithms in PNN and RBS neural network is Dynamic Decay learning algorithm (DDA) (Berthold and Diamond,1995; Berthold and Diamond, 1998).

RBFN-DDA is a dynamically growing neural network that adapts the numbers of neurons to the data space. The main difference of RBFN-DDA to the PNN is the smaller number of utilized neurons in the hidden layer.

While PNN uses one hidden neuron for one available sample, i.e., a kind of memorizing every sample, RBFN-DDA generalizes the data by only inserting a neuron in the network when it becomes necessary (Paetz, 2002)

Topouzelis and Karathanassi & Pavlakis & Rokos, 2004 compared Radial Basis Function (RBF) neural network trained by Dynamic Decay learning algorithm (DDA) and multi-layer neural network to in oil spill detection. The results showed that MLPs appear to be superior to RBFs in detecting oil spills on Synthetic Aperture Radar (SAR) images.

Our purpose of this paper is to introduce the hybrid algorithm to enhance the learning of RBF neural network by DDA.

2. Probabilistic Neural Network (PNN)

Probabilistic neural networks are a special type of feed-forward neural networks that are based on radial basis functions that using Bayes' decision theory to classify input pattern (Donal F.Specht 1988; Donal F.Specht 1991). By doing some modification on RBF neural networks, it is possible to estimate probability density function (PDF) of each class pattern.

Probabilistic neural networks are a form of normalized Radial-Basis Form (RBF) neural networks, where each hidden node is a "kernel" implementing a probability density function (Jones, 2008).

So PNN has three layer fixed structure as RBF. A PNN is a four-layer architecture which consists of input, pattern, summation and output layers (Ding, X., Yeh, C.-H., & Bedingfield, S, 2010). The pattern and summation layer consist the hidden layer. Probabilistic neural network architecture is shown in Figure 1.

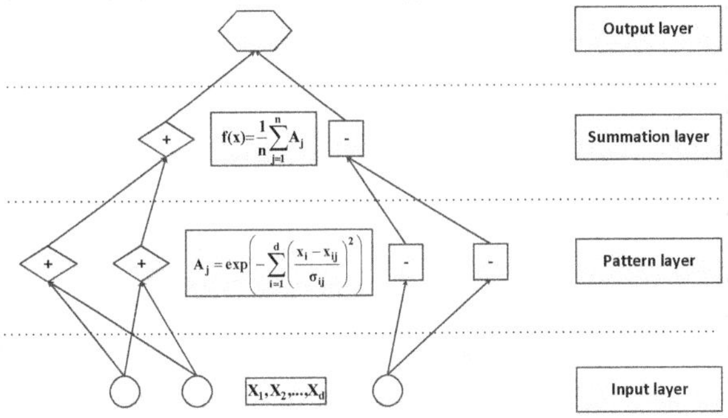

Figure 1. *Probabilistic neural network architecture*

So PNN includes an input layer, a hidden layer and an output layer. Input layer has the responsibility of distribution and it does not implement any process in it. All neurons in the hidden layer have the activation function of radial basis functions usually are selected in the form of Gaussian type. The neurons in the hidden layer receive input vectors and have two parameters: 1) center 2) spread.

Amount of overlap between adjacent neurons in hidden layer neurons determine this spread. Each hidden layer neuron can be said to be simply has larger output when the input vector is closer to the center of the non-linear function of neurons. With increasing the distance of input vector from the center of non-linear functions, the output of neurons is also reduced.

The output layer is a competitive layer. The number of neurons in competitive layer is equal to the number of classes. Activation function of the hidden layer is as follows:

$$a_{ij} = \psi_j\left(\frac{x_i - x_{ij}}{\sigma_{ij}}\right) \tag{1}$$

Where x_i, is ith element of input vector to the jth neuron of the hidden layer. σ_{ij}, x_{ij} and a_{ij} are spread, center and output of jth neuron of hidden layer respectively.

Output of jth neuron of hidden layer (A_j) is obtained by the product of the activation functions that is as follows:

$$A_j = \prod_{i=1}^{d} a_{ij} = \prod_{i=1}^{d} \psi_j \left(\frac{x_i - x_{ij}}{\sigma_{ij}} \right) \tag{2}$$

Where d is the number of dimensions in the input vector. If the activation function (ψ) is selected in the form of the Gaussian:

$$\psi = \exp\left(-\left(\frac{x_i - x_{ij}}{\sigma_{ij}} \right)^2 \right) \tag{3}$$

Eq 2 became as follows:

$$A_j = \prod_{i=1}^{d} a_{ij} = \exp\left(-\sum_{i=1}^{d} \left(\frac{x_i - x_{ij}}{\sigma_{ij}} \right)^2 \right) \tag{4}$$

But the main part of the training of PNN is based on the probabilistic technique that is implemented by using of Bayes strategy and Parzen's non-parametric estimation technique. By using Bayes' decision theory and satisfying following equation, learning sample can be assigned to the category k.

$$h_k l_k f_k(x) > h_q l_q f_q(x) \tag{5}$$

h_k and h_q respectively are prior probability for the class k and q; The probability that sample had been selected from class k or q. l_k and l_q are probability of classification error. On the other hand, Parzen showed that if the kernel function concentrates on some part of a class of samples then it is a good approximation of probability density function for that class. Probability density function for a class can be approximated by the following formula:

$$pdf_k(x) = \left(\frac{1.0}{2\pi^{d/2} \sigma^d} \right) \left(\frac{1.0}{n_k} \right) \sum_{j=1}^{n_k} \left(\exp\left(\frac{-(x - x_{kj})^2}{2\sigma^2} \right) \right) \tag{6}$$

The value of n_k is equal to the number of data in class k, and d is the number of dimensions in the input vector. x_{kj} represents the center of the Gaussian function and is corresponding to the jth sample in the dataset belonging to the class k.

This seemingly complicated formula meaning that at first, the summation of Gaussian functions is calculated then the average of them is gained and then multiplied by weighting factor (the first sentence in the relationship), including fixed terms and nth power of spread. Clearly, the choice of σ has an important effect on $pdf_k(x)$. If σ is too large, the estimate will suffer from too little resolution; if σ is too small, the estimate will suffer from too much statistical variability (Duda, 2001). One way to reach the optimum σ is try and error method.

This algorithm introduces the idea of distinguishing between matching and conflicting neighbors in an area of conflict. Two different thresholds: θ^+ and θ^- as illustrated in Figure 2.

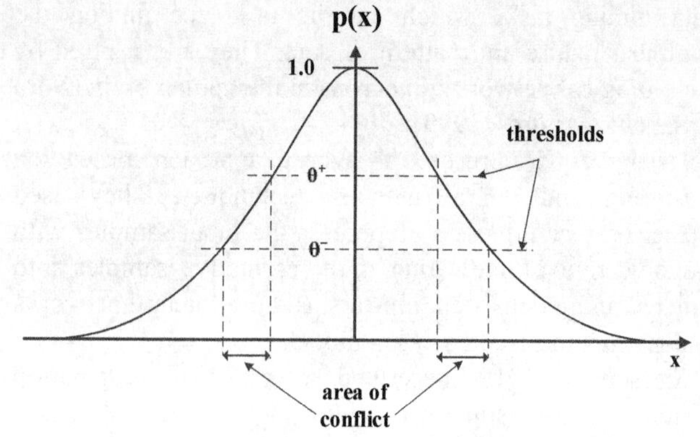

Figure 2. Two different thresholds: θ^+ and θ^- in PNN trained by DDA algorithm

They are used to define intersection of influencing area. θ^+ determines the minimum correct classification probability for training patterns of the correct class. In contrast θ^- is used to avoid misclassifications; that is the probability for an incorrect class for each training pattern is less than or equal to θ^- (M. Berthold, J. Diamond 1998).

In general the DDA algorithm comprises the following three steps (Chiang Tan, 2006):

- Covered. If a new pattern is correctly classified by an already existing prototype, it initiates regional expansion of the winning prototype in the attribute space.
- Commit. If a node of the correct class does not cover a new pattern, a new hidden node will be introduced, and the new pattern is codded as reference vector.
- Shrink. If a new pattern is incorrectly classified by an already existing prototype of conflicting classes, the width of conflicting classes, the width of the prototype will be reduced (i.e. shrunk) for the sake of overcoming the conflict.

These steps are shown in Table 1 and Figure 3.

Table 1. Dynamic Decay learning algorithm (DDA)

FORALL prototypes p_i^k **DO**	// reset weights
$A_i^k = 0.0$	
ENDFOR	
FORALL training pattern $(\vec{x}, c)$ **DO**:	// train one complete epoch
IF $\exists p_i^c : \exp\left(-\frac{\lVert x - r_i^c \rVert^2}{(\sigma_i^c)^2} \right) \geq \theta^+$ **THEN**	
$A_i^c += 1.0$	
ELSE	

add new neuron $p_{m_c+1}^{c}$ with: // commit new neuron

$r_{m_c+1}^{c} = x$

$\sigma_{m_c+1}^{c} = \min_{\substack{k \neq c \\ 1 \leq j \leq m_s}} \sqrt{-\frac{\left\| r_j^k - r_{m_c+1}^c \right\|^2}{\ln \theta^-}};$

$A_{m_c}^{c} = 1.0$

$m_{c^+} = 1$

ENDIF

FORALL k≠c, 1≤j≤ m_k **DO** // shrink radius of conflicting neurons

$\sigma_j^k = \min\left\{ \sigma_j^k, \sqrt{-\sqrt{\frac{\left\| x - r_j^k \right\|^2}{\ln \theta^-}}} \right\};$

ENDFOR

where p_i^c = *i*-the neurons of class *c* ($c \in \{1,\ldots,n\}$, *n* classes), θ_c^+ , θ_c^- : controlling size of overlapping regions of neurons in respect to each class, m_c is the number of neurons for class c, r_i^c is the center of neurons p_i^c , σ_i^c is the radius of neuron p_i^c .

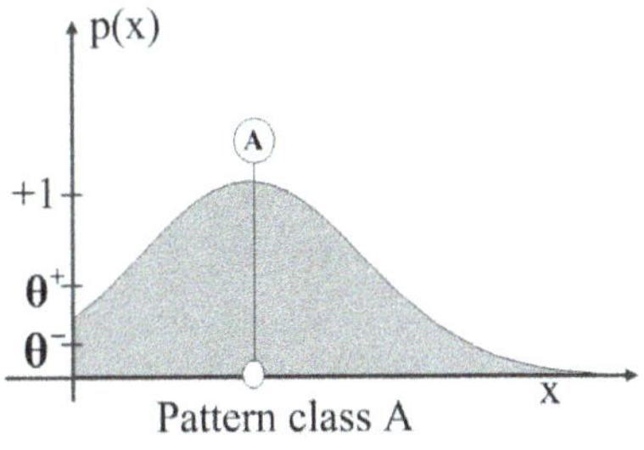

(a)

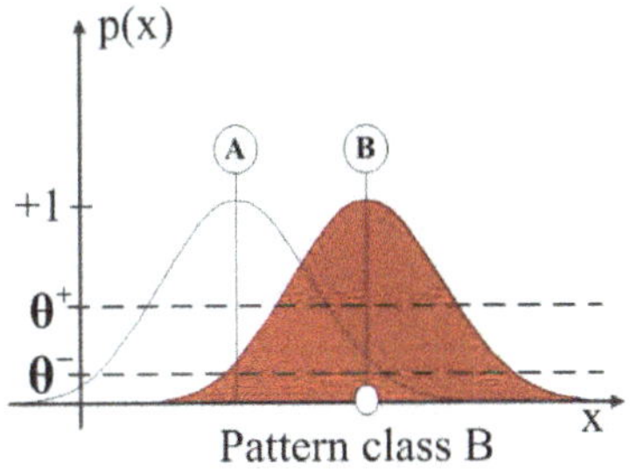

(b)

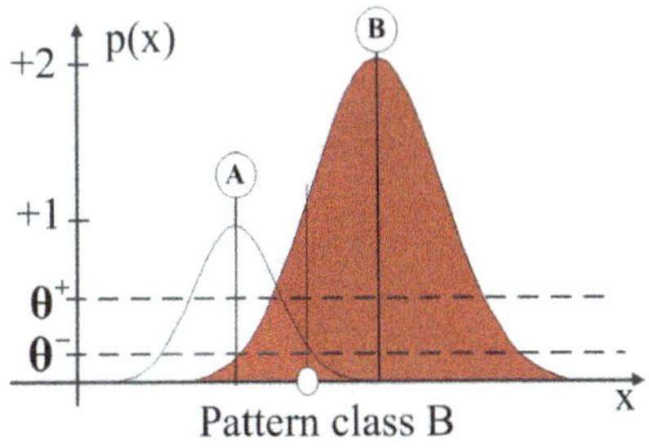

(c)

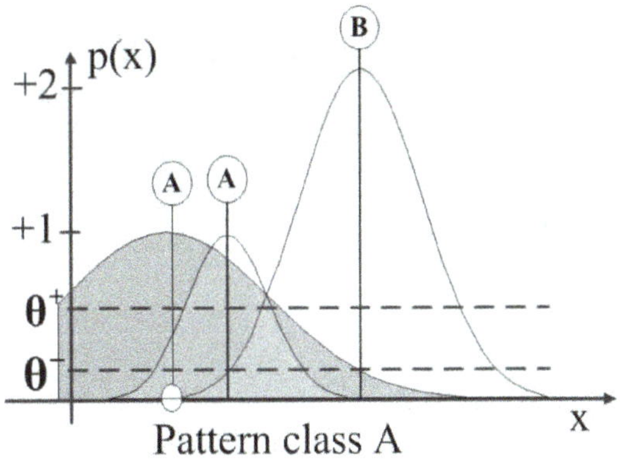

(d)

Figure 3. *The procedure of applying thresholds on data: (a) a pattern of class A is encountered and a new RBF is created (b) a training pattern of class B leads to a new prototype for class B and shrinks the radius of the existing RBF of class A (c) another pattern of class B is classified correctly and shrinks again the prototype of class A (d) a new pattern of class A introduces another prototype of that class.*

3. Particle Swarm Optimization (PSO) Algorithm

Originally, particle swarm optimization was proposed by Kennedy and Eberhart (1995). The main idea of PSO is to mimic social behavior of birds. In PSO algorithm, each particle can move along the linear combination of its personal velocity, towards best global position and towards best local of its personal position in the problem space. The velocity of a particle is updated according to Eqs. 7 and 8.

$$v_{ij}(t+1) = w\, v_{ij}(t) + r_1 C_1 \big(P_{ij}(t) - x_{ij}(t)\big) + r_2 C_2 (g_j(t) - x_{ij}(t)), \quad (7)$$

$$x_{ij}(t+1) = x_{ij}(t) + v_{ij}(t+1), \quad (8)$$

where w is inertia weight that shows the effect of previous velocity on new velocity vector. C_1 and C_2 are positive constant, r_1 and r_2 are random variables with uniform distribution in between *0* and *1*.

4. Empirical Analysis

Credit data Sets

We choose the Australian and German credit data sets which are two real world data sets. They are available from the UCI Repository of Machine Learning Databases (Murphy & Aha, 2001). These data are shown on Table 2. The German credit scoring data are more unbalanced than Australian credit data sets. For each applicant, 24 input variables describe the credit history, account balances, loan purpose, loan amount, employment status, personal information, age, housing, and job title. This data set only consists of numeric attributes. (Cheng and Mu-Chen & Chieh, 2007).

Table 2. Basic information of the two credit datasets

No.	1	2
Names	German	Australian
# classes	2	2
# instances	1000	690
Nominal features	0	6
Numeric features	24	8
Total features	24	14
No. of good instances	700	307
No. of bad instances	300	383

5. Performance Evaluation Criteria

The fundamental issue for rating a classifier's performance is the confusion matrix where the numbers represent the total number of actual classes and predicted classes. (Hassan and Ramamohanarao and Karmakar and Hossain & Bailey, 2010)

A confusion matrix has shown in Table 3.

Table 3. Confusion matrix

	Actual Value P	Actual Value N
Prediction Outcome	True Positive	False Positive
	False Negative	True Negative

Based on the elements in the confusion matrix, following statistics are defined:

$$TPR\, or\, sensitivity = \frac{TP}{TP + FN} \tag{9}$$

$$specificity = \frac{TN}{TP + FN} = 1 - FPR \tag{10}$$

$$classification\, accuracy = \frac{TP + TN}{TP + FP + TN + FN} \tag{11}$$

6. Area under the ROC Curve

Sensitivity and specificity rely on a single cut-point to classify a test result as positive. A more complete description of classification accuracy is given by the area under the ROC (Receiver Operating Characteristic) (Hosmer and Lemeshow, 1989).

The vertical axis of an ROC curve represents TPR. The horizontal axis represents FPR. On the graph, we move right and plot a point. This process is repeated for each of the test tuples in ranked order, each time moving up on the graph for a true positive or toward the right for a false positive (Han and Kamber & Pei, 2012).

The AUC as the fitness function, in the discrete case, can compute with step functions:

$$AUC = \frac{\sum_{i=1}^{n^+} \sum_{j=1}^{n^-} 1_{f(x_i^+) > f(x_j^-)}}{n^+ n^-} \tag{12}$$

Where $f(0)$ is denoted as the scoring function. x^+ and x^- respectively denote the positive and negative samples and n^+ and n^- are respectively the number of positive and negative examples and 1_π is defined to be 1 if the predicate π holds and 0 otherwise (Campbell & Ying 2011; Rakotomamonjy 2004).

7. Fitness Function

There are three common functions to candidate for fitness function in classification problem:

1. Classification accuracy
2. Specificity × sensitivity
3. Area Under ROC

We use Area Under ROC (AUC) as fitness function in PNN-DDA trained by PSO.

8. A Hybrid Algorithm of PSO and PNN-DDA

A PSO algorithm using 5-fold cross-validation is carried out on each training set to find the optimal parameter pair (θ^+, θ^-). Then, the related dataset is trained with the obtained optimal parameter pair (θ^+, θ^-) to get a predictor model. This process is shown on Figure 4:

The parameter used in PSO algorithm is shown on Table 4.

Table 4. The parameter used in PSO

Parameter	Value
Max number of generations	20
Population size	15
inertia weight (w)	0.7298
C1	1.4962
C2	1.4962

9. Results

The classification accuracies and AUC results on the testing data for the 2 datasets are shown in Table 5 and Table 6.

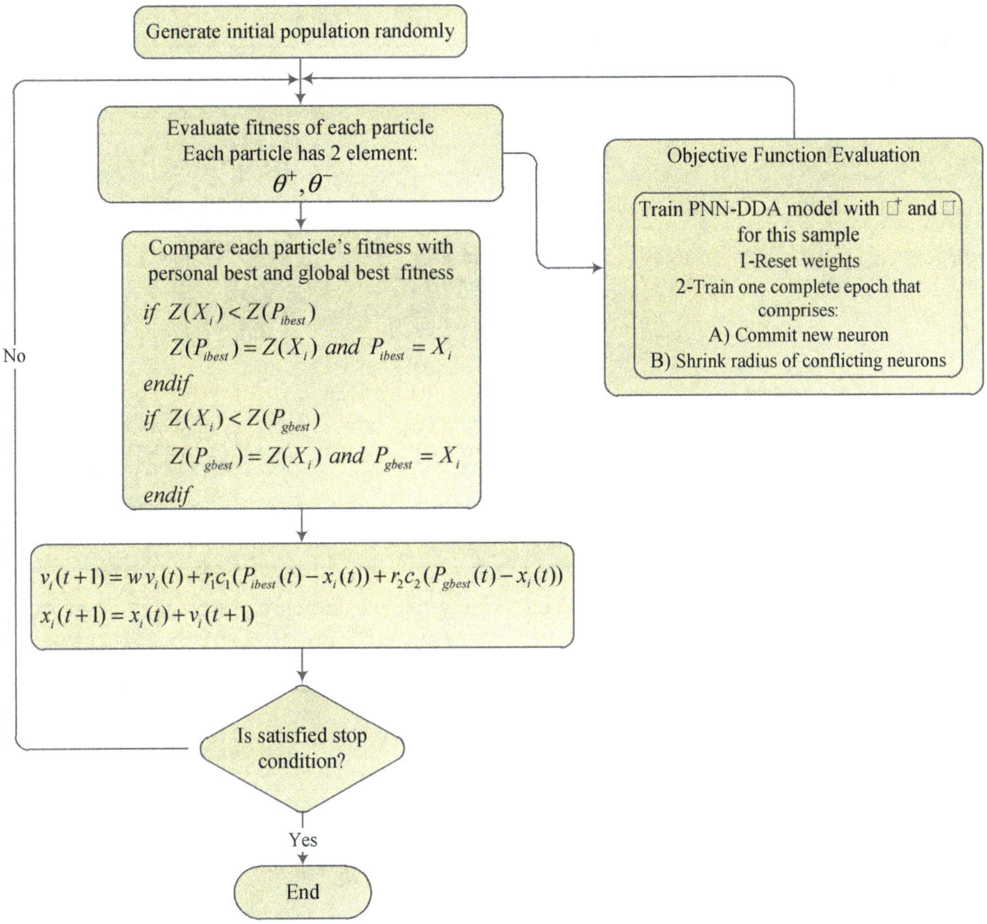

Figure 4. *A hybrid algorithms that PSO optimize the two parameters* (θ^+, θ^-) *in PNN-DDA model*

Table 5. *classification accuracy results of models*

Model	Classification Accuracy (%)							
	70-30% training-test				80-20% training-test			
	German		Australia		German		Australia	
	train	test	train	test	train	test	train	test
ANN	0.6614	0.6400	0.8758	0.8599	0.6975	0.7150	0.8696	0.8188
PSO-PNN	0.8386	**0.7533**	0.9006	**0.8599**	0.8738	**0.7350**	0.9130	**0.8841**

Table 6. *AUC results of models*

Model	Area Under ROC curve (%)							
	70-30% training-test				80-20% training-test			
	German		Australia		German		Australia	
	train	test	train	test	train	test	train	test
ANN	0.8091	0.7522	0.9430	0.9133	0.8134	**0.7920**	0.9497	0.9167
PSO-PNN	0.9317	**0.7860**	0.9632	**0.9200**	0.9258	0.7664	0.9660	**0.9209**

We present values of sensitivity and specificity for the two credit data in Table 7.

***Table 7.** Sensitivity, Specificity for two credit data*

Model	Sensitivity, Specificity for two credit data							
	70-30% training-test				80-20% training-test			
	German		Australia		German		Australia	
	train	test	train	test	train	test	train	test
ANN Threshold= 0.5	[0.9128 0.4928]	[0.8841 0.4624]	[0.8727 0.9028]	[0.8017 0.9011]	[0.8681 0.5858]	[0.8345 0.5738]	[0.8656 0.9150]	[0.7949 0.9000]
PSO-PNN Threshold=0.5	[0.6714 0.9614]	[0.5749 0.7742]	[0.9625 0.7963]	[0.9397 0.6923]	[0.5882 0.9707]	[0.5036 0.8361]	[0.9475 0.8219]	[0.9487 0.7000]
ANN Optimum Threshold	[0.5842 0.8454]	[0.5749 0.7849]	[0.8127 0.9537]	[0.7845 0.9560]	[0.6595 0.7866]	[0.6906 0.7705]	[0.8098 0.9433]	[0.7436 0.9167]
PSO-PNN Optimum Threshold	[0.8377 0.8406]	[0.7874 0.6774]	[0.8989 0.9028]	[0.8793 0.8352]	[0.8414 0.8452]	[0.7626 0.6721]	[0.9115 0.9150]	[0.8846 0.8833]

The best parameter pairs (θ^{+},θ^{-}) of two dataset on each training-test partition are presented in detail in Table 8.

***Table 8.** The optimal parameters for two credit dataset*

Parameters		70-30% training-test		80-20% training-test	
		German	Australia	German	Australia
	θ^{+}	0.8071	0.6109	0.6365	0.6509
	θ^{-}	0.2693	0.2998	0.2608	0.2998
	Optimum threshold	0.5488	0.3488	0.5705	0.3998
ANN	Optimum neuron	4	4	2	5
	Optimum threshold	0.2515	0.3417	0.3302	0.2535

Plotting of ROC curves for the two credit data have shown in Figure 5. The bigger area means better classifier performance.

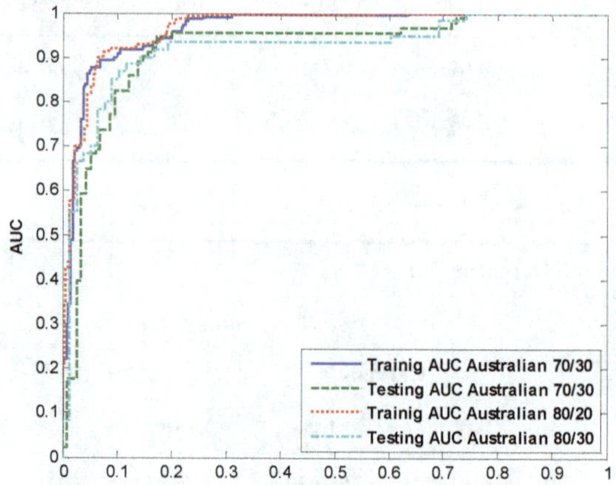

(a)

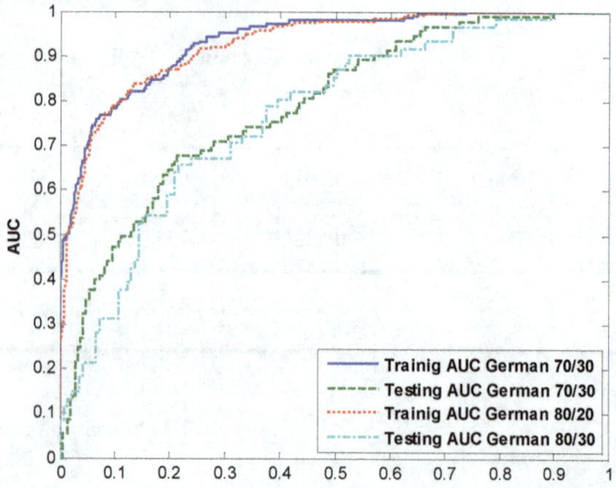

(b)

***Figure 5.** AUC for Australian and German Credit data To observe the evolutionary process in our model, Figure 6 shows the evolution of the best fitness on the two credit dataset.*

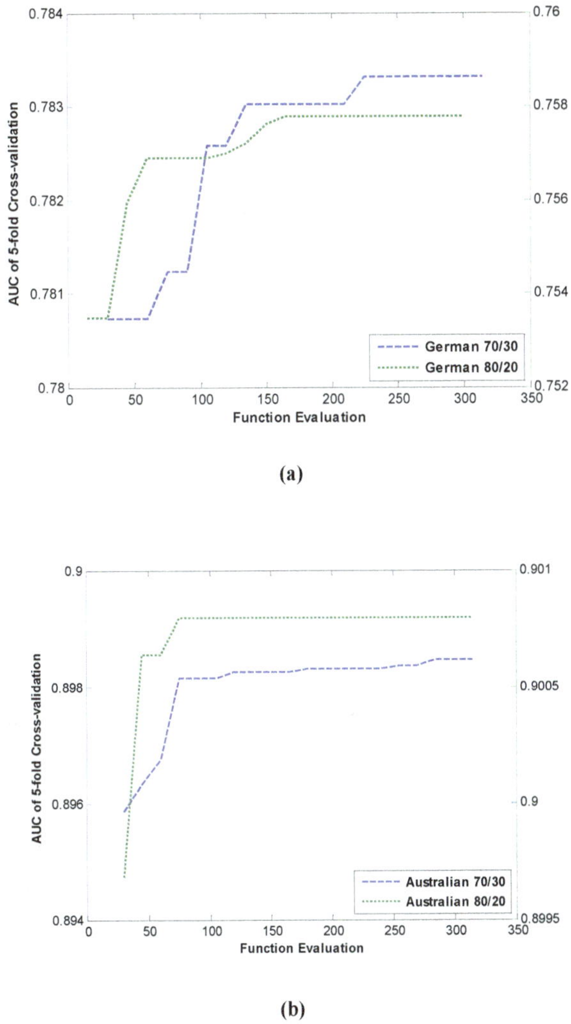

Figure 6. PSO plot for the best fitness during the training phase on the (a) German and (b) Australian credit dataset

It can be observed that in Figure 6.a, the fitness curves gradually improved from 1 to 230 function evaluations and exhibited no significant improvements after that and in Figure 6.b, the fitness curves gradually improved from 1 to 100 function evaluations and exhibited no significant improvements after that. Eventually the optimization stopped at the 350 function evaluations where the generations reached the stopping criterion.

Hence, this hybrid model can converge quickly to global optima and we can enhance its effectiveness for our hybrid model.

10. Conclusion

We have proposed a new effective approach for two class problems in data mining area by applying the PSO algorithm to optimize the two parameter of PNN-DDA. Achieved results show that the proposed model is worthwhile, since it had a reasonable accuracy on two datasets. Furthermore we calculated the best threshold for achieving the best ROC curve. According to the values of classification accuracy, it can be realized that the PSO-PNN has the better performance rather than ANN in two dataset and different training-testing partition. Furthermore, PSO-PNN has considerable performance due to the AUC measure.

References

[1] Altman, E. I. (1968). FINANCIAL RATIOS, DISCRIMINANT ANALYSIS AND THE PREDICTION OF CORPORATE BANKRUPTCY. The Journal of Finance, 23, 589-609.

[2] Beaver, W. H. (1966). Financial Ratios As Predictors of Failure. Journal of Accounting Research, 4, 71-111.

[3] Berthold, M. R., & Diamond, J. (1998). Constructive training of probabilistic neural networks. Neurocomputing, 19, pp. 167-183.

[4] Campbell, C & Ying, Y 2011, 'Learning with Support Vector Machines', in SYNTHESIS LECTURES ON ARTIFICIAL INTELLIGENCE AND MACHINE, Morgan and cLaypool.

[5] Charalambous, C., Charitou, A., & Kaourou, F. (2000). Comparative Analysis of Artificial Neural Network Models: Application in Bankruptcy Prediction. Annals of Operations Research, 99, 403-425.

[6] Chiang Tan, S., Rao, M. V. C., & Lim, C. (2006). An Adaptive Fuzzy Min-Max Conflict-Resolving Classifier. In A. Abraham, B. de Baets, M. Köppen & B. Nickolay (Eds.), Applied Soft Computing Technologies: The Challenge of Complexity (Vol. 34, pp. 65-76): Springer Berlin Heidelberg.

[7] Davis, R. H., Edelman, D. B., & Gammerman, A. J. (1992). Machine learning algorithms for credit-card applications. Journal of Mathematics Applied in Business and Industry, 4, 43–51.

[8] Desai, V. S., Crook, J. N., & Overstreet Jr, G. A. (1996). A comparison of neural networks and linear scoring models in the credit union environment. European Journal of Operational Research, 95, 24-37.

[9] Ding, X., Yeh, C.-H., & Bedingfield, S. (2010). A Probabilistic Neural Network Approach to Modeling the Impact of Tobacco Control Policies by Gender. In Z. Zeng & J. Wang (Eds.), Advances in Neural Network Research and Applications (Vol. 67, pp. 869-876): Springer Berlin Heidelberg.

[10] Duda, R. O., Hart, P. E., & Stork, D. G. (2001). Pattern classification. New York: Wiley.

[11] Efrim Boritz, J., & Kennedy, D. B. (1995). Effectiveness of neural network types for prediction of business failure. Expert Systems with Applications, 9, 503-512.

[12] Elmer, Peter J., and David M. Borowski. (1988). An Expert System and Neural Networks Approach To Financial Analysis, Financial Management, No 12, 66-76.

[13] Hajmeer, M., & Basheer, I. (2002). A probabilistic neural network approach for modeling and classification of bacterial growth/no-growth data. Journal of Microbiological Methods, 51, 217-226.

[14] Han, J., Kamber, M., & Pei, J. (2012). Data mining concepts and techniques, third edition. In. Waltham, Mass.: Morgan Kaufmann Publishers.

[15] Hassan, M. R., Ramamohanarao, K., Karmakar, C., Hossain, M. M., & Bailey, J. (2010). A Novel Scalable Multi-class ROC for Effective Visualization and Computation. In M. Zaki, J. Yu, B. Ravindran & V. Pudi (Eds.), Advances in Knowledge Discovery and Data Mining (Vol. 6118, pp. 107-120): Springer Berlin Heidelberg.

[16] Henley, W. E. (1995). Statistical aspects of credit scoring. Dissertation, The Open University, Milton Keynes, UK.

[17] Henley, W. E., & Hand, D. J. (1996). A k-nearest-neighbor classifier for assessing consumer credit risk. Journal of the Royal Statistical Society Series D: The Statistician, 45, 77-95.

[18] Hosmer, D. W., & Lemeshow, S. (2004). Applied logistic regression. Chichester: Wiley

[19] Hsieh, N. C. (2005). Hybrid mining approach in the design of credit scoring models. Expert Systems with Applications, 28, 655-665.

[20] Huang, C. L., Chen, M. C., & Wang, C. J. (2007). Credit scoring with a data mining approach based on support vector machines. Expert Systems with Applications, 33, 847-856.

[21] J.P. Morgan (April, 1998), Creditmetrics – Technical Document, New York, J.P. Morgan & Co. Incorporated.

[22] Jones, M. T. (2008). Artificial intelligence : a systems approach. Hingham, Mass.: Infinity Science Press.

[23] Kazemi, S. M. R., Hadavandi, E., Mehmanpazir, F., & Nakhostin, M. M. (2013). A hybrid intelligent approach for modeling brand choice and constructing a market response simulator. Knowledge-Based Systems, 40, 101-110.

[24] Kennedy, J. and Eberhart, R. (1995). Particle swarm optimization, Proceeding. IEEE International Conference on Neural Networks (ICNN), Nov./Dec., Australia, Pages 1942–1948.

[25] Khashman, A. (2010). Neural networks for credit risk evaluation: Investigation of different neural models and learning schemes. Expert Systems with Applications, 37, 6233-6239.

[26] Lee, J. J., Kim, D., Chang, S. K., & Nocete, C. F. M. (2009). An improved application technique of the adaptive probabilistic neural network for predicting concrete strength. Computational Materials Science, 44, 988-998.

[27] Lee, K. C., Han, I., & Kwon, Y. (1996). Hybrid neural network models for bankruptcy predictions. Decision Support Systems, 18, 63-72.

[28] M. Berthold, J. Diamond, "Boosting the Performance of RBF Networks with Dynamic Decay Adjustment", Proc. of the Advances in Neural Information Processing Systems (NIPS), Denver, USA, vol. 7, pp. 521-528, 1995.

[29] Malhotra, R., & Malhotra, D. K. (2002). Differentiating between good credits and bad credits using neuro-fuzzy systems. European Journal of Operational Research, 136, 190-211.

[30] Mantzaris, D., Anastassopoulos, G., & Adamopoulos, A. (2011). Genetic algorithm pruning of probabilistic neural networks in medical disease estimation. Neural Networks, 24, 831-835.

[31] Min, J. H., & Lee, Y. C. (2005). Bankruptcy prediction using support vector machine with optimal choice of kernel function parameters. Expert Systems with Applications, 28, 603-614.

[32] Mira, J., Sánchez-Andrés, J. V., Engineering Applications of Bio-Inspired Artificial Neural Networks. Alicante, Spain, vol. 2, 1999.

[33] Murphy, P. M., Aha, D. W. (2001). UCI repository of machine learning databases. Department of Information and Computer Science, University of California Irvine, CA. Available from http://www.ics.uci.edu/mlearn/MLRepository.htmlurlhttp://www.ics.uci.edu/mlearn/MLRepository.html.

[34] Ohlson, & James, A. (1980). Financial Ratios and the Probabilistic Prediction of Bankruptcy. Journal of Accounting Research, 18, 109.

[35] Paetz, J. (2002). Feature selection for RBF networks. In Neural Information Processing, 2002. ICONIP '02. Proceedings of the 9th International Conference on (Vol. 2, pp. 986-990 vol.982).

[36] Rakotomamonjy, A. (2004). Optimizing area under ROC curves with SVMs. In Proceedings of the ECAI-2004 Workshop on ROC Analysis in AI.

[37] Reichert, A. K., Cho, C.-C., & Wagner, G. M. (1983). An Examination of the Conceptual Issues Involved in Developing Credit-Scoring Models. Journal of Business & Economic Statistics, 1, 101-114.

[38] Saunders, A., Allen, L., & Saunders, A. (2002). Credit risk measurement in and out of the financial crisis: new approaches to value at risk and other paradigms. New York: John Wiley and Sons, 2nd edition.

[39] Shin, K. S., & Lee, Y. J. (2002). A genetic algorithm application in bankruptcy prediction modeling. Expert Systems with Applications, 23, 321-328.

[40] Shin, K. S., Lee, T. S., & Kim, H. J. (2005). An application of support vector machines in bankruptcy prediction model. Expert Systems with Applications, 28, 127-135.

[41] Specht, D. F. (1988). Probabilistic neural networks for classification, mapping, or associative memory. In Neural Networks, 1988., IEEE International Conference on (pp. 525-532 vol.521).

[42] Specht, D. F. (1991). A general regression neural network. Neural Networks, IEEE Transactions on, 2, 568-576.

[43] Tam, C. M., Tong, T. K. L., Lau, T. C. T., & Chan, K. K. (2004). Diagnosis of prestressed concrete pile defects using probabilistic neural networks. Engineering Structures, 26, 1155-1162.

[44] Topouzelis K., V. Karathanassi, P. Pavlakis, D. Rokos, 2004. Oil spill detection using RBF Neural Networks and SAR data, XXth ISPRS Congress, Istanbul, Turkey, July 2004

[45] Übeyli, E. D. (2008). Implementing eigenvector methods/probabilistic neural networks for analysis of EEG signals. Neural Networks, 21, 1410-1417.

[46] Varetto, F. (1998). Genetic algorithms applications in the analysis of insolvency risk. Journal of Banking and Finance, 22, 1421-1439.

[47] West, D. (2000). Neural network credit scoring models. Computers and Operations Research, 27, 1131-1152.

Some relationships between left (right) semi-uninorms and implications on a complete lattice

Yuan Wang[1,*], **Keming Tang**[1], **Zhudeng Wang**[2]

[1]College of Information Science and Technology, Yancheng Teachers University, Yancheng 224002, People's Republic of China
[2]School of Mathematical Sciences, Yancheng Teachers University, Jiangsu 224002, People's Republic of China

Email address:
yctuwangyuan@163.com (Yuan Wang), tkmchina@126.com (Keming Tang), zhudengwang2004@163.com (Zhudeng Wang)

Abstract: In this paper, we study the relationships between left (right) semi-uninorms and implications on a complete lattice. We firstly discuss the residual operations of left and right semi-uninorms and show that the right (left) residual operator of a conjunctive right (left) $\vee$-distributive left (right) semi-uninorm is a right $\wedge$-distributive implication that satisfies the neutrality principle. Then, we investigate the left and right semi-uninorms induced by an implication, give some conditions such that two operations induced by an implication constitute left or right semi-uninorms, and demonstrate that the operations induced by a right $\wedge$-distributive implication, which satisfies the order property or neutrality principle, are left (right) $\vee$-distributive left (right) semi-uninorms or right (left) semi-uninorms. Finally, we reveal the relationships between conjunctive right (left) $\vee$-distributive left (right) semi-uninorms and right $\wedge$-distributive implications which satisfy the neutrality principle.

Keywords: Fuzzy Logic, Fuzzy Connective, Left (Right) Semi-Uninorm, Implication, Neutrality Principle, Order Property

1. Introduction

In fuzzy logic, the truth function for a conjunction connective is usually taken as a triangular norm [1] (t-norm for short), which is monotone, associative, commutative and has neutral element 1. However, the t-norms cannot deal with natural interpretations of linguistic words because the axioms of t-norms are strong. To interpret the non-commutative conjunctions, Flondor et al. [2] introduced non-commutative t-norms by throwing away the axiom of commutativity of t-norms and using them to construct pseudo-BL-algebras. About another axiom of t-norms, associativity, Fodor and Keresztfalvi [3] underlined that "if one works with binary conjunctions and there is no need to extend them for three or more arguments, as happens, for example, in the inference pattern called generalized modus ponens, associativity of the conjunction is an unnecessarily restrictive condition". By removing the associativity and commutativity from the axioms of t- norms, weak t-norms [4] and pseudo-t-norms [5] were introduced and discussed.

Uninorms, introduced by Yager and Rybalov [6] and studied by Fodor et al. [7], are special aggregation operators that have proven useful in many fields like fuzzy logic, expert systems, neural networks, aggregation, and fuzzy system modeling [8-11]. This kind of operation is an important generalization of both t-norms and t-conorms and a special combination of t-norms and t-conorms [7]. However, there are real-life situations when truth functions cannot be associative or commutative. By throwing away the commutativity from the axioms of uninorms, Mas et al. introduced the concepts of left and right uninorms on [0, 1] in [12] and later in a finite chain in [13], Wang and Fang [14-15] studied the left and right uninorms on a complete lattice. By removing the associativity and commutativity from the axioms of uninorms, Liu [16] introduced the concept of semi-uninorms and Su et al. [17] discussed the notion of left and right semi-uninorms on a complete lattice. On the other hand, it is well known that a uninorm (semi-uninorm, left and right uninorms) U can be conjunctive or disjunctive whenever $U(0,1)=0$ or 1, respectively. This fact allows to use uninorms in defining fuzzy implications and coimplications [14-16, 18-20].

In this paper, based on [14, 16, 18-20], we study left (right) semi-uninorms and implications on a complete lattice. After recalling some necessary definitions and examples about the left and right semi-uninorms on a complete lattice in Section 2, we discuss the residual operations of left and right semi-uninorms in Section 3 and show that the right (left) residual operator of a conjunctive right (left) $\vee$-distributive left (right) semi-uninorm is a right $\wedge$-distributive implication that satisfies the neutrality principle. In Section 4, we investigate the left and right semi-uninorms induced by an implication and give some conditions such that the operations induced by an implication become either left or right semi-uninorms. In Section 5, we reveal the relationships between conjunctive right (left) $\vee$-distributive left (right) semi-uninorms and right $\wedge$-distributive implications which satisfy the neutrality principle.

The knowledge about lattices required in this paper can be found in [21].

Throughout this paper, unless otherwise stated, L always represents any given complete lattice with maximal element 1 and minimal element 0; J stands for any index set.

2. Left and Right Semi-Uninorms

Noting that the commutativity and associativity are not desired for aggregation operators in a number of cases, Liu [16] introduced the concept of semi-uninorms and Su et al. [17] studied the notions of left and right semi-uninorms. Here, we recall some necessary definitions and examples of the left and right semi- uninorms on a complete lattice.

Definition 2.1 (Su et al. [17]). *A binary operation U on L is called a left (right) semi-uninorm if it satisfies the following two conditions:*

(U1) *there exists a left (right) neutral element, i.e., an element $e_L \in L$ ($e_R \in L$) satisfying $U(e_L, x) = x$ ($U(e_R, x) = x$ for all $x \in L$,*

(U2) *U is non-decreasing in each variable.*

For any left (right) semi-uninorm U on L, U is said to be left-conjunctive and right-conjunctive if $U(0, 1) = 0$ and $U(1, 0) = 0$, respectively. U is said to be conjunctive if both $U(1, 0) = 0$ and $U(1, 0) = 0$ since it satisfies the classical boundary conditions of AND. If $U(1, 0) = 1$ ($U(0, 1) = 1$), then we call U left-disjunctive (right-disjunctive). We call U disjunctive if both $U(1, 0) = 1$ and $U(0, 1) = 1$ by a similar reason.

If a left (right) semi-uninorm U is associative, then U is the left (right) uninorm [14-15] on L.

If a left (right) semi-uninorm U with the left (right) neutral element $e_L \in L$ ($e_R \in L$) has a right (left) neutral element $e_R \in L$ ($e_L \in L$), then $e_L = U(e_L, e_R) = e_R$. Let $e = e_L = e_R$. Here, U is the semi-uninorm [16]. In particular, if the neutral element $e = 1$, then the semi-uninorm U becomes a *t*-seminorm [22] or a semi-copula [23-24]; if the neutral element $e = 0$, then the semi-uninorm U becomes a *t*-semiconorm [25].

Clearly, $U(0, 0) = 0$ and $U(1, 1) = 1$ hold for any left (right) semi-uninorm U on L. Moreover, the left (right) neutral elements need not to be unique. In fact, the projection operator given by $U(x, y) = x$ for all $x, y \in L$ is such that any element in L is a right neutral element. But, left (right) neutral elements are all idempotent [26] because $U(e_L, e_L) = e_L$ ($U(e_R, e_R) = e_R$) for any left (right) neutral element e_L (e_R) of U.

Definition 2.2 (Wang and Fang [15]). *A binary operation U on L is called left (right) $\vee$-distributive if*

$$U(\vee_{j\in J} x_j, y) = \vee_{j\in J} U(x_j, y) \quad \forall x_j, y \in L$$

$$\left(U(x, \vee_{j\in J} y_j) = \vee_{j\in J} U(x, y_j) \quad \forall x, y_j \in L\right);$$

left (right) $\wedge$-distributive if

$$U(\wedge_{j\in J} x_j, y) = \wedge_{j\in J} U(x_j, y) \quad \forall x_j, y \in L$$

$$\left(U(x, \wedge_{j\in J} y_j) = \wedge_{j\in J} U(x, y_j) \quad \forall x, y_j \in L\right).$$

If a binary operation U is left $\vee$-distributive ($\wedge$-distributive) and also right $\vee$-distributive ($\wedge$-distributive), then U is said to be $\vee$-distributive ($\wedge$-distributive).

Noting that the least upper bound of the empty set is 0 and the greatest lower bound of the empty set is 1, we have

$$U(0, y) = U(\vee_{j\in \Phi} x_j, y) = \vee_{j\in \Phi} U(x_j, y) = 0$$

$$\left(U(x, 0) = U(x, \vee_{j\in \Phi} y_j) = \vee_{j\in \Phi} U(x, y_j) = 0\right)$$

for any $x, y \in L$ when U is left (right) $\vee$-distributive,

$$U(1, y) = U(\wedge_{j\in \Phi} x_j, y) = \wedge_{j\in \Phi} U(x_j, y) = 1$$

$$\left(U(x, 1) = U(x, \wedge_{j\in \Phi} y_j) = \wedge_{j\in \Phi} U(x, y_j) = 1\right)$$

for any $x, y \in L$ when U is left (right) $\wedge$-distributive.

For the sake of convenience, we introduce the following symbols:

$U_s^{e_L}(L)$ ($U_s^{e_R}(L)$): the set of all left (right) semi-uninorms with the left (right) neutral element e_L (e_R) on L;

$U_{s\vee}^{e_L}(L)$ ($U_{s\vee}^{e_R}(L)$): the set of all right $\vee$-distributive left (right) semi-uninorms with the left (right) neutral element e_L (e_R) on L;

$U_{\vee s}^{e_L}(L)$ ($U_{\vee s}^{e_R}(L)$): the set of all left $\vee$-distributive left (right) semi-uninorms with the left (right) neutral element e_L (e_R) on L.

Now, we present two examples of left and right semi-uninorms on L.

Example 2.1 (Su et al. [17]). Let $e_L \in L$,

$$U_{sW}^{e_L}(x,y)=\begin{cases} y \text{ if } x\ge e_L, \\ 0 \text{ otherwise,}\end{cases} \quad U_{sM}^{e_L}(x,y)=\begin{cases} y \text{ if } x\ge e_L, \\ 1 \text{ otherwise,}\end{cases}$$

$$U_{sM}^{e_L^*}(x,y)=\begin{cases} 0 & \text{if } y=0, \\ y & \text{if } x\le e_L, y\ne 0, \\ 1 & \text{otherwise,}\end{cases}$$

where x and y are elements of L. Then $U_{sW}^{e_L}$ and $U_{sM}^{e_L}$ are, respectively, the smallest and greatest elements of $U_s^{e_L}(L)$; $U_{sW}^{e_L}$ and $U_{sM}^{e_L^*}$ are, respectively, the smallest and greatest elements of $U_{s\vee}^{e_L}(L)$.

Example 2.2 (Su et al. [17]). Let $e_R \in L$,

$$U_{sW}^{e_R}(x,y)=\begin{cases} x \text{ if } y\ge e_R, \\ 0 \text{ otherwise,}\end{cases} \quad U_{sM}^{e_R}(x,y)=\begin{cases} x \text{ if } y\ge e_R, \\ 1 \text{ otherwise,}\end{cases}$$

$$U_{sM}^{e_R^*}(x,y)=\begin{cases} 0 & \text{if } x=0, \\ x & \text{if } y\le e_L, x\ne 0, \\ 1 & \text{otherwise,}\end{cases}$$

where x and y are elements of L. Then $U_{sW}^{e_R}$ and $U_{sM}^{e_R}$ are, respectively, the smallest and greatest elements of $U_s^{e_R}(L)$; $U_{sW}^{e_R}$ and $U_{sM}^{e_R^*}$ are, respectively, the smallest and greatest elements of $U_{\vee s}^{e_R}(L)$.

3. The Residual Operations of Left and Right Semi-Uninorms

Recently, De Baets and Fodor [18] investigated the residual operators of uninorms on $[0,1]$, Torrens et al. [19-20] studied the implications and coimplications derived from uninorms on $[0,1]$, Wang and Fang [14] discussed the residual implications of left and right uninorms on a complete lattice, and Liu [16] researched semi-uninorms and implications on a complete lattice. In this section, based on [14, 16, 18-20], we consider the residual implications of left and right semi-uninorms on a complete lattice.

First of all, we recall the definition of implications.

Definition 3.1 (De Baets and Fodor [18], Baczynski and Jayaram [27], De Baets [28]). *An implication I on L is a hybrid monotonous (with non-increasing first and non-decreasing second partial mappings) binary operation that satisfies the corner conditions* $I(0,0)=I(1,1)=1$ *and* $I(1,0)=0$.

Implications are extensions of the Boolean implication $\to$ ($P \to Q$ meaning that P is sufficient for Q).

Note that for any implication I on L, due to the monotonicity, the absorption principle holds, i.e., $I(0,x)=I(x,1)=1$ for any $x\in L$.

We denote the set of all implications and the set of all right $\wedge$-distributive implications on L by $I(L)$ and $I_\wedge(L)$, respectively.

Example 3.1. Let

$$I_W(x,y)=\begin{cases} 1 \text{ if } x=0 \text{ or } y=1, \\ 0 \quad \text{otherwise,}\end{cases}$$

$$I_M(x,y)=\begin{cases} 0 \text{ if } (x,y)=(1,0), \\ 1 \quad \text{otherwise,}\end{cases}$$

where x and y are elements of L. It is easy to see that I_W and I_M are, respectively, the smallest and greatest elements of $I(L)$ and I_W is also the smallest element of $I_\wedge(L)$.

Definition 3.1. Let U be a binary operation on L. Define $I_U^L, I_U^R \in L^{L\times L}$ as follows:

$$I_U^L(x,y)=\vee\{z\in L \mid U(z,x)\le y\} \quad \forall x,y\in L,$$

$$I_U^R(x,y)=\vee\{z\in L \mid U(x,z)\le y\} \quad \forall x,y\in L.$$

Here, I_U^L and I_U^R are, respectively, called the left and right residual operators of U.

For any operation U on L and $x,y\in L$, it is straightforward to verify that

(1) $I_U^L(x,1)=I_U^R(x,1)=1$.

(2) $x\le I_U^L(y,U(x,y))$ and $y\le I_U^R(x,U(x,y))$.

(3) If $U(1,0)=0$, then $I_U^L(0,y)=1$ and if $U(0,1)=0$, then $I_U^R(0,y)=1$.

When U is a left (right) semi-uninorm on L, it is easy to see that I_U^L and I_U^R are all non-increasing in the first variable and non-decreasing in the second one.

Example 3.2. For some left and right semi-uninorms in Examples 2.1 and 2.2, a simple computation shows that

$$I_{U_{sW}^{e_R}}^L(x,y)=\begin{cases} y \text{ if } x\ge e_R, \\ 1 \text{ otherwise,}\end{cases} \quad I_{U_{sW}^{e_L}}^R(x,y)=\begin{cases} y \text{ if } x\ge e_L, \\ 1 \text{ otherwise,}\end{cases}$$

$$I_{U_{sM}^{e_R}}^L(x,y)=I_{U_{sM}^{e_R^*}}^L(x,y)=\begin{cases} 1 \text{ if } y=1, \\ y \text{ if } x\ge e_R, \\ 0 \text{ otherwise,}\end{cases}$$

$$I_{U_{sM}^{e_L}}^L(x,y)=\begin{cases} 1 & \text{if } y=1, \\ e_L & \text{if } x\le y<1, \\ 0 & \text{otherwise,}\end{cases}$$

$$I^L_{U^{e_L}_{sM}{}^*}(x,y)=\begin{cases}1 & \text{if } x=0 \text{ or } y=1,\\ e_L & \text{if } 0<x\le y<1,\\ 0 & \text{otherwise,}\end{cases}$$

$$I^R_{U^{e_L}_{sM}}(x,y)=I^R_{U^{e_L}_{sM}{}^*}(x,y)=\begin{cases}1 & \text{if } y=1,\\ y & \text{if } x\le e_L,\\ 0 & \text{otherwise,}\end{cases}$$

$$I^R_{U^{e_R}_{sM}}(x,y)=\begin{cases}1 & \text{if } y=1,\\ e_R & \text{if } x\le y<1,\\ 0 & \text{otherwise,}\end{cases}$$

$$I^R_{U^{e_R}_{sM}{}^*}(x,y)=\begin{cases}1 & \text{if } x=0 \text{ or } y=1,\\ e_R & \text{if } 0<x\le y<1,\\ 0 & \text{otherwise.}\end{cases}$$

When $e_L, e_R\in L\setminus\{0,1\}$, we see that $I^L_{U^{e_L}_{sM}{}^*}$ and $I^R_{U^{e_R}_{sM}{}^*}$ are two implications, $I^L_{U^{e_R}_{sW}}$ and $I^R_{U^{e_L}_{sW}}$ are two right $\wedge$-distributive implications, but $I^L_{U^{e_R}_{sM}}$, $I^L_{U^{e_R}_{sM}{}^*}$, $I^L_{U^{e_L}_{sM}}$, $I^R_{U^{e_L}_{sM}}$, $I^R_{U^{e_L}_{sM}{}^*}$ and $I^R_{U^{e_R}_{sM}}$ are not implications.

Theorem 3.1. Let $U\in U^{e_L}_s(L)$.

(1) *For any* $x,y\in L$, $x\le y\Rightarrow I^L_U(x,y)\ge e_L$.

(2) I^R_U *satisfies the neutrality principle with* e_L, *i.e.,* $I^R_U(e_L,y)=y$ *for any* $y\in L$.

(3) *If* U *is left-conjunctive, then* $I^R_U\in I(L)$.

(4) *If* $U\in U^{e_L}_{s\vee}(L)$ *is left-conjunctive, then*

$$I^R_U\in I_\wedge(L) \text{ and } I^R_U(x,y)=\max\{z\in L\mid U(x,z)\le y\}.$$

Here, I^R_U *is called the right residual implication of the left semi-uninorm* U.

Proof. Clearly, statements (1) and (2) hold.

(3) If U is left-conjunctive, then $U(0,1)=0$ and

$$I^R_U(0,0)=\vee\{z\in L\mid U(0,z)=0\}=1.$$

By the non-decreasingness of U, we see that

$$I^R_U(1,0)=\vee\{z\in L\mid U(1,z)=0\}$$
$$\le\vee\{z\in L\mid z=U(e_L,z)\le U(1,z)=0\}=0.$$

Moreover, it follows from the statements before Example 3.2 that $I^R_U(1,1)=1$ and I^R_U is non-increasing in its first and non-decreasing in its second variable. Therefore, I^R_U is an implication on L.

(4) Assume that U is left-conjunctive right $\vee$-distributive. Then, $I^R_U\in I(L)$ by statement (3). Let $x,y,z\in L$. If $U(x,z)\le y$, then $z\le I^R_U(x,y)$; if $z\le I^R_U(x,y)$, then it follows from the non-decreasingness of U that

$$U(x,z)\le U(x,I^R_U(x,y))=U(x,\vee\{z\in L\mid U(x,z)\le y\}$$
$$=\vee\{U(x,z)\mid z\in L, U(x,z)\le y\}\le y$$

Noting that $U(x,I^R_U(x,y))\le y$, we know that

$$I^R_U(x,y)=\max\{z\in L\mid U(x,z)\le y\}.$$

Moreover, when $J\ne\Phi$, for any $x, y_j\in L\,(j\in J)$, we have that

$$I^R_U(x,\wedge_{j\in J}y_j)=\vee\{z\in L\mid U(x,z)\le\wedge_{j\in J}y_j\}$$
$$=\vee\{z\in L\mid U(x,z)\le y_j\ \forall j\in J\}$$
$$=\vee\{z\in L\mid z\le I^R_U(x,y_j)\ \forall j\in J\}$$
$$=\vee\{z\in L\mid z\le\wedge_{j\in J}I^R_U(x,y_j)\}=\wedge_{j\in J}I^R_U(x,y_j).$$

When $J=\Phi$, we see that

$$I^R_U(x,\wedge_{j\in\Phi}y_j)=I^R_U(x,1)=1=\wedge_{j\in\Phi}I^R_U(x,y_j).$$

Therefore, I^R_U is right $\wedge$-distributive, i.e., $I^R_U\in I_\wedge(L)$. The theorem is proved.

When $e_L<1$, for the right $\vee$-distributive left semi-uninorm $U^{e_L}_{sM}{}^*$, we see that $I^L_{U^{e_L}_{sM}{}^*}$ by Example 3.2, but $I^L_{U^{e_L}_{sM}{}^*}(e_L,y)=e_L\ne y$ when $e_L<y<1$, i.e., $I^L_{U^{e_L}_{sM}{}^*}$ does not satisfy the neutrality principle with e_L. This illustrates Theorem 3.1 doesn't hold for the left residual operator of a left semi-uninorm.

If P and Q are two propositions, then the generalized modus ponens (GMP) [18] gives a lower bound for the truth value of Q when the truth values of propositions P and $P\to Q$ are known. By the proof of Theorem 3.1(4), we know that U and I^R_U satisfy the GMP rule:

$$U(x,I^R_U(x,y))\le y\ \forall x,y\in L$$

and the following right residual principle:

$$U(x,z)\le y\Leftrightarrow z\le I^R_U(x,y))\ \forall x,y,z\in L$$

when a binary operation U is right $\vee$-distributive.

Similarly, U and I^L_U satisfy GMP rule in the form:

$$U(I_U^L(x, y), x) \le y \ \forall x, y \in L$$

and the following left residual principle:

$$U(z, x) \le y \Leftrightarrow z \le I_U^L(x, y)) \ \forall x, y, z \in L$$

when U is left $\vee$-distributive. Thus, for right semi-uninorms on L, we have a similar result.

Theorem 3.2. Let $U \in U_s^{e_R}(L)$.

(1) *For any* $x, y \in L$, $x \le y \Rightarrow I_U^R(x, y) \ge e_R$.

(2) I_U^L *satisfies the neutrality principle with* e_R, *i.e.,* $I_U^L(e_R, y) = y$ *for any* $y \in L$.

(3) *If* U *is right-conjunctive, then* $I_U^L \in I(L)$.

(4) *If* $U \in U_{\vee s}^{e_R}(L)$ *is right-conjunctive, then*

$$I_U^L \in I_\wedge(L) \text{ and } I_U^L(x, y) = \max\{z \in L \mid U(z, x) \le y\}.$$

Here, I_U^L *is called the left residual implication of the left semi-uninorm* U.

Combining Theorems 3.1 and 3.2, we know that both I_U^L and I_U^R are all right $\vee$-distributive implications when U is a $\vee$-distributive left (right) semi-uninorm.

Theorem 3.3. (1) *If* $U \in U_{\vee s}^{e_L}(L)$, *then* I_U^L *is right* $\wedge$ *-distributive and satisfies the left residual principle and the order property with* e_L:

$$x \le y \Leftrightarrow I_U^L(x, y) \ge e_L \ \forall x, y \in L.$$

(2) *If* $U \in U_{s\vee}^{e_R}(L)$, *then* I_U^R *is right* $\wedge$*-distributive and satisfies the right residual principle and the order property with* e_R:

$$x \le y \Leftrightarrow I_U^R(x, y) \ge e_R \ \forall x, y \in L.$$

Proof. Assume that U is a left $\wedge$-distributive left semi-uninorms with the left neutral element e_L. By virtue of the proof of Theorem 3.1(4), we can see that I_U^L is right $\wedge$-distributive and satisfies the left residual principle. Moreover, if $x, y \in L$ and $x \le y$, then it follows from Theorem 3.1(1) that $I_U^L(x, y) \ge e_L$; if $I_U^L(x, y) \ge e_L$, then

$$x = U(e_L, x) \le U(I_U^L(x, y), x) \le y.$$

Thus, I_U^L satisfies the order property with e_L.

Similarly, we can show that I_U^R is right $\wedge$-distributive and satisfies the right residual principle and the order property with e_R when U is a right $\vee$-distributive right semi-uninorms with the right neutral element e_R.

The theorem is proved.

In particular, if U is a $\vee$-distributive semi-uninorm with the neutral element e, then I_U^L and I_U^R satisfy the the residual principle (RP) and the order property (OP) and are all right $\wedge$-distributive implications (see Theorem 3.6 in [16]).

4. The Left and Right Semi-Uninorms Induced by Implications

Liu [16] discussed the semi-uninorms induced by implications and Su and Wang [29] studied the pseudo-uninorms induced by coimplications. In this section, based on these works, we investigate the left and right semi-uninorms induced by implications on a complete lattice.

Definition 4.1. Let I *be a binary operation on* L. *Define two induced operators* U_I^L *and* U_I^R *of* I *as follows:*

$$U_I^L(x, y) = \wedge\{z \in L \mid x \le I(y, z)\} \quad \forall x, y \in L,$$

$$U_I^R(x, y) = \wedge\{z \in L \mid y \le I(y, z)\} \quad \forall x, y \in L.$$

Clearly, $U_I^L(0, x) = U_I^R(x, 0) = 0$, $U_I^L(1, x) = U_I^R(x, 1)$ for any $x \in L$, and $U_I^L = U_I^R$ if I satisfies the condition:

$$x \le I(y, z) \Leftrightarrow y \le I(x, z) \ \forall x, y, z \in L.$$

When I is hybird monotonous, it is easy to see that U_I^L and U_I^R are all non-decreasing in its each variable. Moreover, for any binary operation I, it follows from Definition 4.1 that

$$U_I^L(I(x, y), x) \le y, U_I^R(x, I(x, y)) \le y \quad \forall x, y \in L.$$

These explain that U_I^L and I, U_I^R and I satisfy the GMP rule.

Example 4.1. For two implications I_W and I_M in Example 3.1, we have that

$$U_{I_W}^L(x, y) = U_{I_W}^R(x, y) = \begin{cases} 1 & \text{if } x = 0 \text{ or } y = 0, \\ 0 & \text{otherwise,} \end{cases}$$

$$U_{I_M}^L(x, y) = \begin{cases} \bigwedge_{a \in L \setminus \{0\}} a & \text{if } x > 0 \text{ and } y = 1, \\ 0 & \text{otherwise,} \end{cases}$$

$$U_{I_M}^R(x, y) = \begin{cases} \bigwedge_{a \in L \setminus \{0\}} a & \text{if } x = 1 \text{ and } y > 0, \\ 0 & \text{otherwise.} \end{cases}$$

Thus, four operations induced by implications I_W and I_M are neither left semi-uninorms nor right semi-uninorms on L.

Below, we find some conditions such that these operations induced by implications are left or right semi-uninorms.

Theorem 4.1. Let I be an implication on L.

(1) *If I satisfies the order property with e_L, then $U_I^L \in U_s^{e_L}(L)$; if I satisfies the neutrality principle with e_L, then $U_I^R \in U_s^{e_L}(L)$. Here, U_I^L and U_I^R are called the left semi-uninorms induced by the implication I.*

(2) *If I satisfies the order property with e_R, then $U_I^R \in U_s^{e_R}(L)$; if I satisfies the neutrality principle with e_R, then $U_I^L \in U_s^{e_R}(L)$. Here, U_I^L and U_I^R are called the right semi-uninorms induced by the implication I.*

(3) *If I satisfies the order property with e_L and the neutrality principle with e_R, then U_I^L is a semi-uninorm on L.*

(4) *If I satisfies the order property with e_R and the neutrality principle with e_L, then U_I^R is also a semi-uninorm on L.*

Proof. Assume that $I \in I(L)$. Then U_I^L is non-decreasing in each variable. If I satisfies the order property with e_L, then

$$U_I^L(e_L, y) = \wedge\{z \in L \mid e_L \leq I(y, z)\}$$
$$= \wedge\{z \in L \mid y \leq z\} = y \quad \forall y \in L.$$

Thus, $U_I^L \in U_s^{e_L}(L)$. If I satisfies the neutrality principle with e_L, then

$$U_I^R(e_L, y) = \wedge\{z \in L \mid y \leq I(e_L, z)\}$$
$$= \wedge\{z \in L \mid y \leq z\} = y \quad \forall y \in L.$$

So, $U_I^R \in U_s^{e_L}(L)$.

Similarly, we can show that $U_I^R \in U_s^{e_R}(L)$ when the implication I satisfies the order property with e_R and $U_I^L \in U_s^{e_R}(L)$ when I satisfies the neutrality principle with e_R.

If I satisfies the order property with e_L and the neutrality principle with e_R, then

$$U_I^L(e_L, x) = U_I^L(x, e_R) = x \quad \forall x \in L.$$

Thus, $e_L = e_R$, i.e., U_I^L is a semi-uninorm with the neutral element $e_L (= e_R)$ on L.

In a similar way, we can see that U_I^R is also a semi-uninorm on L when I satisfies the order property with e_R and the neutrality principle with e_L.

The theorem is proved.

When $I \in I(L)$, $I(0, x) = 1$ for any $x \in L$ and hence it follows from Definition 4.1 that

$$U_I^L(1, 0) = U_I^R(0, 1).$$

Thus, U_I^L and U_I^R in Theorem 4.1 are all conjunctive left or right semi-uninorms induced by the implication I.

Theorem 4.2. Let $I \in I_\wedge(L)$.

(1) *If I satisfies the order property with e_L, then $U_I^L \in U_{\vee s}^{e_L}(L)$; if I satisfies the neutrality principle with e_L, then $U_I^R \in U_{s\vee}^{e_L}(L)$.*

(2) *If I satisfies the order property with e_R, then $U_I^R \in U_{s\vee}^{e_R}(L)$; if I satisfies the neutrality principle with e_R, then $U_I^L \in U_{\vee s}^{e_R}(L)$.*

(3) *If I satisfies the order property with e_L and the neutrality principle with e_R, then U_I^L is a left $\vee$-distributive semi-uninorm on L.*

(4) *If I satisfies thet order property with e_R and the neutrality principle with e_L, then U_I^R is also a right $\vee$-distributive semi-uninorm on L.*

Proof. Assume that I is a right $\wedge$-distributive implication. Let $x, y, z \in L$. If $x \leq I(y, z)$, then it follows from Definition 4.1 that $U_I^L(x, y) \leq z$; if $U_I^L(x, y) \leq z$, then

$$I(y, z) \geq I(y, U_I^L(x, y))$$
$$= I(y, \wedge\{ z \in L \mid x \leq I(y, z)\})$$
$$= \wedge\{I(y, z) \mid z \in L, x \leq I(y, z)\} \geq x.$$

Noting that $x \leq I(y, U_I^L(x, y))$, we know that

$$U_I^L(x, y) = \min\{z \in L \mid x \leq I(y, z)\}.$$

Moreover, when $J \neq \Phi$, for any $x_j, y \in L (j \in J)$, we have that

$$U_I^L(\vee_{j\in J}x_j, y) = \wedge\{z\in L \mid \vee_{j\in J}x_j \le I(y,z)\}$$
$$= \wedge\{z\in L \mid x_j \le I(y,z)\,\forall j\in J\}$$
$$= \wedge\{z\in L \mid U_I^L(x_j,y)\le z\,\forall j\in J\}$$
$$= \wedge\{z\in L \mid \vee_{j\in J}U_I^L(x_j,y)\le z\}$$
$$= \vee_{j\in J}U_I^L(x_j,y).$$

When $J=\Phi$, we see that

$$U_I^L(\vee_{j\in\Phi}x_j, y) = U_I^L(0,y) = 0 = \vee_{j\in\Phi}U_I^L(x_j,y).$$

Thus, U_I^L is left $\vee$-distributive. Therefore, by virtue of Theorem 4.1, $U_I^L \in U_{\vee s}^{e_L}(L)$ when I satisfies the order property with e_L and $U_I^L \in U_{\vee s}^{e_R}(L)$ when I satisfies the neutrality principle with e_R.

Similarly, we can show that U_I^R is a right $\vee$-distributive right semi-uninorm and left semi-uninorm when I satisfies the order property with e_R and the neutrality principle with e_L, respectively.

Statements (3) and (4) are the direct consequences of statements (1) and (2) and Theorem 4.1.

The theorem is proved.

By virtue of Theorem 4.2, we see that U_I^L and U_I^R are a left $\vee$-distributive semi-uninorm and a right $\vee$-distributive semi-uninorm on L when $I\in I_\wedge(L)$ satisfies the order property (OP) and neutrality principle (NP), respectively. This explains Theorem 4.2 is a generalization of Theorem 4.5 in [16].

When I is a right $\wedge$-distributive implication on L, by the proof of Theorem 4.2, we know that I, U_I^L and U_I^R satisfy the following adjunction conditions:

$$x\le I(y,z)\Leftrightarrow U_I^L(x,y)\le z,$$
$$y\le I(x,z)\Leftrightarrow U_I^R(x,y)\le z\ \forall x,y,z\in L.$$

Moreover, we have

$$U_I^R(x,y) = \min\{z\in L \mid y\le I(x,z)\}.$$

5. The Relationships between Left (Right) Semi-Uninorms and Implications

In the final section, we reveal the relationships between conjunctive right (left) $\vee$-distributive left (right) semi-uninorms and right $\wedge$-distributive implications which satisfy the neutrality principle on a complete lattice.

Theorem 5.1.

(1) *If $U\in U_{\vee s}^{e_R}(L)$ is right-conjunctive, then $I_U^L\in I_\wedge(L)$ satisfies the neutrality principle with e_R and $U_{I_U^L}^L = U$.*

(2) *If $U\in U_{s\vee}^{e_L}(L)$ is left-conjunctive, then $I_U^R\in I_\wedge(L)$ satisfies the neutrality principle with e_L and $U_{I_U^R}^R = U$.*

(3) *If $I\in I_\wedge(L)$ satisfies the neutrality principle with e_L, then $U_I^R\in U_{s\vee}^{e_L}(L)$ is conjunctive and $I_{U_I^R}^R = I$.*

(4) *If $I\in I_\wedge(L)$ satisfies the neutrality principle with e_R, then $U_I^L\in U_{s\vee}^{e_R}(L)$ is conjunctive and $I_{U_I^L}^L = I$.*

Proof. We only prove the statements (1) and (3) hold.

(1) If U is a right-conjunctive left $\vee$-distributive right semi-uninorm, then $I_U^L\in I_\wedge(L)$ and satisfies the neutrality principle with e_R by Theorem 3.2. Moreover, it follows from the left residual principle that

$$U_{I_U^L}^L(x,y) = \wedge\{z\in L \mid x\le I_U^L(y,z)\}$$
$$= \wedge\{z\in L \mid U(x,y)\le z\} = U(x,y)\quad \forall x,y\in L.$$

Thus, $U_{I_U^L}^L = U$.

(3) If $I\in I_\wedge(L)$ satisfies the neutrality principle with e_L, then U_I^R is a conjunctive right $\vee$-distributive left semi-uninorm by Theorem 4.2. Moreover, it follows from the adjunction conditions that

$$I_{U_I^R}^R(x,y) = \vee\{z\in L \mid U_I^R(y,z)\le y\}$$
$$= \vee\{z\in L \mid z\le I(x,y)\} = I(x,y)\quad \forall x,y\in L.$$

Therefore, $I_{U_I^R}^R = I$.

The theorem is proved.

We denote by $U_{cs\vee}^{e_L}(L)$ and $U_{\vee cs}^{e_R}(L)$, respectively, the set of all conjunctive right $\vee$-distributive left semi-uninorms and the set of all conjunctive left $\vee$-distributive right semi-uninorms; by $I_\wedge^{e_L}(L)$ and $I_\wedge^{e_R}(L)$, respectively, the set of all right $\wedge$-distributive implications which satisfy the neutrality principle with e_L and e_R on a complete lattice.

It is easy to verify that $U_{cs\vee}^{e_L}(L)$ and $U_{\vee cs}^{e_R}(L)$ are two join-semilattices with the smallest elemens $U_{sW}^{e_L}$ and

$U_{sW}^{e_R}$, respectively; and $I_\wedge^{e_L}(L)$ and $I_\wedge^{e_R}(L)$ are two meet-semilattices with the greatest elements $I_{U_{sW}^{e_L}}^R$ and $I_{U_{sW}^{e_R}}^L$, respectively, where

$$(U_1 \vee U_2)(x,y) = U_1(x,y) \vee U_2(x,y),$$
$$(I_1 \wedge I_2)(x,y) = I_1(x,y) \wedge I_2(x,y)\ \forall x,y \in L.$$

Define two mappings $\phi : U_{cs\vee}^{e_L}(L) \to I_\wedge^{e_L}(L)$ and $\varphi : U_{\vee cs}^{e_R}(L) \to I_\wedge^{e_R}(L)$ as follows:

$$\phi(U) = I_U^R\ \forall U \in U_{cs\vee}^{e_L}(L),\ \varphi(U) = I_U^L\ \forall U \in U_{\vee cs}^{e_R}(L).$$

Then it follows from Theorem 5.1 that ϕ and φ are all invertible and

$$\phi^{-1}(I) = U_I^R\ \forall I \in I_\wedge^{e_L}(L),\ \varphi^{-1}(I) = U_I^L\ \forall I \in I_\wedge^{e_{LR}}(L).$$

Moreover, we have the following theorem.

Theorem 5.2.

(1) *If* $U_1, U_2 \in U_{cs\vee}^{e_L}(L)$, *then* $I_{U_1 \vee U_2}^R = I_{U_1}^R \wedge I_{U_2}^R$.

(2) *If* $U_1, U_2 \in U_{\vee cs}^{e_R}(L)$, *then* $I_{U_1 \vee U_2}^L = I_{U_1}^L \wedge I_{U_2}^L$.

(3) *If* $I_1, I_2 \in I_\wedge^{e_L}(L)$, *then* $U_{I_1 \wedge I_2}^R = U_{I_1}^R \vee U_{I_2}^R$.

(4) *If* $I_1, I_2 \in I_\wedge^{e_R}(L)$, *then* $U_{I_1 \wedge I_2}^L = U_{I_1}^L \vee U_{I_2}^L$.

Proof. We only prove the statements (2) and (4) hold.

(2) If $U_1, U_2 \in U_{\vee cs}^{e_R}(L)$, then it follows from the left residual principle that

$$\begin{aligned} I_{U_1 \vee U_2}^L(x,y) &= \vee\{z \in L \mid (U_1 \vee U_2)(z,x) \le y\} \\ &= \vee\{z \in L \mid U_1(z,x) \vee U_2(z,x) \le y\} \\ &= \vee\{z \in L \mid U_1(z,x) \le y, U_2(z,x) \le y\} \\ &= \vee\{z \in L \mid z \le I_{U_1}^L(x,y), z \le I_{U_2}^L(x,y)\} \\ &= \vee\{z \in L \mid z \le I_{U_1}^L(x,y) \wedge I_{U_2}^L(x,y)\} \\ &= (I_{U_1}^L \wedge I_{U_2}^L)(x,y)\ \forall x,y \in L, \end{aligned}$$

i.e., $I_{U_1 \vee U_2}^L = I_{U_1}^L \wedge I_{U_2}^L$.

(4) If $I_1, I_2 \in I_\wedge^{e_R}(L)$, then it follows from the adjunction condition that

$$\begin{aligned} U_{I_1 \wedge I_2}^L(x,y) &= \wedge\{z \in L \mid x \le (I_1 \wedge I_2)(y,z)\} \\ &= \wedge\{z \in L \mid x \le I_1(y,z) \wedge I_2(y,z)\} \\ &= \wedge\{z \in L \mid x \le I_1(y,z), x \le I_2(y,z)\} \\ &= \wedge\{z \in L \mid U_{I_1}^L(x,y) \le z, U_{I_2}^L(x,y) \le z\} \\ &= \wedge\{z \in L \mid U_{I_1}^L(x,y) \vee U_{I_2}^L(x,y) \le z\} \\ &= (U_{I_1}^L \vee U_{I_2}^L)(x,y)\ \forall x,y \in L, \end{aligned}$$

Therefore, $U_{I_1 \wedge I_2}^L = U_{I_1}^L \vee U_{I_2}^L$.

The theorem is proved.

By virtue of Theorem 5.2, we know that ϕ and φ are, respectively, anti-order isomorphisms of $U_{cs\vee}^{e_L}(L)$ onto $I_\wedge^{e_L}(L)$ and $U_{\vee cs}^{e_R}(L)$ onto $I_\wedge^{e_R}(L)$; ϕ^{-1} and φ^{-1} are, respectively, anti-order isomorphisms of $I_\wedge^{e_L}(L)$ onto $U_{cs\vee}^{e_L}(L)$ and $I_\wedge^{e_R}(L)$ onto $U_{\vee cs}^{e_R}(L)$.

6. Conclusions and Future Works

In this paper, we discuss the residual operations of left and right semi-uninorms and the left and right semi-uninorms induced by implications, show that the right (left) residual operator of a conjunctive right (left) $\vee$-distributive left (right) semi-uninorm is a right $\wedge$-distributive implication which satisfies the neutrality principle, give some conditions such that the operations induced by an implication constitute left or right semi-uninorms, demonstrate that the operations induced by a right $\wedge$-distributive implication, which satisfies the order property or the neutrality principle, are left (right) $\vee$-distributive left (right) semi-uninorms or right (left) semi-uninorms, and reveal the relationships between conjunctive right (left) $\vee$-distributive left (right) semi-uninorms and right $\wedge$-distributive implications which satisfy the neutrality principle.

In a forthcoming paper, we will investigate the relationships between left (right) semi-uninorms and coimplications on a complete lattice.

Acknowledgements

This work is supported by the National Natural Science Foundation of China (61379064), Jiangsu Provincial Natural Science Foundation of China (BK2012672) and Science Foundation of Yancheng Teachers University (13YSYJB0108).

References

[1] E. P. Klement, R. Mesiar, and E. Pap, "Triangular Norms", Trends in Logic-Studia Logica Library, Vol. 8, Kluwer Academic Publishers, Dordrecht, 2000.

[2] P. Flondor, G. Georgescu, and A. orgulescu, "Pseudo-t-norms and pseudo-BL-algebras", Soft Computing, 5, 355-371, 2001.

[3] J. Fdor and T. Keresztfalvi, "Nonstandard conjunctions and implications in fuzzy logic", International Journal of Approximate Reasoning, 12, 69-84, 1995.

[4] J. Fdor, "Srict preference relations based on weak t-norms", Fuzzy Sets and Systems, 43, 327-336, 1991.

[5] Z. D. Wang and Y. D. Yu, "Pseudo-t-norms and implication operators on a complete Brouwerian lattice", Fuzzy Sets and Systems, 132, 113-124, 2002.

[6] R. R. Yager and A. Rybalov, "Uninorm aggregation operators", Fuzzy Sets and Systems, 80, 111-120, 1996.

[7] J. Fodor, R. R. Yager, and A. Rybalov, "Structure of uninorms", Internat. J. Uncertainly, Fuzziness and Knowledge-Based Systems, 5, 411-427, 1997.

[8] D. Gabbay and G. Metcalfe, "fuzzy logics based on [0,1)-continuous uninorms", Arch. Math. Logic, 46, 425-449, 2007.

[9] A. K. Tsadiras and K. G. Margaritis, "The MYCIN certainty factor handling function as uninorm operator and its use as a threshold function in artificial neurons", Fuzzy Sets and Systems, 93, 263-274, 1998.

[10] R. R. Yager, "Uninorms in fuzzy system modeling", Fuzzy Sets and Systems, 122, 167-175, 2001.

[11] R. R. Yager, "Defending against strategic manipulation in uninorm-based multi-agent decision making", European J. Oper. Res., 141, 217-232, 2002.

[12] M. Mas, M. Monserrat, and J. Torrens, "On left and right uninorms", Internat. J. Uncertainly, Fuzziness and Knowledge-Based Systems, 9, 491-507, 2001.

[13] M. Mas, M. Monserrat, and J. Torrens, "On left and right uninorms on a finite chain", Fuzzy Sets and Systems, 146, 3-17, 2004.

[14] Z. D. Wang and J. X. Fang, "Residual operators of left and right uninorms on a complete lattice", Fuzzy Sets and Systems, 160, 22-31, 2009.

[15] Z. D. Wang and J. X. Fang, "Residual coimplicators of left and right uninorms on a complete lattice", Fuzzy Sets and Systems, 160, 2086-2096, 2009.

[16] H. W. Liu, "Semi-uninorm and implications on a complete lattice", Fuzzy Sets and Systems, 191, 72-82, 2012.

[17] Y. Su, Z. D. Wang, and K. M. Tang, "Left and right semi-uninorms on a complete lattice", Kybernetika, 49, 948-961, 2013.

[18] B. De Baets and J. Fodor, "Residual operators of uninorms", Soft Computing, 3, 89-100, 1999.

[19] M. Mas, M. Monserrat, and J. Torrens, "Two types of implications derived from uninorms", Fuzzy Sets and Systems, 158, 2612-2626, 2007.

[20] D. Ruiz and J. Torrens, "Residual implications and co-implications from idempotent uninorms", Kybernetika, 40, 21-38, 2004.

[21] G. Birkhoff, "Lattice Theory", American Mathematical Society Colloquium Publishers, Providence, 1967.

[22] F. Suarez Garcia and P. Gil Alvarez, "Two families of fuzzy intergrals", Fuzzy Sets and Systems, 18, 67-81, 1986.

[23] B. Bassan and F. Spizzichino, "Relations among univariate aging, bivariate aging and dependence for exchangeable lifetimes", J. Multivariate Anal., 93, 313-339, 2005.

[24] F. Durante, E. P. Klement, and R. Mesiar et al., "Conjunctors and their residual implicators: characterizations and construct methods", Mediterranean J. Math., 4, 343-356, 2007.

[25] G. De Cooman and E. E. Kerre, "Order norms on bounded partially ordered sets", J. Fuzzy Math., 2, 281-310, 1994.

[26] B. De Baets, "Idempotent uninorms", European J. Oper. Res., 118, 631-642, 1999.

[27] M. Baczynski and B. Jayaram, "Fuzzy Implication", Studies in Fuzziness and Soft Computing, Vol. 231, Springer, Berlin, 2008.

[28] B. De Baets, "Coimplicators, the forgotten connectives", Tatra Mountains Math. Publ., 12, 229-240, 1997.

[29] Y. Su and Z. D. Wang, "Pseudo-uninorms and coimplications on a complete lattice", Fuzzy Sets and Systems, 224, 53-62, 2013.

20

Positioning control system with high-speed starting and stopping for a DC motor using bang-bang control

Hiroki Shibasaki[1,*], Takehito Fujio[1], Ryo Tanaka[1], Hiromitsu Ogawa[1], Yoshihisa Ishida[1,2]

[1]Graduate School of Science and Technology, Meiji University, Kawasaki, JAPAN
[2]School of Science and Technology, Meiji University, Kawasaki, JAPAN

Email address:
shiba@meiji.ac.jp (H. Shibasaki), ce41087@meiji.ac.jp (T. Fujio), rtanaka@meiji.ac.jp (R. Tanaka), h_ogawa@meiji.ac.jp (H. Ogawa), ishida@isc.meiji.ac.jp (Y. Ishida)

Abstract: This paper proposes a positioning control system with high-speed starting and stopping for a DC motor using bang-bang control. The control system of the proposed method continuously operates from the bang-bang control to proportional-integral-derivative (PID) control. The bang-bang control controls the equipment for high-speed starting and stopping. However, torque disturbances in the equipment may prevent the target-value response from stopping at a precise point. Therefore, we introduce a switching control system, which operates on the basis of the velocity signal. The proposed system includes various design factors, such as calculation of the switching time of bang-bang input and a method for continuous operation from the bang-bang control to PID control. Theoretical analysis describes the design details of the proposed method. Simulation and experimental studies show the results of various cases to indicate the effectiveness of the proposed method.

Keywords: Bang-Bang Control, PID Control, Switching Control, Disturbance Observer

1. Introduction

In industry, proportional-integral-derivative (PID) control [1] [2] is used to control the actuator. Furthermore, it is applied to design methods in a wide range of processes in the areas from research and development to production lines. Although an advantage of the PID controller is its general ease of design, the PID controller also has numerous limitations. For instance, tuning of PID controller is determined by the rule of experience. Otherwise, it is difficult to reduce the influence of overshoot. Furthermore, industrial demands require precise starting and stopping in high speed conditions of the actuator.

Bang-bang controllers [3] control the actuator for rapid movement and stopping; This means that it inputs maximum or minimum values to the actuator in open loop, which is known as “Bang-Bang input”.

Research on bang-bang control has been conducted worldwide. Lasalle [4] reported the principle of the bang-bang control. K. Nagakura [5] described time-optimal precise positioning control using a dc servomotor. Rillings. et al. [6] proposed a bang-bang control strategy that uses a tabular adaptive model in a nonlinear system. In addition, K. Furuta et al. [7] designed a bang-bang position control that uses VSS control for the directive drive motor. N. Koreta et al. [8] conducted a study of high accuracy of machine tools with bang-bang control. Further, M. G.Vasek et al. [9] proposed bang-bang control for the double integrator plant in positioning control. Y. Yoshida et al. [10] proposed bang-bang control of a crane in real time. Authors [11] proposed the high speed activation and stopping control system using bang-bang control and simulated for a DC motor.

The present study proposes a position control with high-speed starting and stopping for a DC motor that use bang-bang control. It gradually and continuously converts, a control system started by bang-bang control to PID control. When the system converts, the system generally becomes discontinuous. Therefore, the initial value of the integrator in PID control, which is determined by the whole system is introduced in the control system. In addition, we introduce a disturbance observer for overcoming the variation in torque disturbance. We present the theoretical analysis and design method in detail We show the simulation results and also confirm the responses by using the real system , the DC motor [12].

This paper is constituted as follows. Section 2 shows the DC motor and Section 3 describes the bang-bang control. Section 4 proposes our method and gives design details and theoretical analysis. Section 5 shows the simulation results in various conditions, and Section 6 discusses the experimental study using the DC motor. Finally, we state the conclusion of this paper in Section 7.

2. DC Motor

This section describes the DC motor construction [5]. DC motor electrical characteristic and mechanical characteristic are written as follows:

$$\begin{cases} L\dfrac{di(t)}{dt} + Ri(t) + K_E\omega(t) = u(t) \\ J\dfrac{d\omega(t)}{dt} + D\omega(t) + T_L \operatorname{sgn}\{\omega(t)\} = K_T i(t) \end{cases} \tag{1}$$

which includes the consists of following parameters.

L	inductance of circuit (Henry)
R	resistance of circuit (Ω)
K_E	voltage constant of motor (V/rps)
K_T	torque constant of motor (g・cm/A)
J	inertia of the rotating part (g・cms^2)
D	viscous friction coefficient (g・cm/rps)
T_L	solid friction (g・cm)
i	current (A)
$u(t)$	input (V)
$\omega(t)$	velocity (rad/s)

(1) is rewritten in the Laplace transform as follows:

$$\begin{cases} LsI(s) + RI(s) + K_E\Omega(s) = U(s) \\ Js\Omega(s) + D\Omega(s) + T_L\Omega(s) = K_T I(s) \end{cases} \tag{2}$$

In this equation, the inductance of circuit L is negligible, because it is quite small. From (2), the velocity and the position are as follows:

$$\begin{cases} \Omega(s) = \dfrac{T_m}{T_m s+1}\left(\dfrac{K_T}{JR}U(s) - \dfrac{T_L}{J}\right) \\ \theta(s) = \dfrac{T_m}{s(T_m s+1)}\left(\dfrac{K_T}{JR}U(s) - \dfrac{T_L}{J}\right) \end{cases} \tag{3}$$

where, mechanical time constant T_m is as follows:

$$T_m = \frac{JR}{RD + K_T K_E}. \tag{4}$$

3. Bang-Bang Control

Fig.1 shows a block diagram of the bang-bang control and Fig.2 shows an example response. The maximum input to the DC motor produces the maximum speed. Therefore, the bang-bang control is the control system to stop to the target-value in the minimum time. However, when the plant has load changes or disturbances, controlling a plant using this method is difficult.

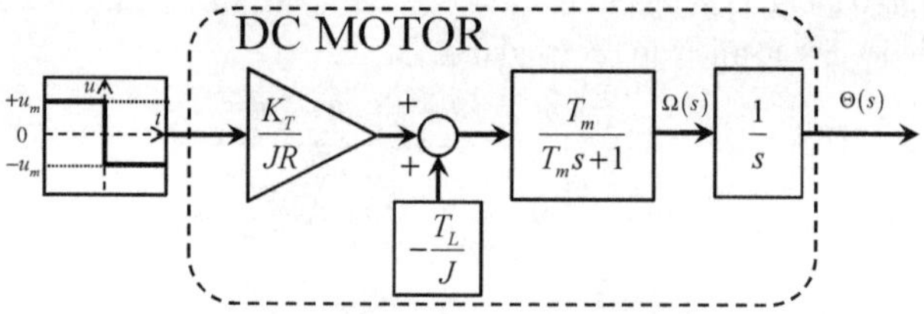

Figure 1. *Block diagram of bang-bang control*

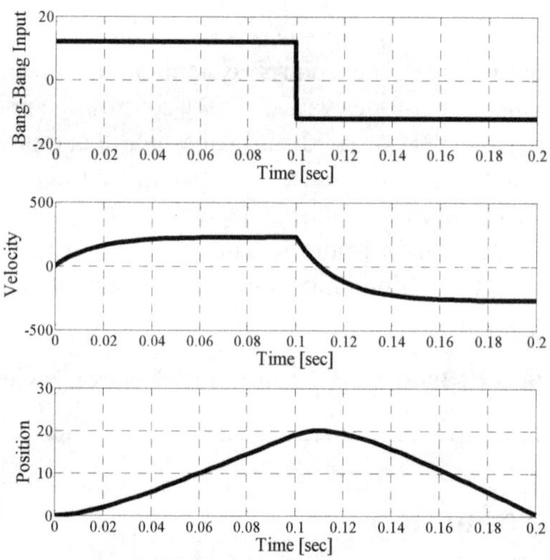

Figure 2. *Example responses of bang-bang Control*

4. Proposed Method

In the proposed method, the motor position is moved near the target value by the bang-bang control and is then converted to PID control. Fig.3 shows a block diagram of the proposed method.

In this section, we discuss the design of the control system and offer a theoretical case. Fig. 4 graphically shows our proposed method.

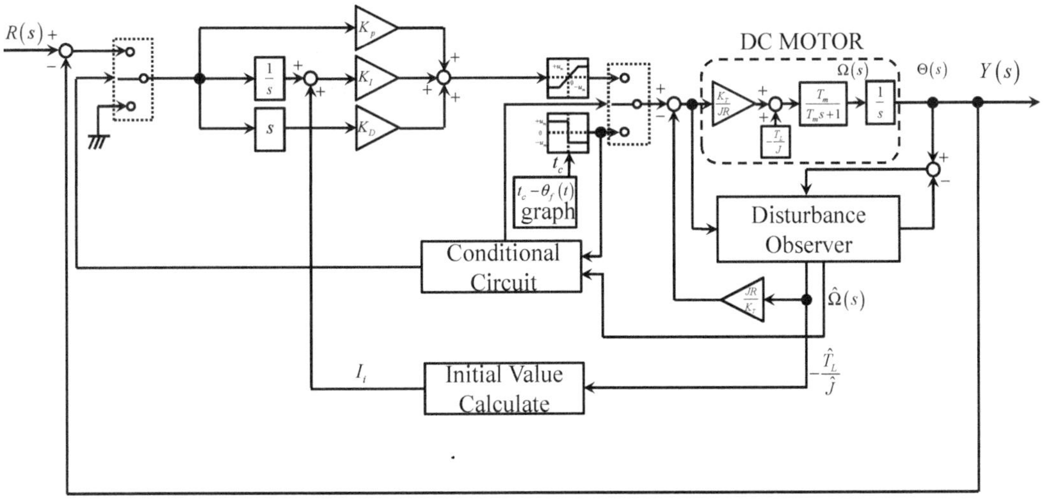

Figure 3. Block diagram of proposed method

4.1. Design of the Control System

First, we build the graph, the relationship of the position switching signal $\theta_f(t)$ and the switching time t_c in bang-bang input using the velocity switching signal $\omega_f(t)$. $\theta_f(t)$ is measured using the bang-bang control as shown in Fig.1. Then, the data are arranged using the least squares method after measurement, where, t_c and $\omega_f(t)$ are arbitrarily determined, e.g., t_c=0.01, or 0.1; $\omega_f(t)$=100 or 200.

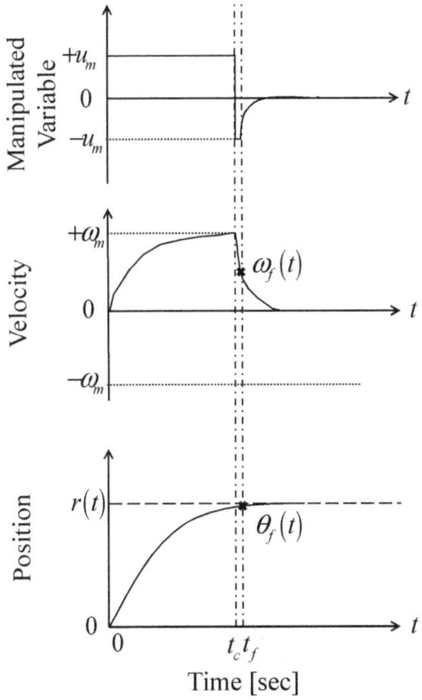

Figure 4. Explanation responses of proposed method

Second, we determine the target-value r (t) and obtain the position switching signal $\theta_f(t)$. From the t_c- $\theta_f(t)$ graph, we get the switching time t_c in the bang-bang input.

Finally, we design PID control using the pole placement method. The system is converted from bang-bang control to PID control on the basis of the t_c- $\theta_f(t)$ graph.

4.2. Theoretical of the Control System

This section describes to derive switching time t_c in bang-bang input and convert time t_f to PID control.

The differential equation about the velocity is written as follows:

$$\frac{d\omega(t)}{dt}+\frac{\omega(t)}{T_m}=-\left(\frac{1}{T_m}-\frac{D}{J}\right)\omega_m-\frac{T_L}{J}, \qquad \omega_m=\frac{u_m}{K_E}, \tag{5}$$

where, ω_m is max velocity

A. *Acceleration Time ($u = u_m$)*

The velocity $\omega(t)$ in the acceleration time is as follows:

$$\begin{aligned}\omega(t)&=T_m\left\{\left(\frac{1}{T_m}-\frac{D}{J}\right)\omega_m-\frac{T_L}{J}\right\}\left(1-e^{-\frac{t}{T_m}}\right)\\&=T_m\left\{\frac{K_T}{JR}u_m-\frac{T_L}{J}\right\}\left(1-e^{-\frac{t}{T_m}}\right)\end{aligned}, \tag{6}$$

B. *Deceleration Time ($u = -u_m$)*

Velocity $\omega(t)$ in deceleration time is follows:

$$\begin{aligned}\omega(t)&=-T_m\left\{\left(\frac{1}{T_m}-\frac{D}{J}\right)\omega_m+\frac{T_L}{J}\right\}\left(1-e^{-\frac{t}{T_m}}\right)\\&=-T_m\left\{2\frac{K_T}{JR}u_m+\frac{T_L}{J}\right\}\left(1-e^{-\frac{t}{T_m}}\right)\end{aligned}. \tag{7}$$

From *A)* and *B)*, the velocity switching signal $\omega_f(t)$ and the position switching signal $\theta_f(t)$ are as follows:

$$\begin{cases} \omega_f(t) = \omega(t_f) = T_m\left\{\dfrac{K_T}{JR}u_m - \dfrac{T_L}{J}\right\}\left(1-e^{-\frac{t_c}{T_m}}\right) \\ \qquad -T_m\left\{2\dfrac{K_T}{JR}u_m + \dfrac{T_L}{J}\right\}\left(1-e^{-\frac{t_f-t_c}{T_m}}\right) \\ \theta_f(t) = \theta(t_f) \\ \quad = T_m\left\{\dfrac{K_T}{JR}u_m - \dfrac{T_L}{J}\right\}\left[t_c - T_m\left(1-e^{-\frac{t_c}{T_m}}\right)\right] \\ \quad \cdots + T_m\left\{\dfrac{K_T}{JR}u_m - \dfrac{T_L}{J}\right\}\left(1-e^{-\frac{t_c}{T_m}}\right)(t_f - t_c) \\ \quad \cdots - T_m\left\{2\dfrac{K_T}{JR}u_m + \dfrac{T_L}{J}\right\}\left[(t_f-t_c) - T_m\left(1-e^{-\frac{t_f-t_c}{T_m}}\right)\right] \end{cases} \tag{8}$$

$\omega_f(t)$ and $\theta_f(t)$ are already specified. By solving the simultaneous equations in (8), the switching time t_c and t_f can be calculated.

Second, we consider the design of the entire system and the method for designing the PID controller.

From Fig. 3, the target-value response is as follows:

$$Y(s) = \frac{K_d K_T T_m s^2 + K_p K_T T_m s + K_i K_T T_m}{JRT_m s^3 + (JR + K_d K_T T_m)s^2 + K_p K_T T_m s + K_i K_T T_m} R(s) \tag{9}$$

From above equation, we can design PID controller by the pole placement method.

Here, a reference model is determined as follows:

$$\frac{\alpha^3}{(s+\alpha)^3} = \frac{\alpha^3}{s^3 + 3\alpha s^2 + 3\alpha^2 s + \alpha^3}, \tag{10}$$

where, α is the model parameter.

The denominator of the target-value response can be compared with denominator of the reference model. Therefore, PID controller gains K_p, K_i, and K_d are written as follows:

$$\begin{cases} \alpha^3 = \dfrac{K_i K_T}{JR} \\ 3\alpha^2 = \dfrac{K_p K_T}{JR} \\ 3\alpha = \dfrac{1}{T_m} + \dfrac{K_d K_T}{JR} \end{cases} \Rightarrow \begin{cases} K_p = 3\alpha^2 \dfrac{JR}{K_T} \\ K_i = \alpha^3 \dfrac{JR}{K_T} \\ K_d = \left(3\alpha - \dfrac{1}{T_m}\right)\dfrac{JR}{K_T} \end{cases} \tag{11}$$

Finally, the system can be converted from bang-bang control to PID control. However, it must maintain continuousness during the conversion process. Therefore, we focus on the integrator of the PID controller and we determine its initial value.

Fig.5 shows a block diagram of the proposed method used to introduce the initial value of the integrator.

First, the transfer function is derived from each signal $R(s)$, I_i, $-T_L/J$, I_v, and I_p, to velocity $Y_v(s)$.

Then, the characteristic of equation (18) is equal to characteristic equation of that (16). Therefore, the numerator of (18) is placed as follows:

$$\begin{aligned} Y_v(s) &= \frac{\begin{array}{c}\{K_d K_T T_m R(s) + JRT_m I_v - K_d K_i T_m I_p\}s^3 \\ + \left\{\begin{array}{l} K_p K_T T_m R(s) - K_i K_T T_m I_i \\ -JR\dfrac{T_m}{J}\left(-\dfrac{T_L}{J}\right) - K_p K_T T_m I_p \end{array}\right\}s^2 \\ + \{K_i K_T T_m R(s) - K_i K_T T_m I_p\}s \end{array}}{JRT_m s^3 + (JR + K_d K_T T_m)s^2 + K_p K_T T_m s + K_i K_T T_m} \\ &= \frac{s(s+\alpha)(s+\beta)}{(s+\alpha)^3} \\ &= \frac{s(s+\beta)}{(s+\alpha)^2} \end{aligned} \tag{12}$$

Furthermore, the initial value of the integrator in the PID controller I_i is

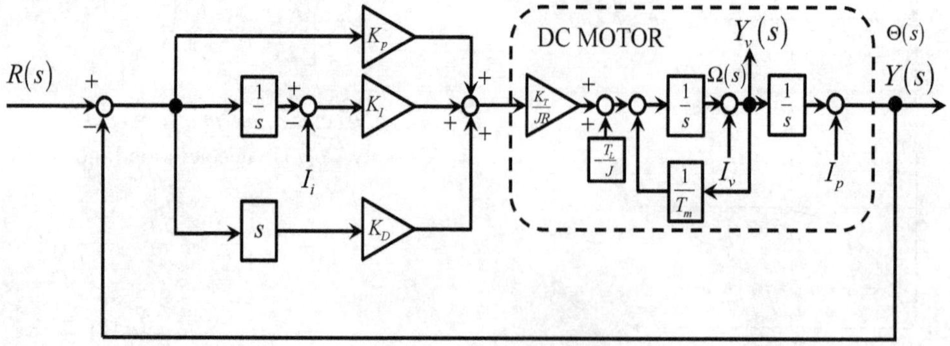

Figure 5. *Block diagram of proposed method to consider initial value*

$$I_i = -\frac{\begin{array}{l}\alpha^2\left\{JRT_m I_v - K_d K_T T_m I_p + K_d K_T T_m R(s)\right\}\\ \alpha\left\{-K_p T_m K_T R(s) - \alpha R T_L T_m + \alpha K_p K_T T_m I_p\right\}\\ +\left\{K_i K_T T_m R(s) - K_i K_T T_m I_p\right\}\end{array}}{\alpha K_i K_T T_m}. \quad (13)$$

Hence, the system cannot be discontinuous from the bang-bang control to the PID control.

Here, the condition of the bang-bang control to the PID control is the velocity $\omega_f(t)$. However, the t_c- $\theta_f(t)$ graph is influenced by the changing of the torque. Therefore, we introduce the disturbance observer, and the system can be estimated as torque-disturbance. Thus, the initial value I_i can be calculated in real-time.

5. Simulation Study

In this section, we show simulation results of the proposed method in several cases, which include variations of target-value, torque-disturbance, and the plant with a modelling error.

Table 1. shows each parameter value of the DC motor [5].

Table 1. *DC motor parameters* [5]

Parameters	Value
L	5.6×10^{-5}
R	7.4×10^{-1}
K_E	4.06×10^{-2}
K_T	4.15×10^{2}
J	4.47×10^{-1}
D	4.30
T_L	5.5×10^{2}
$u(t)$	12

From Fig. 1 and Table 1., $Y(s)$-$U(s)$ and $Y(s)$-Torque-disturbance for the DC motor are written as follows:

$$Y(s) = \theta(s) = \frac{0.0165}{s(0.0165s+1)}\left(1254.6U(s) - 1230.4\right) \quad (14)$$

The disturbance observer is written as follows:

$$\begin{cases} \begin{bmatrix}\dot{x}_1(t)\\ \dot{x}_2(t)\\ \dot{d}(t)\end{bmatrix} = \begin{bmatrix}0 & 1 & 0\\ 0 & -60.56 & 1\\ 0 & 0 & 0\end{bmatrix}\begin{bmatrix}x_1(t)\\ x_2(t)\\ d(t)\end{bmatrix} + \begin{bmatrix}0\\ 1\\ 0\end{bmatrix}u(t) \\ y(t) = \begin{bmatrix}1 & 0 & 0\end{bmatrix}\begin{bmatrix}x_1(t)\\ x_2(t)\\ d(t)\end{bmatrix} \end{cases} \quad (15)$$

The disturbance observer gain L_e is derived by the optimal control method, where, the weight functions Q_e and R_e are as follows:

$$Q_e = \begin{bmatrix}10 & 0 & 0\\ 0 & 10 & 0\\ 0 & 0 & 1\times10^{18}\end{bmatrix}, \; R_e = 1. \quad (16)$$

The disturbance observer gain L_e is as follows:

$$\boldsymbol{L}_e = \begin{bmatrix}1.94\times10^3 & 1.88\times10^6 & 1\times10^8\end{bmatrix}^{\mathrm{T}}. \quad (17)$$

The reference model is as follows:

$$\frac{150^3}{(s+150)^3}. \quad (18)$$

Therefore, the gains of the PID controller K_p, K_i, and K_d are as follows:

$$K_p = 53.80 \quad K_i = 2.69 \quad K_d = 0.31 \quad (19)$$

Fig.6 shows the switching time in bang-bang input t_c- position switching signal $\theta_f(t)$ graph. Here, the measurement points equal 10. The range of the measurement and the velocity switching signal $\omega_f(t)$ are

$$\begin{cases} t_c = 0.01 \sim 0.1 \text{ [sec]}, \; (\textit{increments of } 0.01) \\ \omega_f(t) = 80 \end{cases} \quad (20)$$

The switching of the system operates under the velocity switching signal $\omega_f(t)$

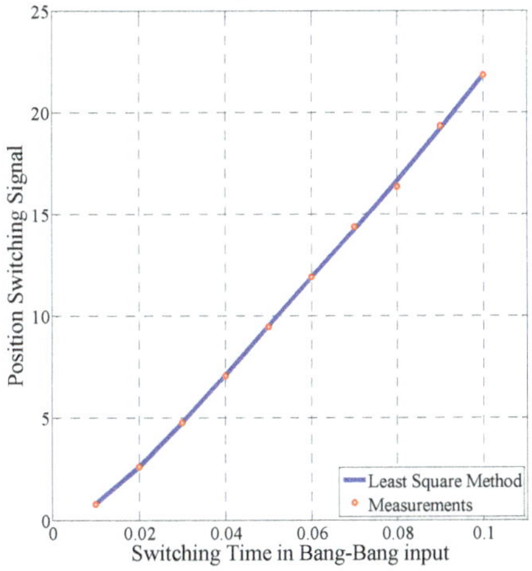

Figure 6. *Measurement result of t_c- $\theta_f(t)$ graph*

5.1. Variations of the Target-Value

The simulation study in this section is performed when the target-value changes.

Here, the position switching signal $\theta_f(t)$ and the velocity switching signal $\omega_f(t)$ are as follows:

$$\begin{cases} \theta_f(t) = r(t) - 0.8 \\ \omega_f(t) = 80 \end{cases} \quad (21)$$

The target values $R(s)$ are 5 and 20. Fig. 7 exhibits the responses in position and velocity.

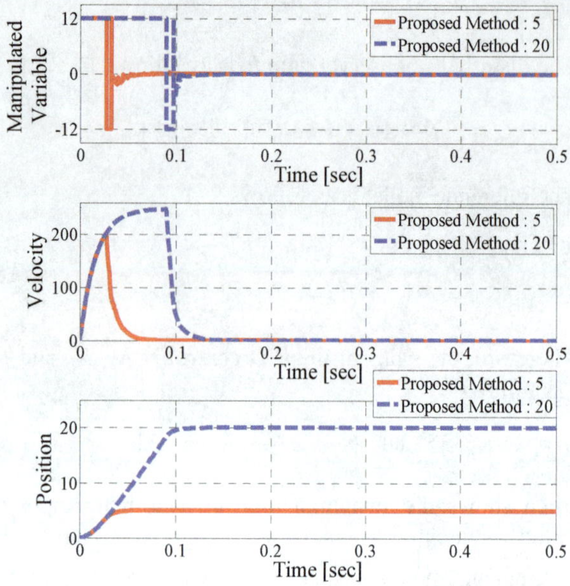

Figure 7. Responses of the target-value variation

The switching time of the bang-bang input from Fig. 6 is as follows:

$$\begin{cases} R(s)=5 \\ R(s)=20 \end{cases} \Rightarrow \begin{cases} t_c=0.028 \\ t_c=0.090 \end{cases} \tag{22}$$

5.2. Variation of the Torque-Disturbance

The simulation study in this section is performed when it changes the torque.

The target-value $R(s)$ is 20, the position switching signal $\theta_f(t)$ is 19.2 and the velocity switching signal $\omega_f(t)$ is 80.

To confirm the simulation results, the torque is changed as follows:

$$-\frac{T_L}{J} \pm 1000 \tag{23}$$

Fig. 8 shows the responses in position and velocity.

5.3. Plant with a Modeling Error

The simulation study in this section includes the plant with a modeling error.

The target value $R(s)$ is 20, the position switching signal $\theta_f(t)$ is 19.2 and the velocity switching signal $\omega_f(t)$ is 80.

The plant with a modeling error from (14) is as follows:

$$\frac{0.0165\times(\pm 20\%)}{0.0165\times(\pm 20\%)s+1} \tag{24}$$

Fig. 9 shows the responses in position and velocity when the plant has a modeling error.

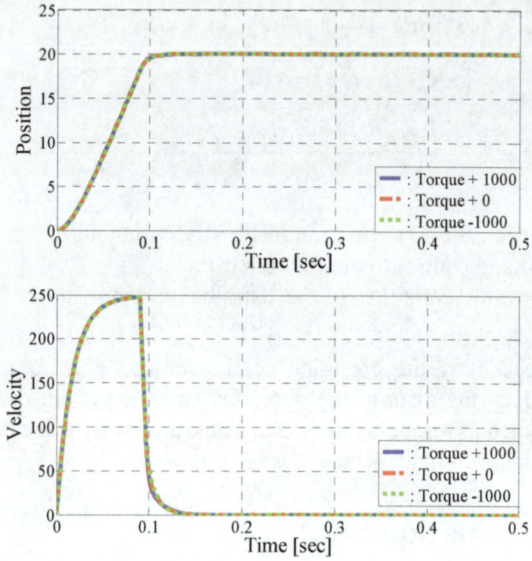

Figure 8. Responses of the torque variation

6. Experimental Study

This section describes the experimental results obtained using a DC motor [12] to show the effectiveness of our proposed method. In industrial applications, electric motors including DC motors are widely employed in equipment such as that used for factory automation, robotics, and hard disk drives.

The experimental control system, shown in Fig. 10, consists of a DC motor, an encoder, a counter that measures the information from an encoder, a host PC that transmits the control information, a target PC that calculates the manipulated variable, and a D/A converter.

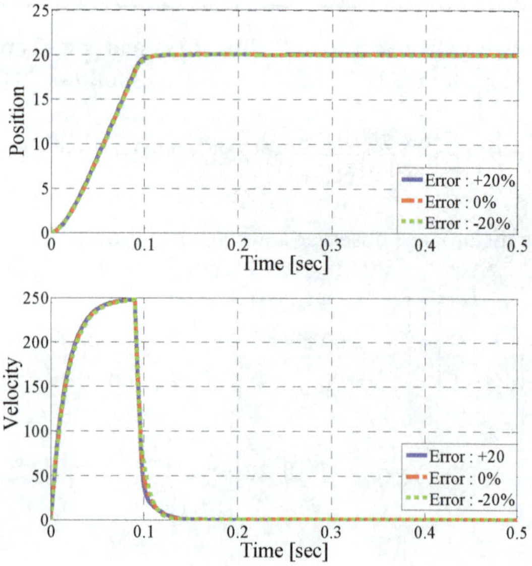

Figure 9. Responses of the plant with a modeling error

First, the control signal used to move the DC motor is transmitted from the host PC to the target PC and then to the D/A converter with a resolution of 12 bits. Next, the output of the motor is transformed to the pulse of Phase A and Phase B by an encoder with a resolution of 0.0879 deg/count. The counter measures the pulses and transmits the positional signal to a host PC through a target PC.

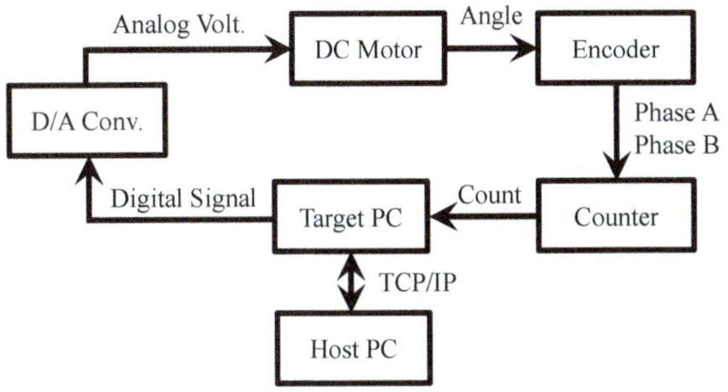

Figure 10. *The experimental Control System*

Fig. 11 shows the DC motor used in the experimental study, which was developed by the PID Corporation (PID-QET ii). The motor is equipped with an inertial load disk that has the radius of 0.0248 m and a weight of 0.068 kg.

Figure 11. *The DC motor*

The transfer function of the DC motor is as follows:

$$\frac{339.6}{s(s+10.78)} \tag{25}$$

The disturbance observer is written as follows:

$$\begin{cases} \begin{bmatrix} \dot{x}_1(t) \\ \dot{x}_2(t) \\ \dot{d}(t) \end{bmatrix} = \begin{bmatrix} 0 & 1 & 0 \\ 0 & -10.78 & 1 \\ 0 & 0 & 0 \end{bmatrix} \begin{bmatrix} x_1(t) \\ x_2(t) \\ d(t) \end{bmatrix} + \begin{bmatrix} 0 \\ 1 \\ 0 \end{bmatrix} u(t) \\ y(t) = \begin{bmatrix} 1 & 0 & 0 \end{bmatrix} \begin{bmatrix} x_1(t) \\ x_2(t) \\ d(t) \end{bmatrix} \end{cases} \tag{26}$$

The disturbance observer gain L_e is derived by the optimal control method, where the weight functions Q_e and R_e are as follows:

$$Q_e = \begin{bmatrix} 1\times10^2 & 0 & 0 \\ 0 & 1\times10^5 & 0 \\ 0 & 0 & 1\times10^{18} \end{bmatrix}, \; R_e = 1 \tag{27}$$

The disturbance observer gain L_e is as follows:

$$L_e = \begin{bmatrix} 1.00\times10^3 & 1.93\times10^5 & 3.16\times10^7 \end{bmatrix}^T \tag{28}$$

The reference model is as follows:

$$\frac{90^3}{(s+90)^3} \tag{29}$$

The gains of the PID controller K_p, K_i, and K_d are as follows:

$$K_p = 71.55 \quad K_i = 214.6 \quad K_d = 0.76 \tag{30}$$

Fig.12 shows the switching time in the bang-bang input t_c-position switching signal $\theta_f(t)$ graph. Here, the measurement points equal 10. The ranges of the measurement and the velocity switching signal $\omega_f(t)$ are as follows:

$$\begin{cases} t_c = 0.01 \sim 0.1 \text{ [sec]}, \; (\textit{increments of } 0.01) \\ \omega_f(t) = 30 \end{cases} \tag{31}$$

The switching of the system operates under the velocity switching signal $\omega_f(t)$

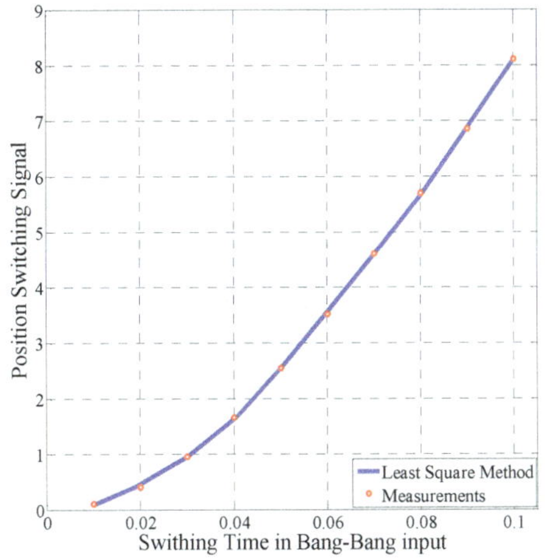

Figure 12. *Measurement result of t_c- $\theta_f(t)$ graph*

6.1. Variations of the Target-Value

In this experimental study, the performance is shown when changing the target-value.

Here, the position switching signal $\theta_f(t)$ and the velocity switching signal $\omega_f(t)$ are as follows:

$$\begin{cases} \theta_f(t) = r(t) - 0.5 \\ \omega_f(t) = 30 \end{cases} \tag{32}$$

The target-values $R(s)$ are 2 and 8. Fig. 13 exhibits the responses in position and velocity. The switching time of the bang-bang input from Fig.12 is as follows:

$$\begin{cases} R(s)=2 \\ R(s)=8 \end{cases} \Rightarrow \begin{cases} t_c=0.0385 \\ t_c=0.0948 \end{cases} \quad (33)$$

6.2. Confirmation of the Robustness

The study in this section shows the robustness of the proposed method.

The target value $R(s)$ is 8, the position switching signal $\theta_f(t)$ is 7.5 and velocity switching signal $\omega_f(t)$ is 30.

Fig. 14 shows the responses in position and velocity when a plant has a weight or a belt in which weight has a radius of 0.0248 m and a weight of 0.001 kg.

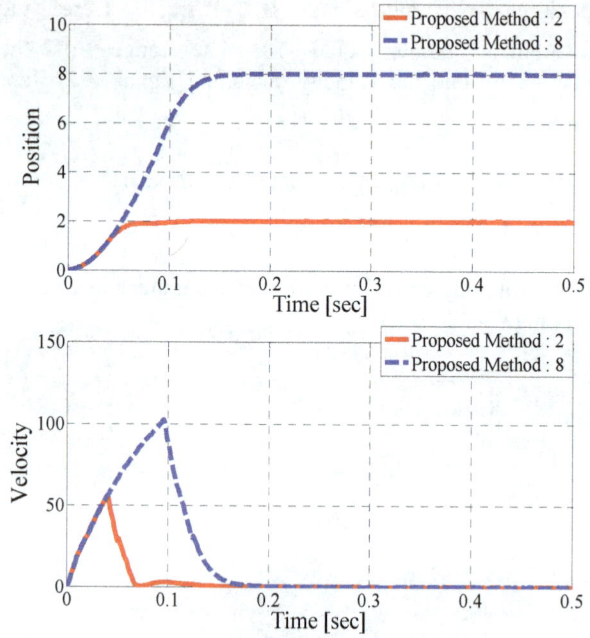

Figure 13. The responses of the torque variation

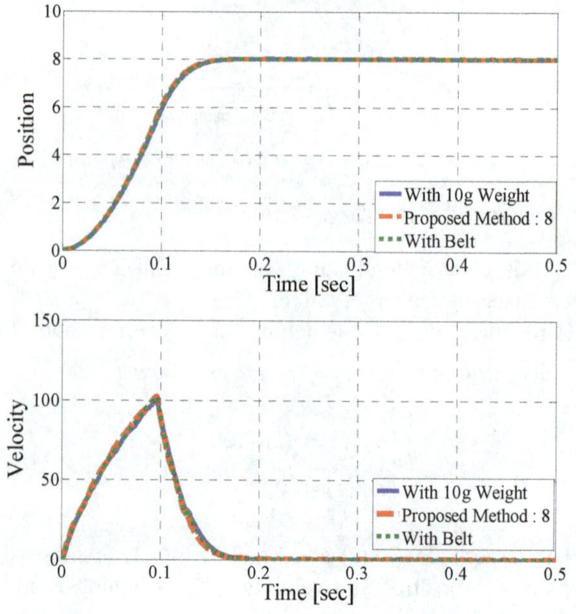

Figure 14. The responses of the torque variation

7. Conclusion

This study proposed a positioning control with high-speed starting and stopping for a DC motor using bang-bang control.

It is easy to determine the switching time of the bang-bang input from graph measurement and theoretical calculations.

When the control system is switched from "Bang-Bang control" to "PID control", it becomes a discontinuous system without the initial value of the integrator of the PID controller. Therefore, we introduced an initial value for the system which is calculated by theoretical analysis. In addition, we obtained the initial value of the integrator in real time by introducing a disturbance observer. Therefore, the system became resistant to changes in torque.

In the simulation and experimental studies, we confirmed the effectiveness of the proposed method in various cases. The responses of the position and the velocity can be obtained when changing cases caused by variations of the target-value, torque-disturbances, and the plant with a modeling error. We can get superior performances in various cases in these studies.

References

[1] N. Suda, System Control and information Society, PID control, Asakura Publication Company , 1992 (In Japanese)

[2] S. Yamamoto, and N. Kato, PID control Foundation and Application, 2nd ed., Asakura Publication Company, 2005 (In Japanese)

[3] S. Miyazaki, and J. Miyazaki, Application of automatic control learning on computer, , CQ Publication Company, 1991 (In Japanese)

[4] J.P. Lasalle, "The 'Bang-Bang' Principle ", Automatic and remote control, Vol.1, Proc. of the First International Congress of the International Federation of Automatic Control, pp.493-497., 1960

[5] K. Nagakura, "Time-Optimal Precise Positioning Control using a DC Servomotor", Journal of The Japan Society for Precision Engineering, Vol. 43, No.9, pp.11056-1062. Sept. 1977(In Japanese)

[6] James. H , Rillings , and Rob J.Roy, "Bang-Bang Control Strategy Using a Tabular Adaptive Model", J IEEE Trans. Automatic Control,Vol.12, Issue 3, pp.310-312, Jun.1967.

[7] K. Furuta, S. Kobayashi, "Bang-Bang Position of Direct Drive Motor", Proc. of IECON'90, 16th Annual Conference of IEEE, Industrial Electronics Society,pp.148-153, Nov. 1990

[8] N. Koreta, H. Okitomo, K. Tsumura, K.Takeuchi, and T.Egawa, "Study of High Accuracy of Machine Tool with Bang-Bang Control", Journal of The Japan Society for Precision Engineering, Vol.6, No.3, pp.427-431, Sept. 1977 (in Japanese)

[9] M. G.Vasek, and R.T. O'Brien. Jr., "Bang-Bang Control of Double Integrator", Proc. The Thirty-Forth Southeastern Symposium on System Theory 2002, pp.275-278, Mar. 2002

[10] Y. Yoshida, and T. Sogo, "Real Time Bang Bang Control of Crane", Departmental Bulletin Paper, Pub. Of Institute of Science and Technology Research, Chubu Univ. Vol.22, 6 page on PDF, Sept. 2010 (in Japanese)

[11] H. Shibasaki, R. Tanaka, H. Ogawa, Y Ishida, "High Speed Activation and Stopping Control System Using the Bang-Bang Control for a DC Motor", Proc. of 2013 IEEE International Symposium on Industrial Electronics (ISIE2013), pp.1-6, May, 2013.

[12] R. Tanaka, H. Shibasaki, H. Ogawa, T. Murakami, and Y. Ishida, "Controller design approach based on linear programming", ISA Trans., vol. 52, pp. 744-751, Nov., 2013.

21

Optimization of goal targeting in 1 on 1 soccer robots using safe points (Experimental Research)

Yahya Hassanzadeh-Nazarabadi[1], Abolfazl Saravani[1], Bahareh Alizadeh[2]

[1]Ferdowsi University, Park Sq, Mashhad, IRAN
[2]Khayam University, Ghasem Abad, Mashhad, IRAN

Email address:
ya_ha_na@ieee.org(Y.H. Nazarabadi), abolfazl.saravani70@gmail.com(A. Saravani), bahareh.alzdh@gmail.com(B. Alizadeh)

Abstract: Due to the fact that targeting the goal is the last part of soccer robot's aggressive decision chains, aiming styles and their lack of proficiency are discussed. Then targeting based on safe points is explained and implemented. Accordingly, this method has perfect accuracy. It brings new standpoint in aiming and it's capable of moving the barrier of its related knowledge.

Keywords: 1 on 1 Soccer Robots, Digital Compass, Binary Tree, Traversing a Binary Tree

---

1. Introduction

Main goal of robocup is to create a robot team fully able to compete against world's best human soccer players until year 2050 [1]. In trying to achieve this purpose, a lot of robots has been made. For instance: Humanoid robots, Small and Middle size robots, 1 to 1 and 2 to 2 Soccer robots.

Humanoid robots' objective is to maintain balance and move like humans. There are agents inside small robots that give orders to others and they control the game. Intentions of the Middle sized ones are single processing and efficiency of their algorithms. In 1 to 1 and 2 to 2, purpose is to go beyond the individual skills to the group skills [2]. Again, the most important responsibility of robots is to score, as a result, many different optimizations have been introduced. Ideally, 1 to 1 model is considered. These intelligent robots play in a dynamic and bounded environment while tracking a special light-emitting ball.

In this paper, goal targeting optimization in 1 to 1 is discussed. First, the problem is defined in details. Then, previous works are studied and their pros and cons are explained using experimental researches with ready samples. In the next part, design of the algorithm is addressed based on artificial intelligence. In this optimization method, a hypothesis named 'safe points' is suggested and its accuracy is analyzed. In the end, conclusion and results are presented.

2. Problem Definition

In this section, properties of the robot, the environment in which it works and the ball are discussed. As illustrated in figure 1, the field has 122cm in width and 183cm in length. Corners are flattened. Team's goals are located at the center of width of the filed at each side. They are 45cm in length and 14cm in height. There are cross-bars at the top of each goal to prevent robots from entering it. Its floor, walls and edges are colored. One side is blue and the other side is yellow also all of other parts of the filed are black. The ground should be positioned in a way that influence of external infrared light is as low as possible and earth magnetic territory is not disturbed in 2 meter depth. Guaranteeing the best conditions is impossible [1, 2].

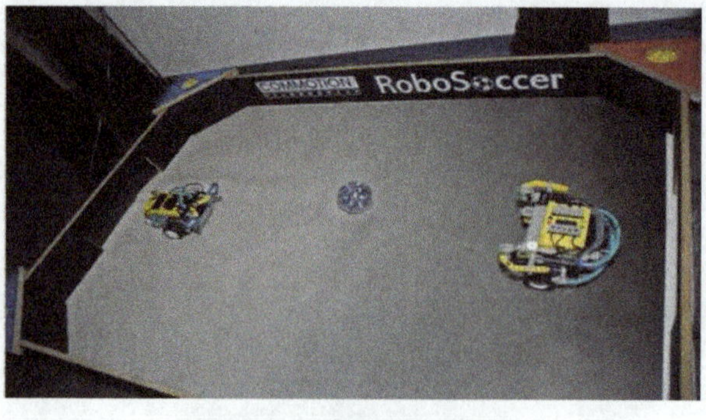

Figure1. *1 to 1 robot game field*

As exhibited in figure 2, 1 to 1 robots are sagacious. They obey specific rules for dimensions and design. These rules are edited by robocup federation every year. By standard, the robot usually can be placed in a circle with a 10cm diameter [1, 2]. Different sensors are installed on the robot. For instance, ultra sound sensor to detect surrounding walls [4] and infrared sensor to detect the ball [5]. There is also a compass attached to it, in order to indicate the angle [3]. An encoder is built in its wheel for positioning. Robot doesn't have any external power source rather than a battery, which it carries [1, 2]. Thus all of decisions are made by the robot itself without any interference of any human being.

Figure 2. 1 to 1 soccer robot

Game ball is an electronic ball which casts out infrared light (IR). In this experiment, functionality of 1 to 1 robot is studied, while assuming one rival robot on the opposite side. Operation of robot is as follows: first it's situated in the field, next it finds the ball using infrared-sensor then it approaches the ball. Ball is obtained by special algorithms. Then it evaluates its next move. In the end, utilizing specific methods, goal is marked and finally it shoots.

3. Study of Previous Work

3.1. Targeting the Goal Using Random Aim Locating

This method is one the most common solution in 1 to 1 and 2 to 2 soccer robots. In thistechnique using confine of opponent's goal, a scale is computed as target. In the moment of the attack, an arbitrary angel, which is in the range of the opponent's goal is chosen by the robot and it sets its shooting system by that angle [4, 5, 6].

Limit of the goal is calculated by the user and it is available to be utilized by algorithms of the robot. It's obvious that processing load is low, due to the fact that user gives the information to the robot. It only chooses a number from the specified range and really doesn't do any particular process at all.

Vital property of this method is haphazard selection of the angle of the objective. This random picking advances the process haste of the robot. Although, quantity of this improvement is scaled by traits of the hardware, implemented data structures and data processing algorithms. Hence, specific measurement is not possible. What is certain is 2 times gain in speed [4].

Granting the haste, random selection decreases accuracy. Because different points of the goal have different fitness values.Witnessed examination of videos of the matches shows center of the goal is more important in 1 to 1 robots and choosing it as aim point gives 98.2% precision in scoring. While for example, points with 2 centimeter distance of pole (boundary between the goal and the rest of the field) of the penalty area have lowest amount of possibility of success, which is about 31.33%. Considering the presented facts, reported accuracy of this method is 41% in average. An optimist approach is 'half of the shots are made through'. Since shooting technique is vital, the amount of correctness is lower than a proper and trusted threshold. Also this method do all of its work without bearing in mind that any opponent can be in the field, which in most cases results in creating conflicts with correctness of the calculated average. For the reason, that opponent is at free will and it can have any kind of movement after shooting, it prevents targeting most of the time.

3.2. Targeting the Goal Using Weighted Random Aim Locating

This method is very similar to the previous technique (3-1), the only difference is that choosing the aim point from the range of possible points of the opponent's goal is adjusted constantly by multiplying it in a coefficient. This factor is 1 at the beginning of the game but it changes according to the points, in which shooting was successful. In other words, range of targeting is the whole goal at the beginning of the game and it gets smaller to the part, where chance of making aneffective shot is higher. But random basis of this method is still in effect, variety from, which target point is chosen and consequently target aim is selected is altering [7].

One of the biggest advantages of this method is high accuracy. This technique is capable of having 79.81% correctness, but it's not practical. Because robot has to be aware whether its shot is successful or not. The stated accuracy is measured, while assuming a user notifies the robot about the result of its shot. But, according to the rules of the robocup, robot has to be independent thus makes user involvement impossible. Implementing an automatic system to recognize the outcome of every shoot produces a big process load, which results in massive reduction in procedure hast. Although this technique has high amount of non-practical accuracy, but it'sunable to find a solution regarding the existence of the opponent and it's completely ignored by this method, which causes in lower correctness of 52.03%.

3.3. Targeting the Goal Using Optimized Angle

In this method, robot is considered to be alone and there

isn't any opponent. After it's situated inside the field, it finds the position of the ball then it's obtained by the robot. So as to target, robot's position is found by calculating the distance from walls of the field making use of ultra-sound sensor. In the end, using mathematical computation accurate angle is determined. At first, robot goes to zero angle then it turns as much as the specified angle and it shoots.

Some of the pros of this algorithm:

- Simple method: presented algorithm doesn't have any specific status and the computations are simple.
- Fast according to time complexity: linear time complexity
- Independent of any special hardware: there isn't any particular hardware needed. It's common to use ultra-sound sensors for determining the distance from the wall and compass for positioning.

Main defect of this method is reduced accuracy when there is an opponent. It's caused by the fact that the robot is assumed to be alone. If any rival is entered into the game, precision is reduced drastically because situating and determining the distance are interfered by rival robot. As a result, wrong angle is calculated. This matter lessens the accuracy nearly to zero percent.

Average practical precision of this method while assuming there isn't any opponent in the field is reported 100%. But, accuracy is reduced to 19.02% by existence of any opponent consequently, make this method less likely to be used.

4. Targeting the Goal Using Safe Points

Corresponding to theprevious experiments, targeting the goal using the optimized angle while assuming the robot to be alone is an efficient method. But any interference from an opponent reduces its accuracy to almost zero percent.

To recognize the existence of the opponent, ultra-sound sensor is utilized. While maintaining zero angles, values of front and back distances are summed. The same happens to left and right gap. If these numbers are equal to length and width of the field respectively, it issues the fact that there is no opponent going to conflict in shooting otherwise preventer is identified.

To solve this problem points, located near to the robot, are used. Before approaching these point using targeting the goal method which utilizes the optimized angle,all of the related calculations are done. After reaching them, the ball is shot. From now on these points are called 'safe points'.

The key to the problem base on safe points is saving the changes of behavior of the opponent (regarding the environment) in a data structure. Afterwards velocity, acceleration andmovement equation of the opponent is worked out. Considering this information, its next move to reach the safe point is settled by binary trees traversal algorithms.

Figure 3 illustrates an example in which using safe points method is possible and figure 4 demonstrates decision making tree considering mentioned example of figure 3.

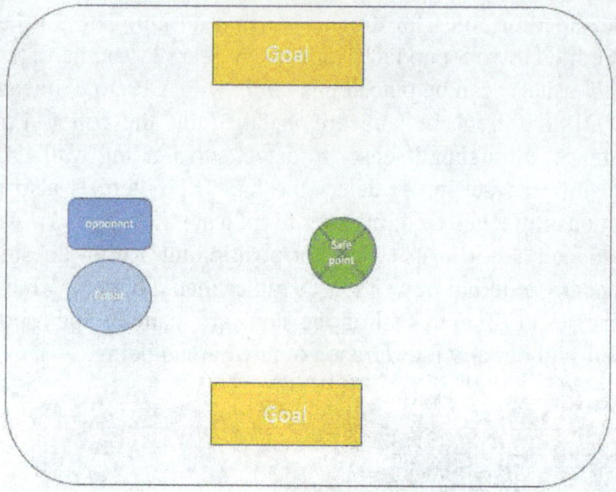

Figure 3. *Positioning opponent and robot*

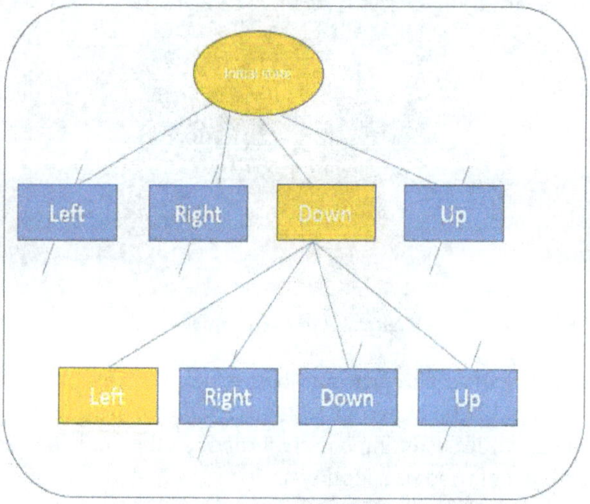

Figure 4.*Decision making tree*

As it can be seen from figure 4 robot can choose 4 different moves, up, down, left and right. At first, opponent is identified using ultra-sound sensor. Next, the wall on the left is recognized. Given the facts, safe point can't be in these 2 directions. For the beginning move, down is chosen so as to create enough space to do the calculations and enough distance from the opponent. After doing so, again 4 different choices appear. To prevent iteration, 'Up' move is not chosen. 'Left' move is not chosen either because of the wall. Although 'Right' and 'Left' have the same amount of computation and same distance from the opponent, 'Right' is selected owing to have less gap to the goal. At every status all of these considerations are made to pick the next move. Search operation is done as soon as robot reaches the safe point. Exclusivity of safe points is when the robot gets to them, according to analyzing the rival's behavior, targeting and shooting procedures are done before opponent reaches them. Accuracy is calculated considering the test that is explained in 'Problem Definition' section and it is shown in table 1.

Table 1.Results of inducted experiments using safe point algorithms

Number	X1	Y1	Accuracy
1	31.6	59.3	92.33%
2	42.1	96.7	89.17%
3	132.9	100.2	97.90%
4	91.6	90.2	91.32%
5	60.0	88.2	96.67%

In table 1, average precision is calculated using 5 points as representation of entire field. Each point is evaluated 20 times. According to technical repots of participants of Robocup 2013 and videos related to the games, accuracy of this tentative algorithm is magnificent up to this day [4, 5, 6, 7, 8, 9, 10, 11].

5. Conclusion

In this tentative experiment, 1 to 1 robots and their operational environment are presented and examined. Then, Importance of accuracy of targeting is discussed. Next, concerned practices are appraised. It ended that maximum practical accuracy of these methods is 50% which reduces drastically when existence of an opponent is considered and conflicts are created by it. In other words, all of these techniques think through the environment as a static one.

After that, targeting the goal algorithm using safe points, which is designed by the authors, is suggested. This method considers opponent making conflicts with targeting process, meaning environment is considered to be dynamic. So, aiming isn't affected by it. Average preciseness is 93.47%.

References

[1] Robocup Organization, "Junior Soccer Robots Rules", Robocup, Mexico City, Mexico, 2012

[2] Iranian Robocup Organization, "Junior Soccer Robots Rules", Robocup Iran Open, Tehran, Iran, 2012

[3] Hassanzadeh Nazarabadi Y, Saghlatoon H, Sharif Shazileh A., "A method to create the most accurate goal targeting in 1 on 1 soccer robots", 5thENASE, Athens, Greece, 2010.

[4] GOROGORO, Technical Document Paper, Robocup, Mexico City, Mexico 2012.

[5] Submarine, Technical Document Paper, Robocup, Mexico City, Mexico 2012.

[6] CATASTROPHY, Technical Document Paper, Robocup, Mexico City, Mexico 2012.

[7] FRF, Technical Document Paper, Robocup, Mexico City, Mexico 2012.

[8] Fast and Smart, Technical Document Paper, Robocup, Mexico City, Mexico 2012.

[9] XYZ, Technical Document Paper, Robocup, Mexico City, Mexico 2012.

[10] FTA, Technical Document Paper, Robocup, Mexico City, Mexico 2012.

[11] A2Z, Technical Document Paper, Robocup, Mexico City, Mexico 2012.

Washing machine using fuzzy logic

Mustafa Demetgul[1], Osman Ulkir[2], Tayyab Waqar[2]

[1]Marmara University, Technology Faculty, Department of Mechatronics Engineering, Istanbul, TURKEY
[2]Marmara University, Institute of Pure and Applied Sciences, Department of Mechatronics, Istanbul, TURKEY

Email address:
mdemetgul@marmara.edu.tr (M. Demetgul)

Abstract: For the past few years, different types of control techniques are being used in various fields of industry. Fuzzy logic based control system is one of them. Fuzzy logic uses statements instead of mathematical model for solving a given problem. In this paper, a normal household washing machine, which is used very often, is modeled with the help of Fuzzy logic. Both the simulation and the control of the aforementioned device have been done by using MATLAB's fuzzy logic toolbox.

Keywords: Fuzzy Logic, Washing Machine, MatLab, Optimization, Automatic Sensing

1. Introduction

Fuzzy logic is a concept which helps computers in making decisions in a way which resembles human behaviors. It helps industry in increasing productivity, creates the opportunity to make production more convenient and most importantly it helps industries in economical terms.

The concept of fuzzy logic was first proposed by Professor Lotfi A. Zadeh in 1965. It was presented in one of his research papers under the name Fuzzy logic or Fuzzy sets [1, 2].

The first fuzzy logic based control experiment was conducted by Mamdani [3] in 1974 who designed the fuzzy logic for a steam engine. With his experiment, Mamdani showed how easy it is for a computer to process linguistic statements which was proposed by Zadeh.

After 1980 the use of fuzzy logic based control system becomes common in vacuum cleaners, washing machines, elevators, metro and company operations. Advancements in engineering in recent years have allowed fuzzy logic to be used in many fields [4].

Nowadays fuzzy logic has found its application in several fields like electronic control systems, automotive industries, breaking systems and home electronics etc.

Everyday many home appliances are being upgraded using fuzzy logic to save time and to conserve electricity [5].

Many necessary home appliances like washing machines, dish washers; vacuum cleaners etc. are based on fuzzy logic nowadays. Tiryaki and Kazan's dish washer using fuzzy logic and Alhanjouri and Alhaddad's optimize wash time of washing machine using fuzzy logic are one of few studies which are based on fuzzy logic [5,6].

Alhanjouri and Alhaddad's washing machine takes dirt type and degree of dirtiness as inputs while wash time is the only output of the system [6].

Agarwal controlling the washing time using fuzzy logic control. Type of dirt, dirtness of clothes were selected as input [7].

Also aim to reduce wasting of electric and water. The washing machine fuzzy controller neural network is researched deeply, which is based on fuzzy logic, neural network and its learning algorithm [8-9].

Kumar and Haider aim to reduce washing time. Quantity and dirtiness were selected as input [10].

The device which is being presented in this study has 4 different inputs depending on which 5 different outputs are being controlled by using fuzzy logic.

In contrast to the previous studies, the washing machine discussed in this paper has more inputs and outputs for example depending on the sensitivity and the quantity of cloths, washing machine will automatically adjust its washing speed, amount of detergent, hotness of water and water level. It will help us in conserving water and detergent while washing our cloths.

In this paper, fuzzy logic control for a very common household, i.e. washing machine, has been developed. The modeling of washing machine has been based on the parameters regarding washing of cloths.

2. Fuzzy Logic

In crisp logic, classifications are definite, i.e. a member is either a part of group or not. It can't be both. In short, crisp logic has two values 0 and 1. In contrast, fuzzy logic can work in complex conditions similar to humans.

If we consider solving a problem regarding human age by using both the crisp and fuzzy logic we can understand the difference between them. Figure 1 shows crisp logic while fuzzy logic can be seen in figure 2.

According to figure 1, people in between 0-30 years will be considered young; 30-50 years old will be considered middle aged and above 50 will be considered old.

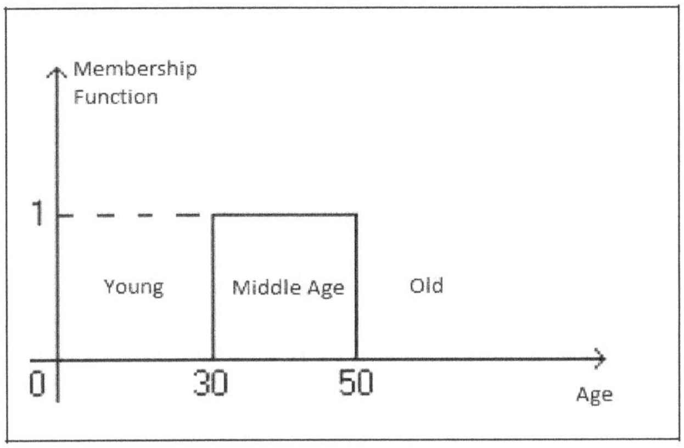

Figure 1. *Crisp Logic [11]*

According to those laws, a 31 years old person will be considered middle age while a 29 years old will be considered young.

If we examine this condition using fuzzy logic then a 30 years old person, in an appropriate proportion, will be considered both young and middle age (figure 2). Unlike Crisp logic, Fuzzy logic doesn't have only 0 and 1. It is more flexible thus a more realistic approach can be applied [12].

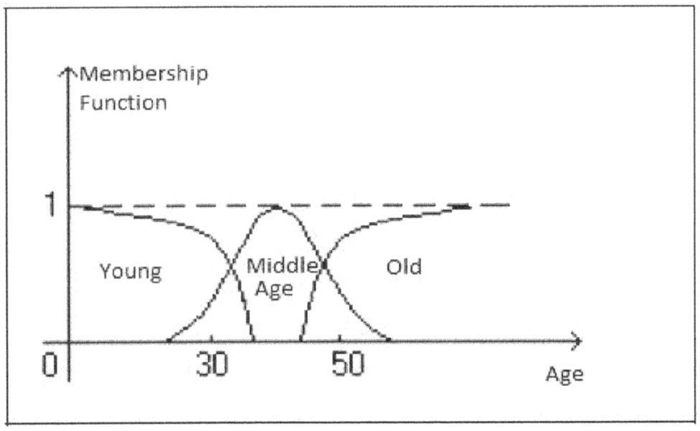

Figure 2. *Fuzzy Logic [11]*

2.1. Fuzzy Logic System Structure

The basic elements of fuzzy logic; Fuzzy inputs, outputs, rules and defuzzification; are shown in figure 3. [13-14]

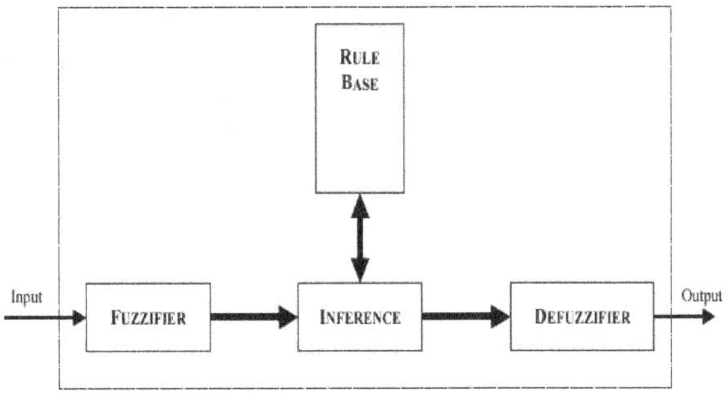

Figure 3. *Structure of a fuzzy logic system [12, 13]*

Fuzzy take the information from a system; which is in normal language; and converts it to values. The values of input quantities, which are associated with of membership functions, are given in form of words such as small, smallest [2].

Input and output variables of fuzzy system are determined and their values i.e. less, much, hot and cold etc. are being selected. Afterwards, rules are being developed and by using those rules input and output relationship is developed.

Outputs are being produced by using fuzzified inputs and rules which are being determined. Those fuzzified outputs must be converted to real values so that they can be used in real systems. This whole process is known is defuzzification. [13].

3. Fuzzy Logic Modeling of Washing Machine

To model a system using fuzzy logic, the first step is to determine the inputs and outputs of it. A washing machine's most important duty is to clean the cloths without damaging them. In order to achieve it, the output parameters of fuzzy logic, which are related to washing, must be paid importance. Inputs and outputs of fuzzy logic system are shown in figure 4.

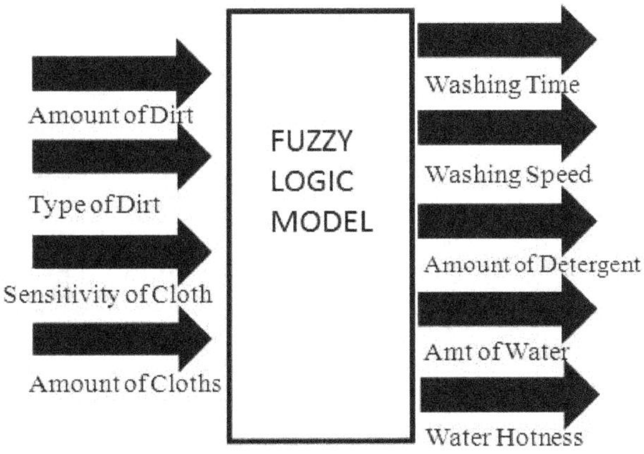

Figure 4. *Inputs and Outputs of the System*

Using fuzzy logic, input parameters such as Amount of dirt, type of dirt, sensitivity of cloth and amount of cloths, will help washing machine to achieve economical wash.

The input and output parameter's membership function values, names, upper and lower limits are being set based on a given problem. The membership functions, with upper and lower limits, of input and output parameter's are shown in figure 5 and 6 respectively. Figure 5 and figure 6 show membership functions of input and output respectively and also their upper and lower limits.

After determining the membership functions and their upper and lower limits required for the modeling of necessary parameters, a total of 81 rules have been established to define relationship among those parameters.

In order to apply fuzzy logic to washing, it is necessary to establish fuzzy logic rules. These rules can be seen in Table 1.

Input Parameters:

1. Amount of Dirt
2. Type of Dirt
3. Sensitivity of Cloth
4. Amount of Cloths

Output Parameters:

1. Washing Time
2. Washing Speed
3. Amount of Detergent
4. Amount of Water
5. Water Hotness

Table 1. *Fuzzy Logic Rule Table*

	INPUTS				OUTPUTS				
	1	2	3	4	1	2	3	4	5
1	S	NG	LS	S	S	LW	LT	LT	LW
2	S	G	LS	L	LG	M	MN	MN	H
3	S	NG	NS	M	M	M	LT	LT	LW
4	M	M	NS	M	M	M	N	N	N
5	M	M	LS	S	S	M	N	N	N
6	M	M	LS	L	LG	H	MN	MN	N
7	M	M	VS	M	S	M	N	N	N
8	L	G	VS	L	VL	VH	MN	MN	H
9	L	G	LS	M	LG	H	MN	MN	H
10	L	NG	NS	L	VL	H	N	N	LW

S = Small, M = Medium, L = Large, NG = Not Greasy, G = Greasy, LS = Less Sensitive, NS = Normal Sensitive, VS = Very Sensitive, S = Short, LG = Long, VL = Very Long, LW = Low, H = High, VH = Very High, LT = Little, N = Normal, MN = Many.

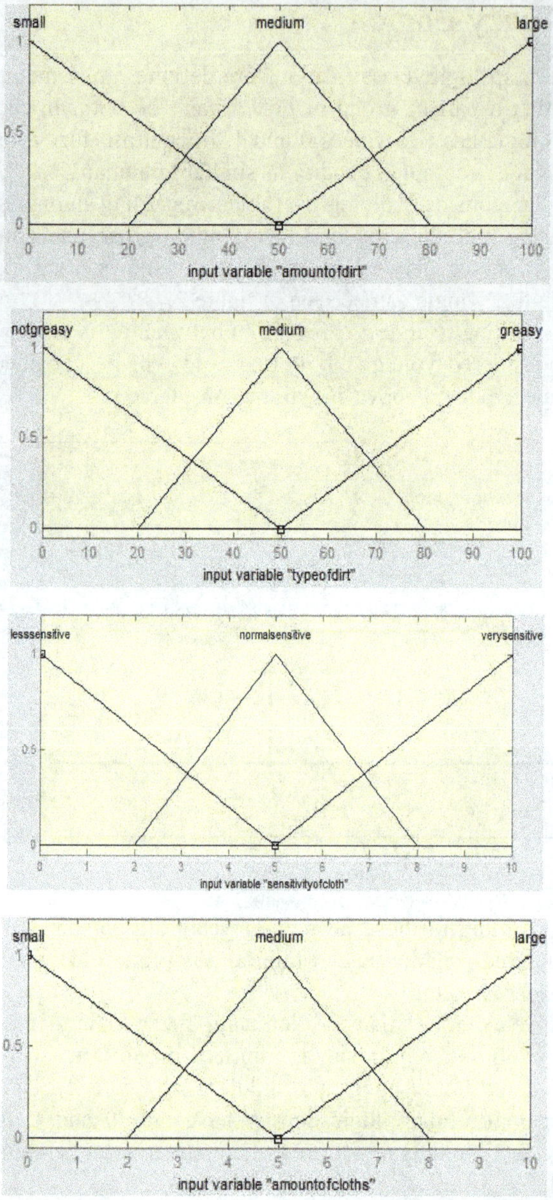

Figure 5. *Fuzzy logic input membership functions*

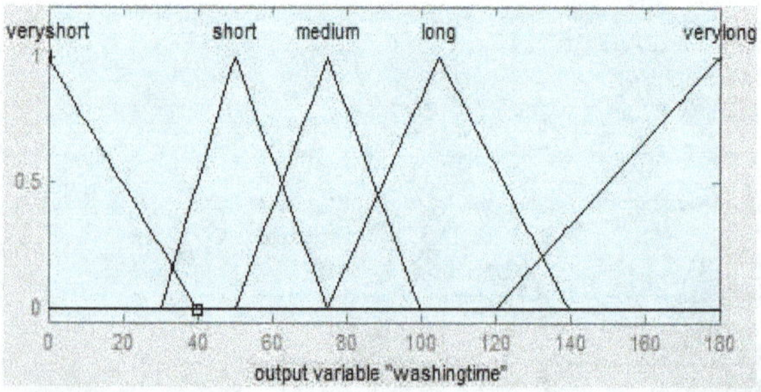

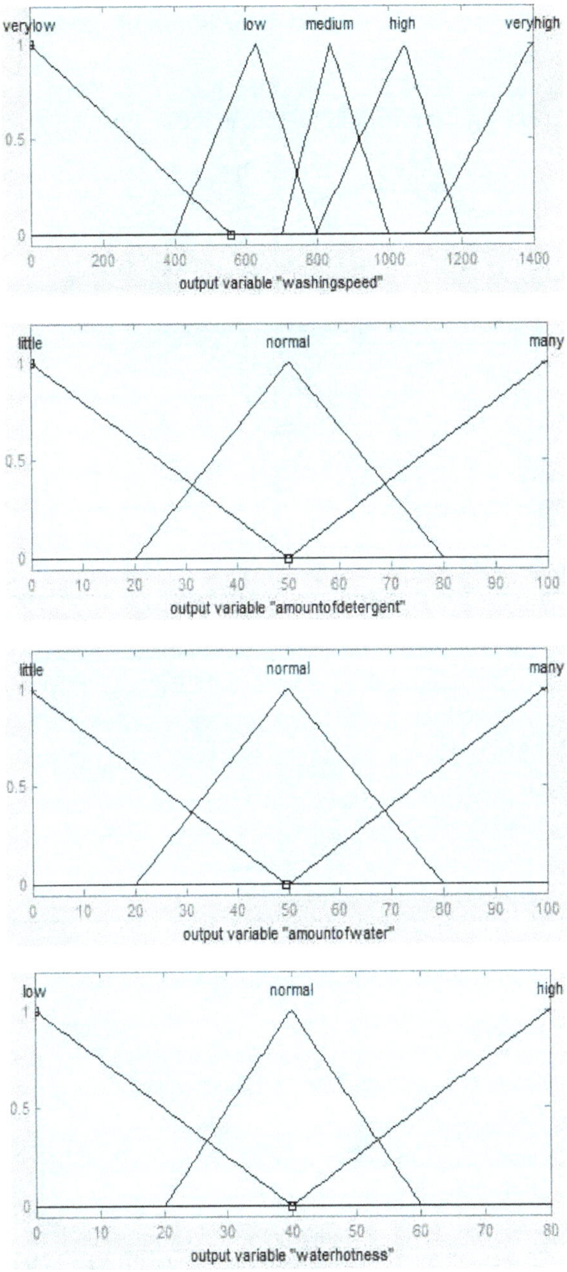

Figure 6. Fuzzy logic output membership functions

Fuzzy rules have been established for the modeling of washing machine. The whole system has been developed by using MATLAB's fuzzy logic toolbox.

The results of those rules, which have been determined by applying Min-Max operator, are illustrated in the form of 3D graphs in figure 7, 8, 9, 10, 11, 12. These figures show the relationship between input and output parameters.

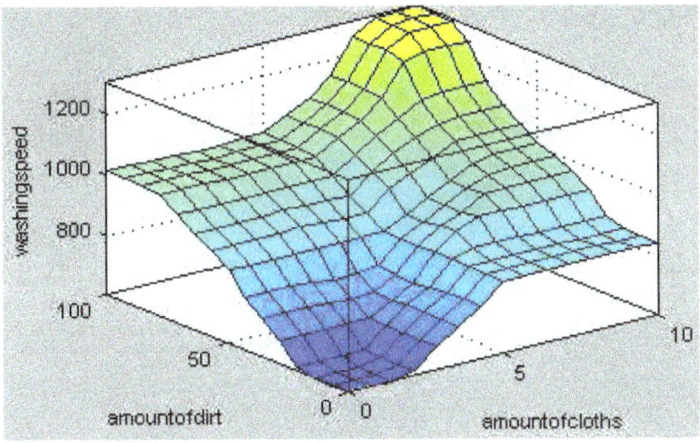

Figure 7. Amount of cloths and amount of dirt affects the washing speed

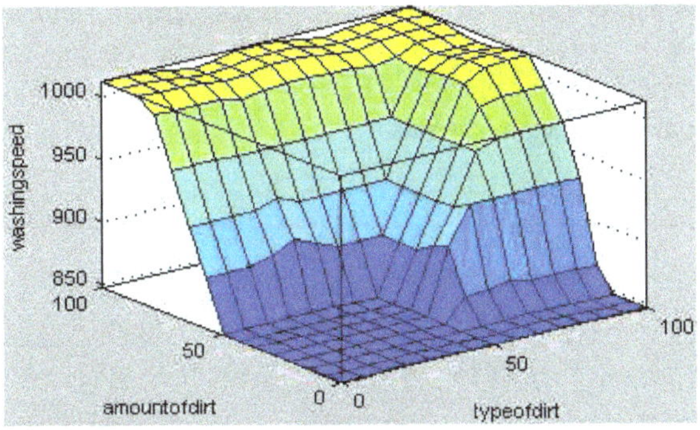

Figure 8. Type of dirt and amount of dirt affects the washing speed

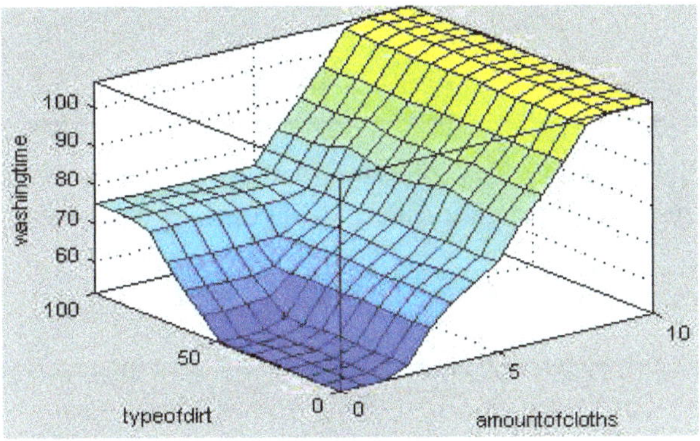

Figure 9. Amount of cloths and type of dirt affects the washing time

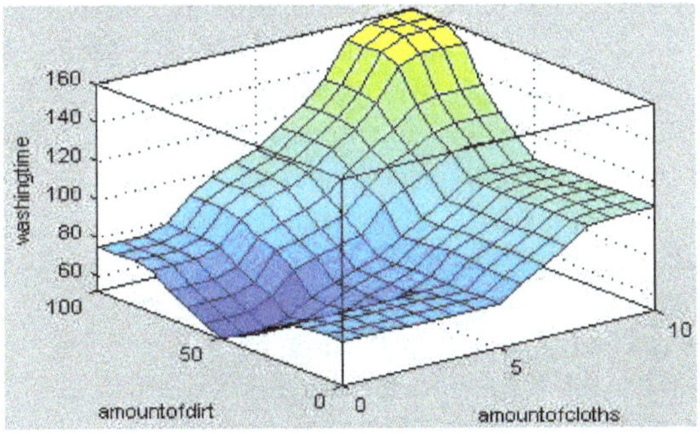

Figure 10. Amount of dirt and amount of cloths affects the washing time

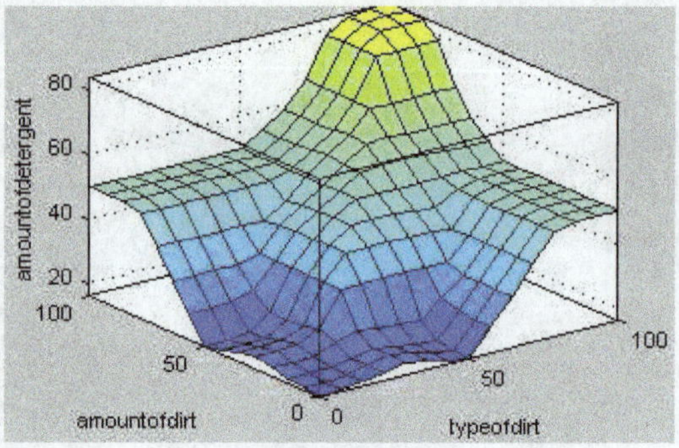

Figure 11. Type of dirt and amount of dirt affects the amount of detergent

As can be seen in figure 7, washing speed isn't affected much by the amount of cloths. Amount of dirt and sensitivity of cloth are the most important factors which regulate the washing speed.

Figure 9 shows that washing time is directly proportional to the amount of cloths.

Figure 12 tells us that water hotness is very much affected by the type of dirt present in the cloths.

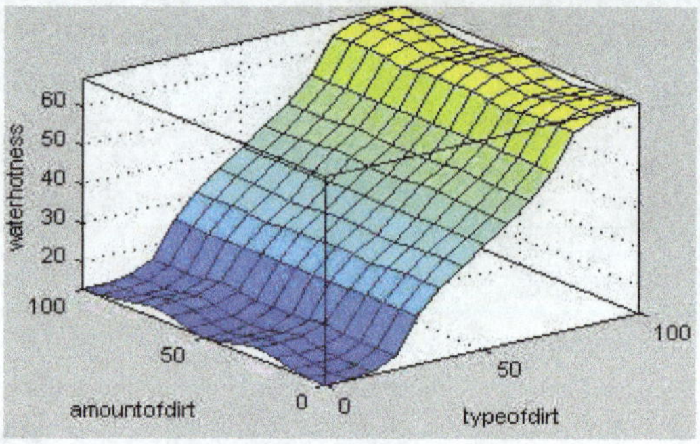

Figure 12. Type of dirt and amount of dirt affects the water hotness

Table 2. Fuzzy logic outputs in response to the inputs

INPUTS				OUTPUTS				
Amount of Dirt	Type of Dirt	Sensitivity of cloth	Amount of Cloths	Washing Time	Washing Speed	Amount of detergent	Amount of Water	Water Hotness
%50	%50	5	5	95 d	1000d/d	% 45,8	% 48,5	60 °C
%88.7	%82.2	1.8	7.67	140 d	1200d/d	% 60,5	% 60	70°C
%62	%66.4	3.67	6.2	128 d	1100d/d	% 52	% 55,8	65°C
%24.7	%20.4	7.27	3.27	60 d	700d/d	% 30,5	% 35	40°C
%11.3	%11.2	8.73	1.93	40 d	400d/d	% 20,5	% 30	30°C

Table 2 describes the variation in the output parameters, in response to the given inputs, of the washing machine.

4. Conclusion

In this paper, rule based fuzzy logic for washing machine has been developed. It will help in achieving economical washing procedure by sensing amount of dirt, type of dirt, sensitivity of cloth and amount of cloths. Based on input parameters, i.e. amount of dirt, type of dirt, sensitivity of cloth and amount of cloths; washing speed, washing time, water hotness and amount of detergent will be regulated on the output. Due to this adjustment of output parameters, cloths will come out cleaner and it will also make the whole washing process economical by reducing the amount of water, detergent, electricity and time.

Results of this simulation based study are pretty good which is clearly shown in the graphs. When we compare the results we have gotten with the expected results, it shows that this model, which has been developed in this paper, is extremely useable.

MATLAB/Fuzzy logic toolbox has been used to materialize this study. It can be practically implemented by using necessary mechanical and electronics engineering concepts.

References

[1] Zadeh, L.A., "Fuzzy Sets", Information and Control,8,338-353, 1965.

[2] Elmas, C., "Bulanik Mantik Denetleyicileri", ISBN 975-347-613-2, 2003

[3] Mamdani, E.H., "Application of Fuzzy Algorithms for Control of Simple Dynamic Plant", Proc. IEEE, 121(12), 1585-1588, 1974.

[4] Şen Z., "Bulanık (Fuzzy) Mantık Ve Modelleme İlkeleri", Bilge Sanat Yapım Yayınevi, İstanbul, 2001.

[5] Tiryaki, A.E., Kazan, R., "Bulasik Makinesinin Bulanik Mantik ile Nodellenmesi", Muhendis ve Makina Dergisi, Cilt:48, Sayi:565, Sakarya.

[6] Alhanjouri, M. and A. Alhaddad, A., "Optimize Wash Time of Washing Machine Using Fuzzy Logic", Islamic University of Gaza.

[7] Agarwal, M. (2007). Fuzzy logic control of washing machines.URL:http://softcomputing.tripod.com/sample_termpater:pdf.

[8] Zhen. A. and Feng R. G., "The design of neural network fuzzy controller in washing machine," in Proc. 2012 International Conference on Computing, Measurement, Control and Sensor Network (CMCSN), Shanxi, China, 136-139, 2012.

[9] Virkhare N., Jasutkar R.W., Neuro-Fuzzy Controller Based Washing Machine, International Journal of Engineering Science Invention, 3(1), 48-51, 2014.

[10] Kumar D., Haider Y., Fuzzy Logic Based Control System for Washing Machines, International Journal of Computer Science and Technology, 4(2), 198-200, 2013.

[11] Ozek, A., Sinecen, M., "Klima Sistem Kontrolunun Bulanik Mantik ile Modellenmesi", Pamukkale Universitesi Muhendislik Fak. Muhendislik Bilimleri Dergisi, 10(3), 353-358, 2004.

[12] Tuncer, S. 1999. "Değişken Hızlı Sürücü Sistemleri İçin Fuzzy Denetleyicili Yeni Bir Algoritmanın Geliştirilmesi ve Uygulaması" , F. Ü., Fen Bilimleri Enstitüsü, Yüksek Lisans Tezi.

[13] Dadone, P., "Design Optimization of Fuzzy Logic Systems", Doctor of Philosophy in Electrical Engineering, Virginia Polytechnic Institute and State University, 2001.

[14] Mendel, J.M., "Fuzzy Logic Systems for Engineering: A Tutorial", Proceedings of the IEEE, 83(3), 1995.

Design of magnetorheological fluid dynamometer which electric current and resisting moment have corresponding relationship

Luo Yiping, Xu Biao, Ren Hongjuan, Chen Fuzhi

College of Automobile Engineering, Shanghai University of Engineering Science, Shanghai, China

Email address:

lyp777@sina.com (Luo Yiping), xubiao0813@163.com (Xu Biao), ren-hongjuan@163.com (Ren Hongjuan), 284774881@qq.com (Chen Fuzhi)

Abstract: Based on the research of the new material magnetorheological fluid, magnetorheological fluid dynamometer is designed. Under the premise of certain structure size and material, there is a one-to-one correspondence between MRF dynamometer loading current and load torque provided with the machine electricity and the theoretical calculation. This papergives the design method and specific geometric parameters of magnetorheological fluid dynamometer. The process of magnetorheological fluid dynamometer theory design is obtained by taking a specific model motor as an example, which provides a theoretical basis for the application of MRF in the field of dynamometer.

Keywords: MRF, Dynamometer, Load Torque, Design, Corresponding Relationship

1. Introduction

Magnetorheological Fluid, called MFR, is an active branch in the research of intelligent material field. MFR is made of tiny soft magnetic particles and non-conducting magnetic mother liquid with high permeability and low hysteresis of magnetic and mixed with an emulsifier[1]. Under the action of external magnetic field, MRF can instantly (in milliseconds) achieve consecutive reversible transition between the low viscosity easy flow of Newtonian fluid and high viscosity hard plastic Bingham[2]. It has a wide range of applications in many fields because of its "liquid", "solid" state transition reversible, controllable and rapid and other outstanding technological features[3]. In the current technology, Bossis and Cutillas et al from French University of Nice has done a lot of work in the MRF mechanism research, especially in the aspect of microstructure analysis[4]. Kormann et al from Germany BASF G have developed a stable of nanoscale MRF[5]. The engineering and technical personnel Lord company has developed a vehicle seat suspension damper[6]. Gm Foister and Gopalswamy developed magnetorheological Fluid and magnetorheological clutch[7]. Jianhua Ni[8] from Xi'an Jiaotong University had carried out a research on the application of magnetorheological damper in the vehicle suspension control.

The existing measurement devices for rotating parts (engines, motors, etc.) power are hydraulic dynamometers , electricity dynamometer , eddy current dynamometer , etc. which can relatively accurately measure the power of the rotary member to a certain extent but also have such defects as large size, slow response, energy consumption, expensive. Magnetorheological fluid power measuring device has broad prospects with its compact structure, rapid response and low power consumption.

2. Principle of Work

Rotating parts power calculated formula is:

$$P = \frac{nT}{9550} \tag{1}$$

Where P is the power of rotating parts(kW), n is the rotation speed ($r \cdot min^{-1}$), T is the torque ($N \cdot m$).

Magnetorheological fluid dynamometer bench as shown in Fig. 1, viscosity of magnetorheological Fluid increases dramatically under the action of external magnetic field and imposes resistance moment on the moving object and we can get the value of the torque, namely output torque of the rotating parts, based on excitation coil size and structure size, rotating speed based on speed sensor, so that we can get the output power of the rotating parts by the formula (1).

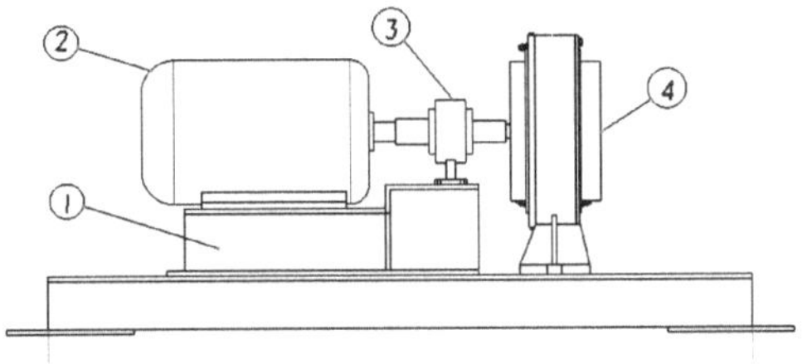

1. chassis ; 2.rotating parts ; 3. speed sensor ; 4. MRF dynamometer

Fig 1. *Magnetorheological fluid dynamometer test bench.*

3. Torque Calculation

Currently, the most common view about the constitutive nature of MRF is to treat magnetorheological fluid as a Bingham fluid. The constitutive equation[9] is:

$$\tau = \tau_B + \eta\gamma \quad (2)$$

Where τ is the shear stress (P_a); τ_B is the dynamic yield stress(P_a), η is the liquid viscosity ($P_a{\cdot}s$); γ is the shear rate[10].

$$\gamma = \omega \cdot \frac{r}{h} \quad (3)$$

Where ω is the MRF angular velocity at any position ($rad{\cdot}s^{-1}$), r and h are radius and thickness of disk.

Through the study of magneto-rheological fluid existing products, we chose disk magnetorheological fluid power absorption device, torque analysis as shown in Fig. 2

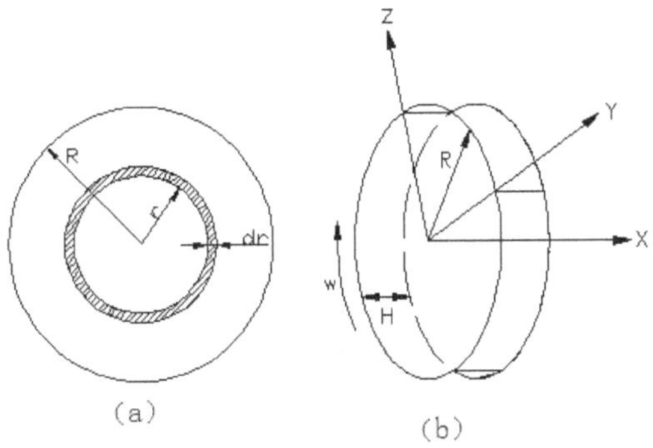

Fig 2. *Magnetorheological Fluid Torque Analytical Diagram.*

The shaft rotates around x–axis in angular velocity (ω), shears with MRF that contacts the disk and forms a disc shape by the magnetorheological fluid that has been through magnetorheological effect. Assumed the outer diameter of disc shape is R, we take the micro annular disc-shaped magneto-rheological fluid whose width is d_r to study. At the position of radius r, the micro torque of micro annular disc-shaped magnetorheological fluid is:

$$d_T = r d_{Fr} \quad (4)$$

Where r is the radius of micro disk; d_{Fr} is the micro shear stress, in addition

$$d_{Fr} = \tau d_s \quad (5)$$

$$d_s = 2\pi r d_r \quad (6)$$

We can get load torque that single disk can offer from formula (2)~(6)

$$T = \frac{2}{3}\pi\tau_B R^3 + \frac{1}{2h}\pi\eta\omega R^4 \quad (7)$$

As is shown in formula (7),load torque consists of two parts. First part relates to magnetic induction intensity and structure size $T_B = \frac{2}{3}\pi\tau_B R^3$ and second part is viscous resistance moment of magnetorheological fluid $T_\eta = \frac{1}{2h}\pi\eta\omega R^4$. Total load torque equals the sum of two parts: $T = T_B + T_\eta$. Volume of magnetorheological fluid:

$$V = \pi R^2 H \quad (8)$$

From formula (7) and (8), we can get:

$$\frac{H}{R} = \frac{3}{4}\cdot\frac{T_B}{T_\tau}\cdot\frac{\eta\omega}{\tau_B} \quad (9)$$

Formula shows that geometric constraints of disk magnetorheological fluid power measurement device relates to the following parameters: the material properties of magnetorheological fluid (η / τ_B), torque component ratio expected value (T_B / T_τ), input rotational speed ω.

4. Magnetic Field Analysis

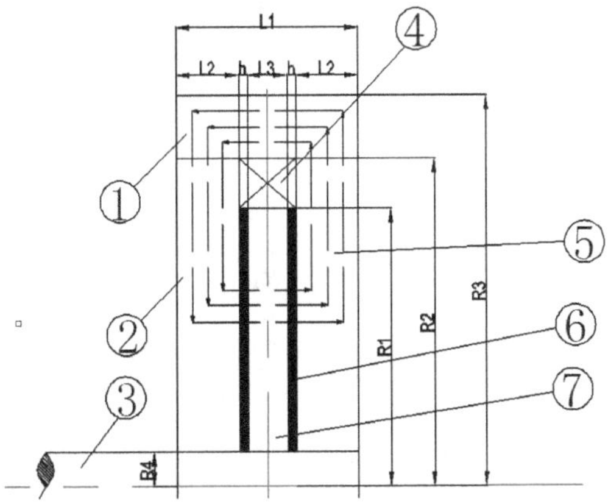

1: outer race 2: left shell 3: rotating shaft 4: excitation coil 5: right shell 6: magnetorheological fluid, 7: turntable

Fig 3. *Double disk type magnetorheological fluid magnetic circuit diagram.*

Double disk type magnetorheological fluid magnetic circuit diagram as shown in Fig. 3, the rotating shaft drives the rotation of the turntable. There is disk type MRF on both sides of turntable respectively, with thickness of h. When the turntable rotates, the excitation coil is energized, and imposes magnetic field on magnetorheological fluid and magnetorheological Fluid produces magnetorheological effect thus providing the load torque. The dotted line with an arrow in the Fig. is the trend of virtual magnetic induction line: outer race→ left shell→ magnetorheological fluid→turntable→ magnetorheological fluid→ right shell→ outer race forms a loop. Thus each disk needs to provide the load torque of half shaft torque, namely:

$$\frac{T}{2} = \frac{2}{3}\pi\tau_B R^3 + \frac{1}{2h}\pi\eta\omega R^4 \tag{10}$$

Reluctance calculation is similar to the calculation of the resistance, so it is also called the magnetic Ohm 's Law[11]. Reluctance calculation formula:

$$R_M = \frac{L}{\mu \cdot s} = \frac{L}{\mu_R \mu_0 s} \tag{11}$$

Where L is the circuit length, μ is the material permeability, s is the magnetic circuit cross section, μ_R is the material relative magnetic permeability, μ_0 is the permeability of vacuum, $\mu_0 = 4\pi \times 10^{-7} N \cdot A^{-2}$.

As is shown in formula (11), reluctance value is proportional to magnetic circuit length and inversely proportional to magnetic permeability and magnetic circuit cross section. Reluctance value can be changed by these three parameters in the design. Fig. 3 loop is series arrangement.

When magnetic circuit passes turntable 7, magnetorheological fluid 6,and outer race 1,substituted formula (11) and get:

$$R_{M7} = \frac{L_3}{\pi(R_1^2 - R_4^2)\mu_7} \tag{12}$$

$$R_{M6} = \frac{h}{\pi(R_1^2 - R_4^2)\mu_6} \times 2 \tag{13}$$

$$R_{M1} = \frac{L_1 - L_2}{\pi(R_3^2 - R_2^2)\mu_1} \tag{14}$$

When magnetic circuit passes left and right shell, because they are symmetrical, substituted formula (11) and get:

$$R_{M2} = \frac{\ln(\frac{R_2 + R_3}{2}) - \ln R_4}{2\pi L_2 \mu_2} \tag{15}$$

$$R_{M5} = R_{M2} \tag{16}$$

So the total reluctance of reluctance loop is:

$$R_M = R_{M1} + R_{M2} + R_{M5} + R_{M6} + R_{M7} \tag{17}$$

According to the Ampere circuital theorem in magnetic circuit[12], magnetomotive force of magnetic circuit:

$$V_m = \Phi R_M = NI \tag{18}$$

Here Φ is the magnetic flux, N is number of coil; I is the coil current.

When determining the material and structure parameters of the magnetic circuit, the total reluctance and action area under the corresponding magnetic induction intensity of magnetic circuit are uniquely determined, load torque and magnetic induction intensity exist one-to-one relationships.

5. Determination of Current I

After getting the magnetomotive force of magnetic circuit, we need a further determination of specific parameters of magnetic excitation coil, mainly including coil conductor diameter (coil current) and turns per coil. Magnetic excitation coil conductor diameter is determined by the formula[13]:

$$d = \sqrt{\frac{4\rho_c d_{cp} NI}{U}} \tag{19}$$

Where ρ_c is the resistivity of the wire material ($\Omega \cdot mm^2 \cdot m^{-1}$), d_{cp} is the coil average diameter(m), NI is the coil ampere turns, U is the supply voltage.

6. Motor Dynamometer Device Design

Take QABP-160M2A motor as example. Parameters are as follows:

Table 1. *QABP-160M2A technical data.*

Rated Power P(kW)	Rated Torque T(N·m)	Rated Rotational Speed n(r·min^{-1})	Max Torque/Rated Torque(times)
11	35	2930	3

We choose Tider MFR270/50 type magnetorheological fluid, whose technical data is as table 2. The constitutive equation is[14]:

$$\tau_B = 64.72 \times [1 - \exp(-1.63B)] \tag{20}$$

Table 2. *Magnetorheological Fluid Technical Data.*

Magneto-rheological Fluid Type	Relative Permeability	Viscosity at 20^{0}C(cP)	Magnetic Saturation Intensity (T)
Tider MFR270/50	8	0.27	1~2

Permeability material used for the magnetic structure(including left and right shell, outer race and turntable) around coil is A_3 steel whose relative permeability can increase from the initial several hundreds to thousands under different magnetic induction intensity, but it has no change in magnitude, so we ignore the change and choose relative permeability $\mu_{rA}=800$, magnetic saturation intensity B_A=2.14T[15].

Structure of magnetorheologicalfluid dynamometer is shown in Fig. 4.

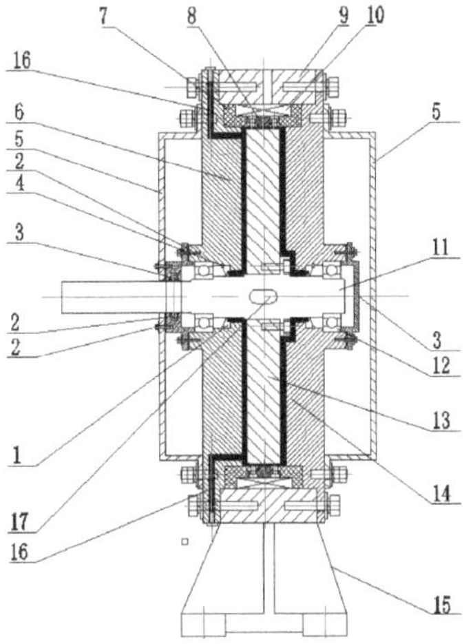

1: framework oil seals; 2: felt ring seal;3 bearing end cover;4: clip ring for hole;5:cooling water-jacket;6:left shell;7:coil support ;8:isolation magnetic ring;9:outer race;10:coil;11:rotatingshaft;12:bearing;13:turntable;14:right shell;15:base;16: magnetorheological fluid injection hole;17: A type key

***Fig 4.** Magnetorheological Fluid Dynamometer.*

Considering structure size against Fig. 3, magnetorheological fluid effective outer diameter is R_1=0.1m,coil outer diameter R_2=0.11m,rotating shaft diameter R_4=0.015m. To make full use of magnetorheological fluid, magnetorheological fluid should saturate first and from table 2, we chose magnetic saturation intensity limit B_M=1T, so the when magnetorheological fluid saturates, the magnetic flux is $\Phi=3.07\times10^{-2}Wb$. Because magnetic flux in every magnetic circuit loop is same, substitute data and get: R_3=0.13m, L_2=0.022m.

As for the turntable, because its section is the same as magnetorheological fluid and magnetic saturation intensity is bigger than magnetorheological fluid's, its thickness is ok if slightly bigger than magnetorheological fluid's. We chose magnetorheological fluid thickness h=0.0015m[16]and considering the strength requirement of turntable, its thickness chooses L_3=0.008mand substituted formula (12)~(16),we get reluctance of each part: R_{M7}=259H^{-1}, R_{M6}=9728H^{-1}, R_{M2}=R_{M5}=31565H^{-1}, R_{M1}=2179H^{-1}. Therefore, the total reluctance is: R_M=75296H^{-1}.

When the dynamometer works, the torque increases as the current increases, so the coil should have the ability to provide maximum torque current. From table 1, we know T_{max}=105N·mand the rotation speed is n=1000r·min^{-1}, namely $\omega=105rad\cdot s^{-1}$. According to formula (10)、(20) and (18),we can get: dynamic yield stress $\tau_B=21.96\times10^3\,p_a$, magnetic induction B=0.25T, magnetomotive force V_m=577.78A.

Coil wire is copper enameled wire with good conductivity, whose standard specification as table 3. Voltage is common voltage U=12V. From (18) and (19),we know conductor diameter d=0.865mm. According to the experience of each square millimeter of wire can pass through the 5A current, namely coil current I=2.9A. From (18), we get coil turns: N=199, the coil resistance $R_{coil}=4.19\Omega$.

Thus, we finish the calculation process of motor power →rotational speed →torque → current. As for the device structure size that has been known, if rotational speed measured by speed sensors, the power given by the output current, one-to-one correspondence between current and torque, by calculating the corresponding load torque calculated from current, we can get power motor of rotating parts.

***Table 3.** Copper Enameled Wire Specification.*

Copper Core Nominal Diameter (mm)	Copper Enameled Wire Max Diameter (mm)	Copper Core Cross Section (mm^2)	No. (cm)	Resistivity ($\Omega\cdot mm^2\cdot m^{-1}$)
0.83	0.89	0.5411	11.2	
0.86	0.92	0.5809	10.9	0.01851
0.9	0.96	0.6362	10.4	

Both engine and motor rotating parts have the character of decreasing and increasing torque, namely torque will change when rotational speed changes. As for the above device, we can increase or decrease the current to control the load torque, and further control the rotational speed of rotating parts.

7. Conclusion

(1) Put forward a new theoretical dynamometer design method, the calculation process of torque and reluctance during the design of magnetorheological fluid dynamometer and a determination method of current, turns, diameter and other parameters.

(2) Taking some motor as example into practice proves its feasibility of this method, enlarges dynamometer range and provides a theoretical basis for the design of magnetorheological fluid dynamometer.

(3) Magnetorheological fluid dynamometer design process still ha some problems, such as sealing, cooling and control problems. If these problems are solved, the design of the dynamometer will be more perfect.

Acknowledgements

This research was supported by the modern automobile service engineering subject platform construction in Shanghai City (A-0507-13-0226).

The first author Luo Yiping was born in 1966. He gained a master's degree of Industrial Engineering in Shanghai Jiao Tong University in 1987. Now he is a professor in Shanghai University of Engineering Science.

The second author Xu Biao was born in septemper.1990 in Anhui province. He got bachelor's degree of vehicle engineering in Anhui Polytechnic University in 2012 and now is a graduate student in Shanghai University of Engineering Science.

The third author Ren Hongjuan was born in 1978 in Shandong province. She gained a master's degree of Power Machinery and engineering in Shandong University in 2002. Now she is an associate professor in Shanghai University of Engineering Science.

References

[1] Meng Li, Jiangang Lv, Yong Wei. Study on the System of Tracked Vehicles Magnetorheological Fluid Damper Suspension[J]. Mechanical Design, 2004, 21(12): 52-55.

[2] Yanrong Yang, Huiyong Shan, Yong Wei. The Theoretical Analysis and Design of a Cylindrical Magnetorheological Brake [J]. Electromechanical Engineering Technology , 2005, 34(10): 15-16.

[3] Lin Zhang. Research and Design of Disk Type Magnetorheological Transmission Mechanism [J]. Mechanical Design, 2009, 26(1): 31-32.

[4] Bossis G, Mathis C, Mimouni Z. Magneto rheolgical suspensions[J]. Euro phys Lett, 1990, 11(2): 133-137.

[5] Kormann C, Laun H. M, Richter H J. MR Fluid with Nano-Sized Magnetic Particles Technology [J]. International Journal of Modern Physics, 1996, 10(23): 3167-3172.

[6] Cutillas S. Bossis G CebersA [J]. Vhys Rev E, 1998, 57(1): 804-811.

[7] Gopalswamy S, Linzell S M, Jones G L. MR fluid clutch with minimized reluctance [P]. USA:US Patent: 5896965,1999.

[8] Jianhua Ni, Zhiqian Zhang, Ke Zhang. A New Type of Magnetorheological Damper and Its Application in Semi-active Control of Vehicle Suspension [J]. Mechanical Science and Technology , 2004, 23(1): 4-6.

[9] O. Ashour, A. Craig. Magnetorheological Fluid: materials, characterization, and devices [J]. Int. Mater. Syst. struct, 1996, 7(2): 123-130.

[10] Jian Chang, Yunmin Yang, Xianghe Peng. Research on testing device for magnetorheological fluid properties [J]. Chinese Journal of scientific instrument, 2001, 22(4): 354-358.

[11] Wu Ai, Cheng Li. Circuit and Magnetic Circuit [M]. Wuhan City: Huazhong University of Science and Technology press , 2002.

[12] Canbin Liang, Guangrong Qin, Zhujiang Liang. Electromagnetics [M]. Beijing: Higher Education Press, 1980.

[13] Jinming Tian. Design and Application of Electromagnetic Clutch [M]. Suzhou: Jiangsu Science and Technology Publishing House, 1982.

[14] Kejun Jiang, Chengye Liu. Design and Performance Simulation of Magnetorheological Fluid Clutch [J]. Computer Sumulation, 2011, 28(8): 337-341.

[15] Jun Zheng. Magnetorheological Transmission Theory and Experimental Study [D]. Chongqing: Chongqing University, 2008.

[16] W. H. Li and, H. Du. Design and Experimental Evaluation of a Magnetorheological Brake [J]. The International Journal of Advanced Manufacturing Technology, 2003, 21(7): 508-515.

Application of fuzzy logic to multi-objective scheduling problems in robotic flexible assembly cells

Khalid Abd[1, 2, *], Kazem Abhary[1], Romeo Marian[1]

[1]School of Engineering, University of South Australia, Mawson Lakes 5095, South Australia
[2]School of Industrial Engineering, University of Technology, Baghdad, Iraq

Email address:
abdkk001@mymail.unisa.edu.au(K. Abd), kazem.abhary@unisa.edu.au(K. Abhary), romeo.marian@unisa.edu.au(R. Marian)

Abstract: This paper is aimed at developing a methodology to solve a multi-objective problem in robotic flexible assembly cells. The proposed methodology is based on three main steps: (1) scheduling of the RFACs using different common rules, (2) normalisation of the scheduling outcomes, and (3) selection of the optimal scheduling rules, using a fuzzy inference system. In this paper, four rules, namely short processing time, long processing time, earlier due date and random, are examined. Four objectives are considered simultaneously: scheduling length, total transportation time, utilisation rate and workload rate. A realistic case study is provided for demonstrating applicability of the suggested methodology. The results show that the methodology is practical and works in RFACs settings.

Keywords: Assembly Cells, Scheduling Rules, Fuzzy Logic, Robotics

1. Introduction

Flexible manufacturing systems have attracted significant attention in recent years, due to their flexibility and dexterity in dealing with unexpected events. One class of such systems is called robotic flexible assembly cells (RFACs). RFACs are highly modern systems, structured with industrial robot(s), assembly stations and an automated material handling system, all monitored by computer numerical control [1-3].

The design of RFACs with multi robots leads to increased productivity in a shorter cycle time and with lower production costs [4]. However, there are certain difficulties that have arisen with this design concept. For example, more than one robot operating simultaneously in the same work environment requires a complex control system to prevent collisions between robots [5], and also to prevent deadlock problems [6]. Moreover, industrial robots must be employed as effectively as possible due to the high cost of the robots [4]. To overcome the above difficulties, efficient scheduling of RFACs is required.

Few studies have been devoted to scheduling RFACs [7]. These studies may be categorised according to the approaches adopted. In the first category are those studies which applied heuristic approaches to solve scheduling problems such as Lee and Lee [6], Nof and Drezner [8], Lin et al. [9], Pelagagge et al. [10], Sawik [11], Jiang et al. [12] and Rabinowitz et al. [13]. The studies in the second category investigated simulation as an approach to scheduling RFACs, for instance, Gilbert et al. [14], Hsu and Fu [15] and Barral et al. [16]. There are only two studies in the third category, by Brussel et al. [17] and Dell Valle and Camacho [18], which implemented expert systems approaches to solve scheduling problems. The major limitation of the above studies is that they concentrated on assembly of only one type of product at a time.

In our previous study [19], scheduling RFACs for concurrent assembly of multi-products was proposed. Different scheduling rules were implemented. The results showed that making a decision on the best scheduling rule based on multi criteria is a considerably complex task. This study does not address how to select the most suitable scheduling rule.

The problem of rule selection implies that a set of rules should be evaluated and ranked according to different criteria, which are conflicting with each other. Accordingly, rule selection is considered as a multi criteria decision-making (MCDM) problem [20]. The aim of this paper is to propose a methodology to select the best scheduling rule for RFACs using Fuzzy logic.

2. Fuzzy Logic

Fuzzy Logic (FL) was first introduced by Zadeh in 1965 [21]. FL is a nonlinear mapping of an input data vector into a scalar output. A fuzzy inference system (FIS) consists of four components [22, 23]:

- Fuzzification: In this interface, the real world variables (crisp input data) are converting into linguistic variables (fuzzy values). This step can be done using the membership functions of input variables.
- Knowledge base: In this component, the membership functions are determined. These membership functions reflect a human reasoning mechanism.
- Inference engine: In this component, fuzzy input values convert to fuzzy output using IF-THEN type fuzzy rules. These rules reflect human experts' knowledge of the system. The number of decision rules depends on the number of input variables and their linguistic values.
- Defuzzification: In this interface, the fuzzy outputs are translates into a crisp value. The defuzzification process can be achieved using the membership functions of output variable. Several methods have been proposed for defuzzification process. The well-known method for defuzzification process, named centre of gravity (COG) [23, 24] is used to transform the fuzzy inference output into non-fuzzy value.

The important component in a FIS is the knowledge base. This component stores both the membership functions and the IF-THEN rules base provided by experts. Three steps, linguistic variables, membership functions and fuzzy rules, are prepared to establish a knowledge base [25, 26]. The next sub section will describe these three steps.

2.1. Linguistic Variables

A linguistic variable is the procedure to describe variables in terms of words instead of their values. A linguistic variable consists of a set of terms called linguistic terms, denoted by *T*. For example, if processing time is interpreted as a linguistic variable, terms such as "*Low*", "*Medium*" and "*High*" are used in a real industry context. Hence, a linguistic variable of processing time could be *T* [processing time] = [*Low, Medium, High*].

2.2. Membership Functions

A membership function (MF) embodies a fuzzy set Ã graphically. The values of the membership functions are between 0 and 1, denoted by μÃ (x) where x is an element of Ã; these values are called degrees of membership.

2.3. Fuzzy Rule

A fuzzy rule is structured to control the output variable. These rules can be provided by experts or may be extracted from numerical data. A fuzzy rule has two parts, the antecedent and the consequent: IF <antecedent> THEN <consequent>. For instance, IF *x* is *A* THEN *y* is *B*; where *x* and *y* are variables and *A* and *B* are linguistic variables.

3. Proposed Methodology

In this section, the architecture of the proposed methodology is presented. The methodology has three main steps: scheduling, normalisation and evaluation, as shown in Figure 1. The next sub sections will present these steps in detail.

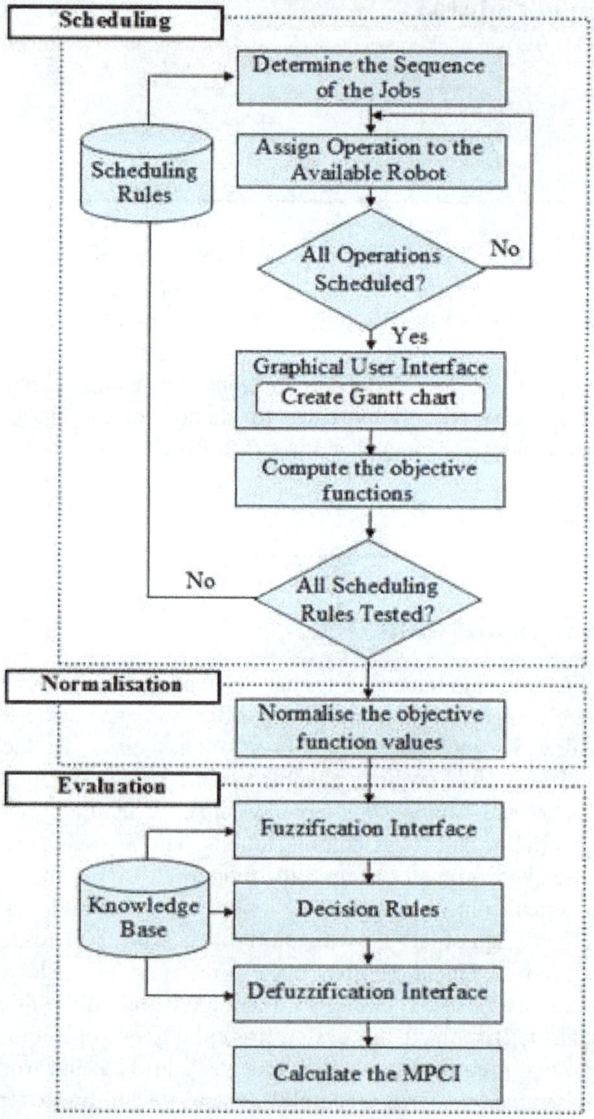

Figure 1.*Proposed methodology for scheduling problems in RFACs*

3.1. Scheduling

In scheduling RFACs, when a robot becomes free and more than one job is waiting for processing, the jobs will be scheduled, from the highest priority to the lowest priority. This can be done using scheduling rules. These rules are used to generate the sequence of job flow to the system. The following is a list of the common rules used in this study.

- Short Processing Time *(SPT)*: select job with minimum processing time first.
- Long Processing Time *(LPT)*: select job with maximum processing time first.
- Random *(RAND)*: jobs are sequenced randomly.
- Earlier Due Date *(EDD)*: jobs are sequenced according to their due dates.

Four objective functions are considered to evaluate the scheduling. The mathematical expressions of these objectives were formulated in our previous study [19].

- Scheduling length (T_{max}): One of the common objective functions in scheduling is called scheduling length or cycle time. T_{max} represents the maximum total completion time performed by the robot. Total Transportation Time (T_{tran}): The sum time required to travel the robot between cell resources to finish assembly of products. T_{tran} aims at measuring the amount of movement of each robot in RFACs, during one cycle time.
- Utilisation rate (U_R): Another important objective function, that gives a clear perception as to whether the robots are used efficiently.
- Workload rate (W_R): W_R is a measure of RFAC balance.

In this study, six assumptions are considered. First, the optimum assembly sequence of each product is given in advance. Second, each product uses some or all of the cell resources. Third, each robot can perform only one task at a time. Fourth, no interruptions such as resources breakdown occur in the cell. Fifth, the processing time of each task is deterministic and is known in advance. Sixth, the set-up times are assumed done when the cell is off-line, so not considered.

3.1. Normalisation of the Objective Values

This is done by converting the objectives functions values to range between 0 and 1. This process is called normalisation. In this study, two objective functions, T_{max} and T_{tran}, are to be minimised, and the other objective functions, U_R and W_R, are to be maximised. Therefore, the normalisation can be done using the following two equations, where $\square_{i(k)}$ denotes the value of objective function k in scheduling rule i. The equation (1) is for the objective functions to be maximised, while the equation (2) is for the objective functions to be minimised.

$$\mu_k^i = \frac{\left[\text{Max}_{i(k)} - {}_{i(k)}\right]}{\left[\text{Max}_{i(k)} - \text{Min}_{i(k)}\right]}, 0 \leq \mu_k^i \leq 1 \qquad (1)$$

According to the equation (1), it can be seen that the rule i with minimum value of objective function k has a normalised value of 1 and the rule i with maximum value of objective function k has a normalised value equal to 0.

$$\mu_k^i = \frac{\left[{}_{i(k)} - \text{Min}_{i(k)}\right]}{\left[\text{Max}_{i(k)} - \text{Min}_{i(k)}\right]}, 0 \leq \mu_k^i \leq 1 \qquad (2)$$

From the equation (2), it can be concluded that the rule i with minimum value of objective function k has a normalised value of 0 and the rule i with maximum value of objective function k has a normalised value of 1.

Therefore, the scheduling rule with low T_{max}, low T_{tran}, high U_R and high W_R will take high rank.

3.3. Evaluation of Scheduling Rules using FIS

In order to find out the optimum scheduling rule for multiobjective problems in RFACs, FIS is used. For a single objective, the optimum scheduling rule is the one having the highest normalisation value. Multi-objective function optimisation is not as straightforward as that of a single objective function optimisation. To overcome this problem, a multiple performance characteristics index (MPCI) based FIS is developed to derive the optimal solution.

In this study, FIS is implemented using the MATLAB fuzzy toolbox. The fuzzy logic toolbox consists of five graphical user interface tools (GUIs) for building, editing and observing any fuzzy inference system [27, 28]. These tools are: the fuzzy inference system (FIS) editor, the membership function editor, the rule editor, the rule viewer, and the surface viewer, as shown in Figure 2. The GUIs are dynamically connected, and the altering of any GUI can affect the other GUIs.

Figure 2. *Fuzzy inference system and its integral components in MATLAB software*

3.3.1. FIS Editor

The FIS editor handles the information related to the variables of inputs and output, such as variables' names and their numbers. In the present study, The FIS contains four input parameters and one output parameter. The input parameters are the normalisation of T_{max}, T_{tran}, U_R and W_R, while MPCI is the output parameter.

The T_{max}, T_{tran}, U_R and W_R are break down into a set of linguistic values for the inputs: low (L), medium (M), and high (H); while the output is set into seven linguistic values: tiny (T), very small (VS), small (S), medium (M), large (L), very large (VL) and huge (H). Table 1 shows the different linguistic values of the inputs/output and their numerical range.

3.3.2. Membership Function Editor

The membership function editor is used to construct the shapes of all the input/output parameters. There are different types of membership functions' shapes such as triangular, trapezoidal, Gaussian, singleton, etc. The triangular shape is the common membership functions shape and a powerful way to approach the convex function [29]. In this study, the membership functions for inputs/output are plotted using the triangular shape, shown in Figures 3 and 4 respectively.

Table 1. *Input and output variables with their fuzzy values.*

System variables	Linguistic variables	Linguistic Value	Numerical Range
Inputs	T_{max}, T_{tran}, U_R and W_R	Low	[0– 0.25]
		Medium	[0.25–0.75]
		High	[0.75–1]
Outputs	MPCI	Tiny	[0– 0.167]
		Very Small	[0– 0.334]
		Small	[0.167 – 0.5]
		Medium	[0.334 – 0.66]
		Large	[0.5 – 0.834]
		Very Large	[0.667 – 1]
		Huge	[0.834 –1]

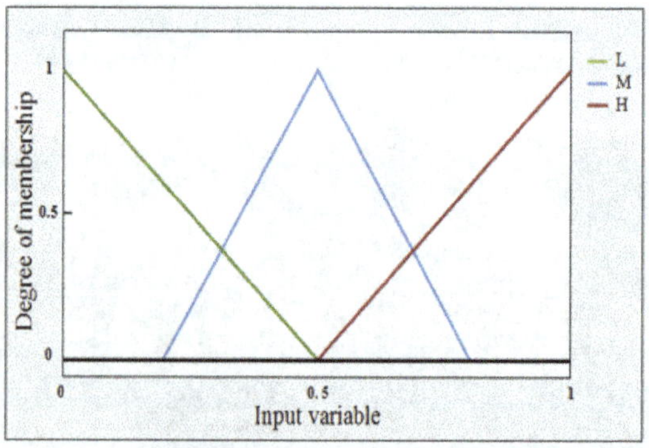

Figure 3. *Membership functions for fuzzy input variables*

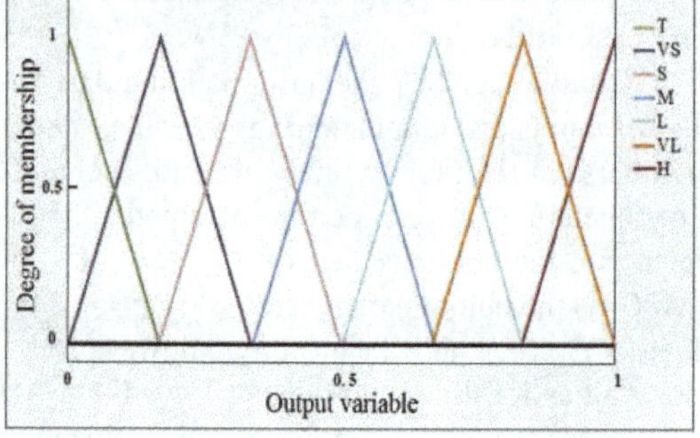

Figure 4. *Membership functions for fuzzy output variable*

3.3.3. Rule Editor

The rule editor is for editing the list of fuzzy rules that are used to control the output variable. The fuzzy rule is constructed based on the number of linguistic variables for inputs/output. In the present study, to control the output parameter (MPCI), fuzzy rules are structured. Fuzzy rules are derived directly based on the formula (n^m), where $n$and m denote input parameters and their linguistic values. Thus, the number of fuzzy rules is $4^3 = 64$. Each rule is mathematically evaluated through a process named implication. In this study, Mamdani implication is applied [30].

The generic form of a fuzzy rule can be stated in the following form: IF (T_{max} is ■) and (T_{tran} is ■) and (U_R is ■) and (W_R is ■) THEN (*MPCI* is ■). The black boxes represent the linguistic variables for each of the fuzzy variables. Example of the fuzzy rules derived is shown in the examples below.

1. IF (T_{max} *is Low*) and (T_{tran} *is Low*) and (U_R *is High*) and (W_R *is High*) THEN (*MPCI is Huge*).
2. IF (T_{max} *is Medium*) and (T_{tran} *is Low*) and (U_R *is High*) and (W_R *is High*) THEN (*MPCI is Very Large*).
3. IF (T_{max} *is High*) and (T_{tran} *is Low*) and (U_R *is High*) and (W_R *is High*) THEN (*MPCI is Large*).

.

.

64. IF (T_{max} *is High*) and (T_{tran} *is High*) and (U_R *is Low*) and (W_R *is Low*) THEN (*MPCI is Tiny*).

3.3.4. Surface Viewer

The surface viewer allows the user to visualize the relation between input fuzzy variables and the output of a fuzzy system in a three-dimensional graph, the X-axis and Y-axis in the 3D graph represent any two selected input variables, and the Z-axis represents the output of a fuzzy system. In this study, since the number of input variables is four, the number of generated 3D graphs is six.

3.3.5. Rule Viewer

The rule viewer displays a graphical representation of the values of the input variables and the output of a fuzzy system, through all the fuzzy rules.

4. Case Study

The illustration of the proposed approach is demonstrated using a realistic case study. The multi robot assembly cell consist of two robots (R_1 and R_2) that can use a number of tools that can be changed in a tool magazine (GC), assembly stations (AS_1, AS_2, AS_3, AS_4& AS_5) where components are assembled, transfer table (TT) to transfer partial assemblies from one robot to another. There are also two conveyors. The first one (IC) supplies components to the cell and the second one (OC) is for conveying out a final product when assembly processes are completed, as shown in Figure 5.

Three constraints have been taken into account. First,

robot arms cannot move from one place to another directly. The reason for this is to avoid collision with the other robot arms. This is achieved by assigning control points $\{C_1, C_2, \dots, C_4\}$ to simplify path planning and avoid collision. Second, to prevent collisions between robots in a shared area, more than one robot cannot access the same resource simultaneously. Third, to fetch and assemble parts, the hand of each robot should be equipped with the right tool; however, a specific tool may not be available for the two robots concurrently, due to the restricted number of available tools [19].

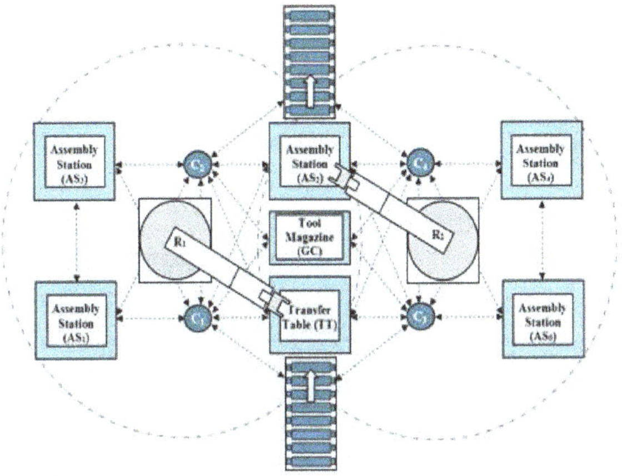

Figure 5. Robotic flexible assembly cells

Assume that three sizes of supply tanks are to be assembled in the RFACs presented as shown in Figure 6. The tanks are, essentially, chambers of constant level which supply, through gravity, liquid fuel for burners. They are composed of 13, 15 and 31 parts, respectively.

For manipulation within the cell, the tanks are loaded on a pallet that requires a gripper G1 (to be used for T_1, T_2 and T_3), whereas for the lids, a second pallet is utilised, requiring another gripper G2 (for L_1, L_2 and L_3). The grippers are interchangeable. Each robot can work with either G1 or G2, as required. The times needed to perform `tool change operations` are shown in Table 3. Table 4 shows the time required to move the robot between two positions in the cell.

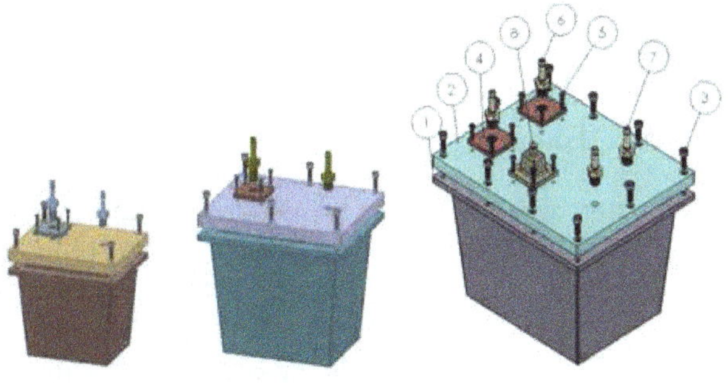

Figure 6. Exploded view of supply tanks 11, 21 and 51

Further, assume that RFACs do not have to be stopped to load the AS with consumables like fasteners, parts and sealants (a realistic assumption for modern assembly machines with magazines of parts and components). The Assembly stations in the RFACs are dedicated to the specific assembly operations, as shown in Table 2. This table also gives details of description for each assembly operations.

Table 2. Assembly operations requirements

Description	Station	Operation times		
		Tank 11	Tank21	Tank51
Sealant on Tank	AS_1	5	10	18
Assemble Lid	AS_2	5	5	5
Insert screws M6x16	AS_2	8	12	20
Sealant on Lid	AS_1	5	5	10
Fit the Level Control unit	AS_3	5	5	10
Fit the Safety Valve	AS_4	0	0	5
Insert screws M3x12	AS_2	8	8	24
Assemble Inlet & Outlet	AS_5	10	10	20
Total processing time (Sec.)		46	55	112
Due date (Sec.)		250	200	300

Table 3. Tool change requirements

Tool name	Number of available tools	Part assignment	Tool change time (s)
Gripper 1	2	T_1, T_2, T_3	3
Gripper 2	1	L_1, L_2, L_3	3

Table 4. Transportation time for robots between cell resources

Path description	Position	Time (s)
Robot move from resource to control point	AS1,2,3/GC/TT→C1,2 AS2,4,5/GC/TT→ C3,4	1
Robot move from control point to resource	C1,2→AS1,2,3/GC/TT C3,4→AS2,4,5/GC/TT	2
Robot move between control point and conveyor	C1, C3←→IC C2, C4←→OC	1.5
Robot move between two control points	C1←→C2 C3←→C4	1
Robot move directly from station to another	AS1←→AS3 AS4←→AS5	2

5. Results and Discussion

In order to examine the effectiveness of the proposed methodology, four experiments are executed. Each experiment is performed with different scheduling rules. These rules are used to generate the sequence of job flow to the RFACs. Table 5 shows the list of scheduling rules adopted in this study.

Table 5. List of dispatching rules and the priority of the jobs

NO	Scheduling Rule	Sequence
A	Short Processing Time (SPT)	$Tank_{11}$→ $Tank_{21}$ →$Tank_{51}$
B	Long Processing Time (LPT)	$Tank_{51}$→ $Tank_{21}$ → $Tank_{11}$
C	Due date (DD)	$Tank_{21}$→ $Tank_{11}$→ $Tank_{51}$
D	Random (RAND)	$Tank_{11}$ → $Tank_{51}$→ $Tank_{21}$

Figure 7 illustrates the Gantt chart of the final schedule produced by different rules. The results of the overall objective functions of the four scheduling rules are presented in Figure 8. Figure 8(a) shows the T_{max} and T_{tran} results of the scheduling rules. From this figure it can be seen that the *SPT* and *RAND* obtain the best results for minimising the T_{tran} and T_{max} respectively compared with the other scheduling rules. *LPT* and *EDD* are the worst in minimising the T_{tran} and T_{max} respectively.

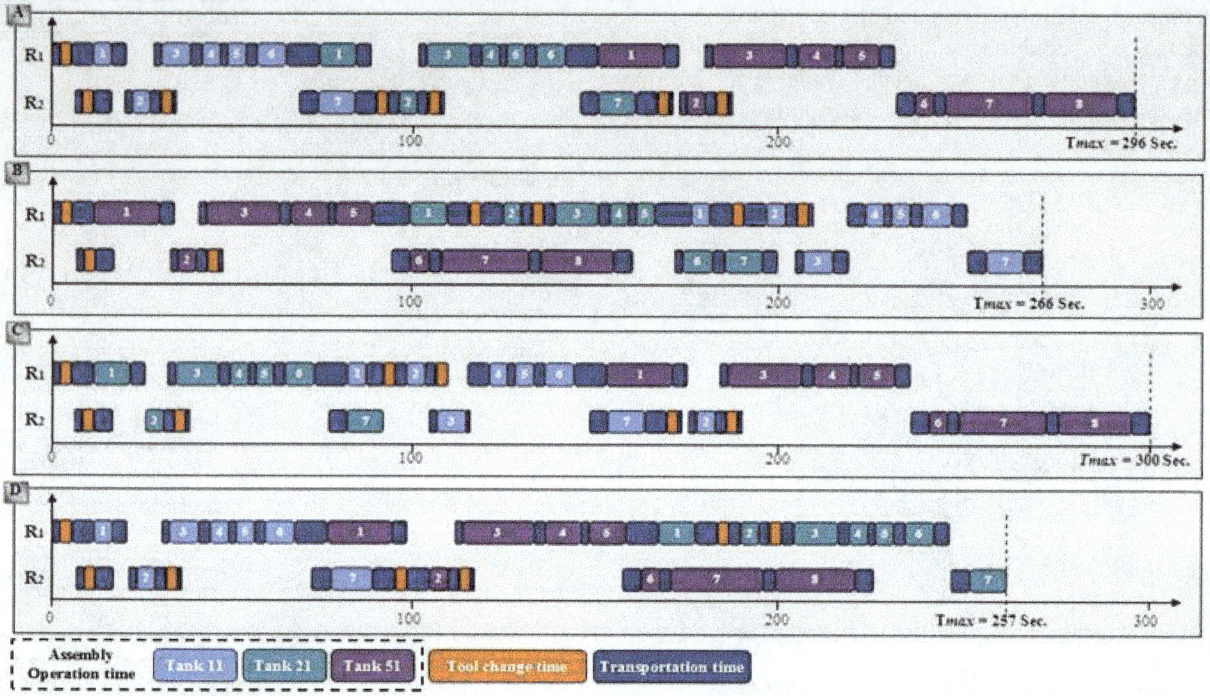

Figure 7. *Gantt chart of experimental studies*

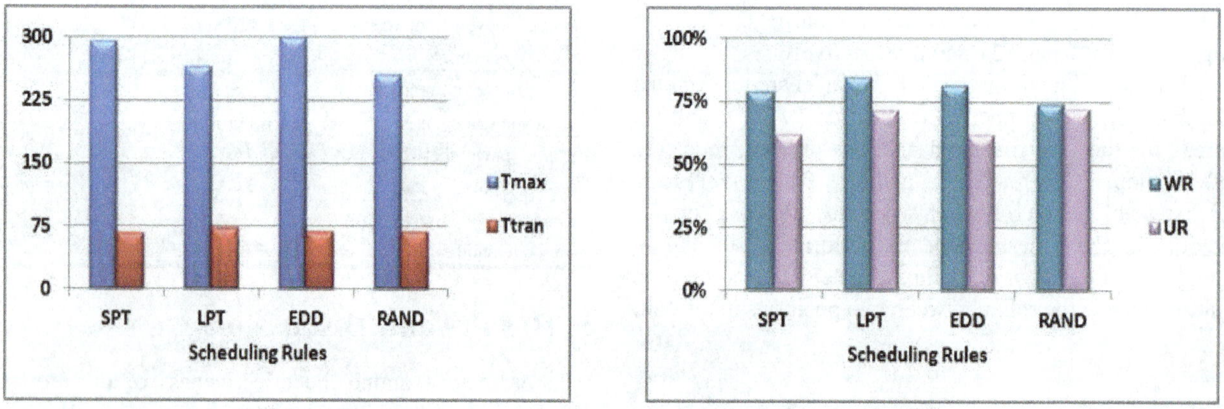

(a) (b)

Figure 8. *Experimental results for scheduling rules*

Figure 8(b) shows the U_R and W_R results of scheduling rules. In this figure, *LPT* emerges as the best rule among all four scheduling rules for maximising the W_R; EDD ranks second, and still obtains good results among all the other selected rules from the previous literature. RAND appears to be the worst rule for maximising the W_R. RAND emerges as the best rule among all four rules for maximising the U_R, followed by LPT, EDD and SPT. Depending on the previous results, making a decision of the best scheduling rule based on one objective function is simple; nevertheless determination of the optimal scheduling through consideration of multi objective functions is a considerably more complex task. The next step of the proposed methodology is normalisation of the objective functions' values. The ranges of values for the four objective functions, T_{max}, T_{tran}, U_R and W_R are all different. To avoid the different ranges, the values must be normalised to values between 0 and 1. In this case, since the objective of T_{max} and T_{tran} is minimising, and the objective of U_R and W_R is maximising, therefore the normalisation is determined using the equations 1 and 2. Figure 9 shows the normalisation values of the experimental results.

After the objective functions' values are normalised, MPCI is calculated to derive the optimal scheduling rule. The MPCI calculation can be done using FIS. The fuzzy logic toolbox in MATLAB is used to construct the FIS of the MPCI. In this study, since the number of input parameters is four, the number of generated 3D graphs is six. Figure10 illustrates one example of a 3D graph. In this Figure, the MPCI resulting from the interaction of T_{max} and T_{tran} is shown. It can be seen that the low T_{max} and T_{tran} values give a high score of MPCI. Moreover, it can be concluded that the T_{max} has a higher influence than the T_{tran} on the MPCI.

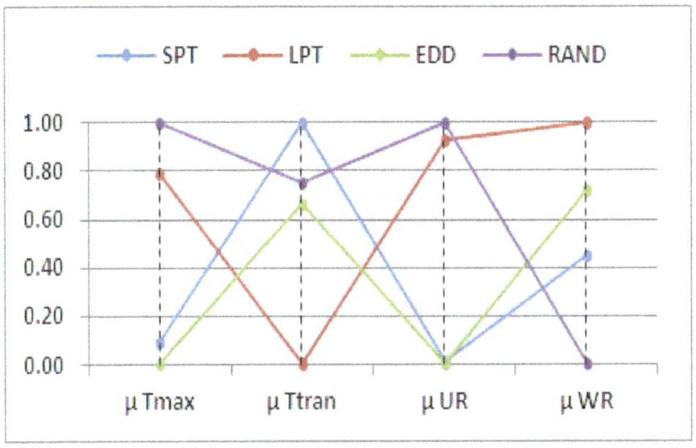

Figure 9. *Normalisation of the objective functions' values*

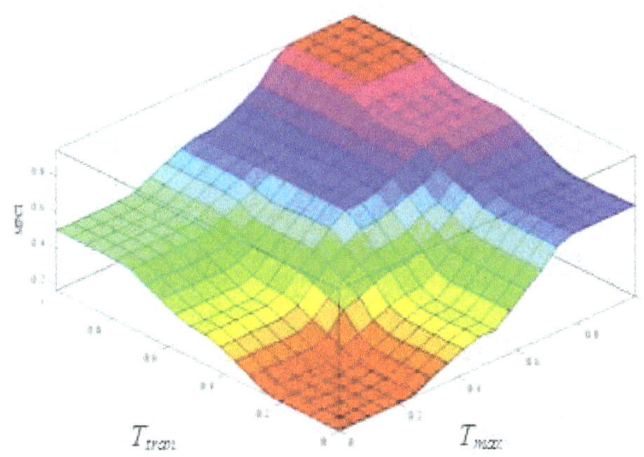

Figure 10. *Surface analysis between inputs/output combinations*

The rule viewer, which displays a graphical representation of the values of the input parameters and the output of a fuzzy system through all the fuzzy rules, is shown as an example in Figure 11. This figure shows the rule viewer for MPCI, which can accept any value of four input parameters: T_{max}, T_{tran}, U_R and W_R. The output (MPCI) in this Figure can be interpreted easily, as for example, as in the following: IF T_{max} is (0.09), T_{tran} is (1.00), the U_R is (0.01) and W_R is (0.45) THEN MCPI will be (0.36).

Table 6 shows the MPCI values corresponding to each experimental run obtained by using the FIS. LPT gives the highest MPCI value among the four rules; *EDD* is the worst one on this numerical example. The final ranking of the scheduling rules is *LPT – RAND – SPT – EDD* in descending order of preference.

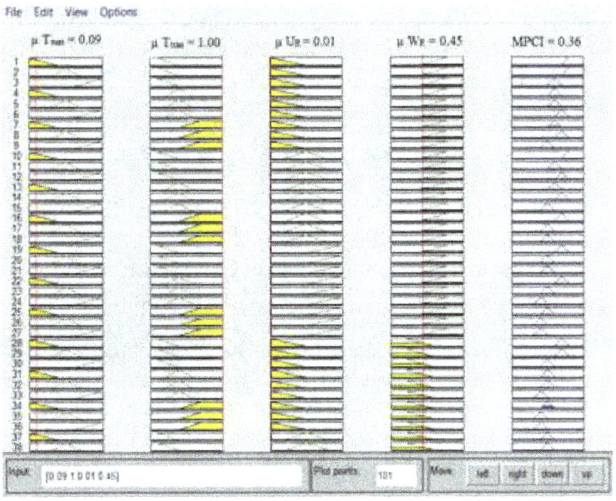

Figure 11. *Final output of fuzzy rules*

Table 6. *List of dispatching rules and the priority of the jobs*

Scheduling Rule	$\mu\, T_{max}$	$\mu\, T_{tran}$	$\mu\, U_R$	$\mu\, W_R$	MPCI
SPT	0.09	1.00	0.01	0.45	0.36
LPT	0.79	0.00	0.93	1.00	0.71
EDD	0.00	0.67	0.00	0.72	0.33
RAND	1.00	0.75	1.00	0.00	0.61

6. Conclusion

This paper has dealt with the problem of scheduling RFACs with consideration of assemble multi-product, under different experiments. These experiments were performed using four scheduling rules: *SPT*, *LPT*, *EDD* and *RAND*. The scheduling results showed that the decision making of selecting the best rule based on one objective function is a simple way; nevertheless determination of the optimal scheduling rule through consideration of multi-objective functions is a considerably more complex task. Consequently, in this paper, FIS is developed to select the optimal rule for scheduling RFACs; to minimise the scheduling length (T_{max}) and total transportation Time (T_{tran}); and to maximise of the utilisation rate (U_R) and workload rate (W_R). The FIS results showed that fuzzy logic is a powerful technique and easy to use for handling the multi objective optimisation problem. The proposed methodology is implemented using only four scheduling rules. This study could be extended to include other types of rules such as critical ratio (CR) and minimise slack time (MST). A possible extension would be development of a new rule for scheduling RFACs, by combine all input variables such as processing time, due date and batch size in one rule; and comparing the results that are obtained by the new rule with the common rules.

References

[1] S. Manivannan, "Robotic collision avoidance in a flexible

assembly cell using a dynamic knowledge base", IEEE transactions on systems, man, and cybernetics, vol. 23, pp. 766-782, 1993.

[2] T. Sawik, "Production planning and scheduling in flexible assembly systems" Springer -Verlag, Berlin, 1999.

[3] R. M. Marian, A. Kargas, L. H. S. Luong, and K. Abhary, "A framework to planning robotic flexible assembly cells", presented at the 32nd International Conference on Computers and Industrial Engineering, Limerick, Ireland, 2003.

[4] E. K. Xidias, P. T. Zacharia, and N. A. Aspragathos, "Time optimal task scheduling for two-robotic manipulators operating in a three-dimensional environments", Journal of Systems and Control Engineering, vol. 224, pp. 845-855, 2010.

[5] S. Y. Nof and J. Chen, "Assembly and disassembly: an overview and framework for cooperation requirement planning with conflict resolution", Journal of Intelligent and Robotic Systems vol. 37, pp. 307–320, 2003.

[6] J.-K. Lee and T.-E. Lee, "Automata-based supervisory control logic design for a multi-robot assembly cell", International Journal Computer Integrated Manufacturing, vol. 15, pp. 319-334, 2002.

[7] K. Abd, K. Abhary, and R. Marian, "A scheduling framework for robotic flexible assembly cells", AIJSTPME-Asian International Journal of Science and Technology in Production and Manufacturing Engineering, vol. 4, pp. 30-37, 2011.

[8] S. Y. Nof and Z. Drezner, "The multiple-robot assembly plan problem", Journal of Intelligent and Robotic Systems vol. 7, pp. 57-71, 1993.

[9] H. C. Lin, P. J. Egbelu, and C. T. Wu, "A two-robot printed circuit board assembly system", International Journal of Computer Integrated Manufacturing, vol. 8, 1995.

[10] P. M. Pelagagge, G. Cardarelli, and M. Palumbo, "Design criteria for cooperating robots assembly cells", Journal of Manufacturing Systems, vol. 14, pp. 219-229, 1995.

[11] T. Sawik, "Integer programming models for the design and balancing of flexible assembly systems", Mathematical and Computer Modelling vol. 21, pp. 1-12, 1995.

[12] K. Jiang, L. D. Seneviratne, and S. W. E. Earles, "Scheduling and compression for a multiple robot assembly work cell", production Planning & Control, vol. 9, pp. 143-154, 1998.

[13] G. Rabinowitz, A. Mehrez, and S. Samaddar., "A scheduling model for multi-robot assembly cells", International Journal of Flexible Manufacturing Systems vol. 3, pp. 149-180 1991.

[14] P. R. Glibert, D. Coupez, Y. M. Peng, and A. Delchambre, "Scheduling of a multi-robot assembly cell", Computer Integrated Manufacturing Systems, vol. 3, pp. 236-245, 1990.

[15] H. Hsu and L. C. Fu, "Fully automated robotic assembly cell: scheduling and simulation", presented at the IEEE International Conference on Robotics and Automation, National Taiwan University, 1995.

[16] D. Barral, J.-P. Perrin, and E. Dombre, "Flexible agent-based robotic assembly cell", New Mexico, 1997.

[17] H. Van Brussel, F. Cottrez, and P. Valckenaers, "SESFAC: A scheduling expert system for flexible assembly cell", Annals of The CIRP, vol. 39, pp. 19-23, 1990.

[18] C. Del Valle and E. F. Camacho, "Automatic assembly task assignment for a multi robot environment", Control engineering practice, vol. 4, pp. 915-921, 1996.

[19] K. Abd, K. Abhary, and R. Marian, "Scheduling and performance evaluation of robotic flexible assembly cells under different dispatching rules", Advances in Mechanical Engineering, vol. 1, 2011.

[20] K. Abd, K. Abhary, and R. Marian, "An MCDM Approach to Selection Scheduling Rule in Robotic Flexible Assembly Cells", presented at the International Conference on Mechanical, Aeronautical and Manufacturing Engineering, Venice, Italy, 2011.

[21] L. A. Zadeh, "Fuzzy sets", Information and Control, vol. 8, pp. 338-353, 1965.

[22] L. A. Zadeh, "Fuzzy-algorithm approach to the definition of complex or imprecise concept", International Journal of Manachines Studies, vol. 8, pp. 249–291, 1976.

[23] J. M. Mendel, "Fuzzy logic systems for engineering: A tutorial", Proceedings of the IEEE 83 pp. 345–377, 1992.

[24] J. Yen, R. Langari, "Fuzzy logic intelligence, Control, and Information", Prentice Hall Publishing Company, 1999.

[25] K. Abd, K. Abhary, and R. Marian, "Intelligent modeling of scheduling robotic flexible assembly cells using fuzzy logic", presented at the 12th WSEAS International Conference on Robotics, Control and Manufacturing Technology, Rovaniemi, Finland, 2012.

[26] K. Abd, K. Abhary, and R. Marian, "Efficient scheduling rule for robotic flexible assembly cells based on fuzzy approach", presented at the 45th CIRP Conference on Manufacturing Systems, 2102.

[27] Mathworks 2009. "Fuzzy logic toolbox user's guide", http://www.mathworks.com/access/helpdesk/help/pdf_doc/fuzzy/fuzzy.pdf.

[28] S. N., Sivanandam, S. Sumathi, and S. N. Deepa, "Introduction to fuzzy logic using MATLAB", New York, Springer, 2007.

[29] W. Pedrycz, "Why triangular membership functions?" Fuzzy Sets and Systems pp. 21–30, 1994.

[30] G. Klir and B. Yuan, "Fuzzy sets and fuzzy logic: theory and applications", Prentice-Hall, Englewood Cliffs, Upper Saddle River, NJ 1995.

25

Interfacing with adaptive systems

René Ahn, Emilia Barakova, Loe Feijs, Mathias Funk, Jun Hu, Matthias Rauterberg

Designed Intelligence Group, Industrial Design Department, Eindhoven University of Technology, Eindhoven, the Netherlands

Email address:

r.ahn@tue.nl (R. Ahn), e.barakova@tue.nl (E. Barakova), l.m.g.feijs@tue.nl (L. Feijs), m.funk@tue.nl (M. Funk), j.hu@tue.nl (J. Hu), g.w.m.rauterberg@tue.nl (M. Rauterberg)

Abstract: We cast a design view on the interactions that occur when humans face (interconnected) adaptive systems. As humans are also adaptive, the combined behavior of such systems and humans can exhibit various phenomena that are especially of concern to designers of adaptive systems dealing with the inherent complexity of systems, systems' interfaces, interconnectivity, and other design factors. Based on examples of interactions between humans and systems at different levels of complexity, we propose a hierarchical taxonomy of increasingly complex challenges that system engineers will encounter when designing adaptive systems. Among adaptive systems, the taxonomy distinguishes *closed* and *open* systems, embodying processes that are *unaware* or *aware*, and finally, *friendly* and *hostile*. This taxonomy can be of use in designing these systems and their interfaces, as it helps to categorize the information needs of users. In fact, systems at various levels in the hierarchy need to offer certain cognitive affordances for users to operate these systems successfully. We illustrate how complex the information needs of users in these different situations can be, and formulate emerging design research questions. These could be of particular interest to designers who create intelligent systems, products, and related services in a societal context.

Keywords: Adaptive Systems, Human Behavior, Interfaces, Systems Design, Taxonomy

1. Introduction

Scientific and technological progress enables us to create new types of products, systems, and services. In particular, miniaturization, affordability, and the advancement in design of interactive and intelligent systems extends the range of situations and applications where we can now create products or systems that can react on changes in their environment without explicit prompting from their direct users. These systems often have a repertoire of actuators and sensors through which they perceive certain characteristics of their environment, and a 'policy' which makes them 'choose' an 'appropriate' action adapted to a given situation.

From a user point of view, these systems act independently, adapting to changes in the environment or in the situation, e.g., for traffic situations as will be described below. When such systems are introduced in a given environment, they may bring positive changes and even save users time and trouble. But they also will introduce changes in this environment, changes to which the user needs to adapt, for instance, uncertainty, ambiguous signals, and behavior that changes and adapts over time. Aarts and de Ruyter propose and rank (from low to high adaptivity) the following ambient system intelligence levels: (1) context aware, (2) personalized, (3) adaptive, and (4) anticipatory [1]. Although this specific level system is useful for intelligent environments, we try to address a general niveau by following the definition of complex adaptive systems [2]: A complex adaptive system consists of inhomogeneous, interacting adaptive agents; adaptive means capable of learning; an emergent property of a complex adaptive system is a property of the system as a whole which does not exist at the individual elements (agents) level.

As the number of systems keeps growing, their effect on the environment accumulates, and the user's position in his environment changes, first gradually, but later more extensively [3]. This happens because the user will, eventually, interact more and more with his environment through these systems, so his interaction with his environment becomes more and more indirect and mediated through electronic systems. The range of the environment tends to expand, and the interaction with the environment may become more varied, wide-ranging, sophisticated, and extensive, but also more incomprehensible, ambiguous, conditional, complex, and artificial.

We assume that there are clear trends to be observed here, and that designers should investigate these trends closely to achieve a better fundamental understanding of the various trade-offs and design issues that play a role in them. We can see that there are specific problems resulting from these trends, problems that users are currently already facing [4]. Clearly, these problems relate to new design decisions that need a novel framework of thinking which can result from a careful analysis [5].

2. Interaction Patterns and Societal Trends in Technological Realities

As humans, we find ourselves involved in complicated causal chains of interaction, where we juggle a large number of external demands and internal objectives. Furthermore, these causal chains are shaped by the distribution of power and the exertion of social pressure. Therefore, this network is seemingly becoming increasingly large, fast, dense, and sophisticated.

Naturally, as social creatures, humans have always been a part of a large network of interactions [6]. Accordingly, our societies have already developed cultural conventions, as well as legal rules that supported our functioning within these networks. However, many of the regulating features that have been developed and which are currently embedded in our habits, institutions, laws and culture, have all evolved under circumstances that may now no longer be warranted. The existing culture, institutions, and laws are still largely based on certain outdated assumptions about the flows of information in the aforementioned causal chains of interaction: Firstly, interaction patterns are still assumed to work at a human pace. This reflects the historical fact that changes in the causal networks surrounding us were largely driven by humans. Secondly, the topology of the causal network was considered to be more or less static or only subject to very slow changes [7]. This reflects the historical fact that the creation of connections in the causal network was costly, labor intensive, and time consuming. Thirdly, our culture still assumes that the timeliness, cost, and frequency of interactions largely depend on proximity, which is indeed still dominant in face-to-face interaction. This reflects the fact that most interactions were direct physical and mechanical interactions, e.g., pushing buttons on the dashboard of a car or operating a screwdriver, and these interactions were therefore clearly localized. Fourthly, another assumption that may have been valid for most civil interactions throughout history, but which is now no longer warranted, is the explicit presence or absence of anonymity, the reciprocity or symmetry (at least in terms of available information) in certain types of interactions between different parties. Finally, our assumptions are still limited to expecting contradictions, ambiguity, and uncertainty in human counterparts of interaction [8]. This is and will be changing drastically: complex intelligent devices, systems of such (remote) devices, and services will involve more uncertainty than ever before [9]. They will operate on more uncertainty, but also create more uncertainty themselves. A good example are automatic stock trading agents, which utilize limited models about the behavior of the market, and by trading, they let the market slowly deviate from the models as they exploit the model with tailored heuristics.

We ought to explain why we think this matters to designers and system engineers [10] [11]. Whereas Industrial Design, until recently, used to be about form giving, efficient manufacturing, product aesthetics, and ergonomics, nowadays it encompasses much more [12]: first, classical products such as coffeemakers and shavers have embedded electronics, so industrial designers also had to take responsibility for the behavior of the embedded software. Then all similar products, and of course also mobile phones and televisions etc., became Internet enabled. So now they are part of complex socio-technical systems and the designers are co-responsible for the efficiency, the trust, and the aesthetics of the interaction with these systems [13]. Consequently, the interactions and considerations addressed in this paper are of utmost importance to the design community [7].

As designers, we now need to acknowledge that many of the basic circumstances that shape human-product interactions are changing rapidly. Our design practice has not yet been adapted to these changes, and we need to make a conscious effort to avoid the threats that these developments may pose and grasp the opportunities that they may offer [14]. To identify these more clearly, we need to thoroughly study how adaptive systems present themselves from a human viewpoint, and how human users deal with such systems, either by operating on the assumptions of causality, or by finding new strategies to cope with the inherent uncertainty of adaptive systems [4].

3. The Human Point of View in Design

As the result of actions always lies in the future, purposeful human action is ultimately rooted in expectations [15]. More precisely speaking, what often motivates human action is some change that is (consciously or unconsciously; see [16]) expected as a result of this action. Actions are motivated by the likely changes in the environment that are expected as a result of these actions [17]. The expectations that humans have are dependent on their ideas, reflexes, skills, prejudice, and knowledge as well as on their current assessment of the given situation [18].

Human environments are extremely complex, and stretch far beyond human mental capacities, so humans are never able to fully predict the consequences of their actions [5]. Yet, in many natural situations, in which humans are competent, they are capable of predicting quite adequately how their actions can affect a certain environment. They understand which aspects of an environment tend to remain static, and which will change as a result of their direct actions. But that is not all. They are also able to foresee (and deal with) the various spontaneous changes that may arise in the different elements of a given environment. So, as a rule, humans cannot only estimate the potential of their own actions in a given situation, but can also anticipate to a certain extent the spontaneous changes that may happen in various elements of their environment. Humans can have quite detailed expectations about the behavior of their environment, and they may also understand how various independent actors in their environment are interconnected and can influence each other.

For instance, an experienced cook that prepares various dishes simultaneously does not only know that dishes and spoons will remain where he puts them, but he will be able to foresee when to turn the meat or stir the sauce in order to keep it from burning, he knows how to initialize and control the many variables that determine the end result, and may know special tricks that enable him to deal with exceptions: He knows when to add a drop of water to save the mayonnaise. What is crucial for his competence is that he has reliable expectations about the current state of the system 'kitchen' and about the way in which different aspects of the system will evolve, and that he knows where and when to inspect the system to track its development. Last but not least he also knows how to intervene in order to steer the system in the desired direction and to avoid unwanted outcomes.

So, humans are able to predict where spontaneous (i.e. unprovoked) changes in their environment may occur, and how these events can influence the systems the user is currently interacting with. Humans know which elements in the environment they need to (consciously or subconsciously) monitor to manage changes or to react to them. Thus, in the following we will abstract from humans in more theoretical terms as smart adaptive controllers or responders with a limited foresight [19] [20].

To better understand how the human ability to take care of (complex) situations based on limited foresight may be affected (and perhaps even challenged) by some of the trends that we have identified, we will consult a model that can describe a human that 'takes care' of a *situation*, i.e., a system within an environment. Such a *control situation* is an environment with actors and products, in which a control task takes place [21]. This definition holds for more traditional 'static' systems as well as for adaptive systems. Such systems are specializations of control situations. The model is inspired by control theory and cybernetics and reframes control theory for 'systems in contexts' (see for 'context' [22]). In this discipline, 'taking care' of a control situation is described as a 'control task'. The following short excursion into cybernetics will help us understand why different control situations lead to different types of control tasks which may be more or less challenging to humans, and it may also help us to identify the designable aspects of a control situation that may make these tasks more or less demanding.

3.1. Human Control as a Cybernetic Task

The field of Cybernetics (from the Greek word κυβερνητική [kyvernitikí = "government"]) casts a structured view on control tasks [23]. The essence of a control task is that a 'controller' (for instance a human) who performs actions (provides inputs) that keep 'essential variables' of those objects and processes (the system within its environment) that matter to the controller (which may include the controller itself) within desired bounds. The controller protects or shields these variables from the possible influences of external disturbances: "In general, then, an essential feature of the good regulator is that it blocks the flow of variety from disturbances to essential variables" [23, p. 201]. A well-known classical example is the temperature controller in a room: when the (variable) temperature becomes too low, the heater is switched on. When too high, the heater is switched off. When there is a cold wind blowing from outside, that's an external disturbance. In the rest of this paper we focus on one control task chosen for practical reasons: *driving a car*. This example allows us to illustrate a number of important points of increasing complexity. At the same time we assume most readers are familiar with what it means to drive a car. An example for a user who is interacting with processes involving essential variables that need to be regulated is a person actually driving a car. We can assume that the driving process which involves interactions between the car, the driver, and the environment, is characterized by some state that is changing over time. To drive successfully, the driver needs to undertake many actions, which influence how the state of the process changes over time. These actions, the inputs from the driver (controller) to the process, need to ascertain that essential variables like distances, (relative) velocities, fuel level, engine temperature, or even his own fatigue remain within certain limits.

Such a control process can be modeled as an optimization process. We can assume that the driver tries to minimize some 'error' quantity that can be calculated from the values of certain essential variables. It is important for the driver to keep this error within certain bounds. The driver needs to keep the system state within a 'safe' region. If the driver is not able to do this, an accident may occur.

How the state of the process develops over time does not only depend on the inputs that the controller provides. There will also be external factors that influence the time development of the process. From the viewpoint of the controller these 'external' inputs constitute disturbances over which he has no direct control but for which s/he may have to compensate. In the example of driving a car, external inputs or 'disturbances' can be other drivers on the same road, animals, road conditions, weather conditions, the state of the brakes and the engine, etc. It is important to realize that the state of the process being controlled may not always be completely observable to the controller. Only certain aspects of that process may be observable. The collection of these aspects constitutes a certain (limited) view of the controller on the process of interacting in the system.

The human now faces a 'control task'. This task consists of 'steering' the process through the manipulation of the inputs, based on the observations that the view on the process allows. Despite the external disturbances, the controller needs to keep the 'error' of the process within the prescribed bounds. The very simplest example of this notion of being 'within bounds' could be keeping the car on the road. Secondly, respecting the min/max speed limits and thirdly keeping a safe distance from obstacles, but other types of bounds, not just literal bounds, are included as well.

The multitude of different influencing factors of a control task needs to be defined structurally. Figure 1 shows a special control situation, namely an adaptive system, consisting of one User A, the process to be controlled, other users (User B and C), and the environment. Embedded in the adaptive system are control tasks, which express the dynamics of the control situation, not just between primary user and process, but also between other users or the environment and the process. Finally, the process state is depicted as being a floating 'dot' that needs to be maintained in a 'safe region' to keep the process and potentially the entire adaptive system alive, by preventing fatal states.

Note that while the human user and the control task are depicted here 'outside' the controlled process, it can and will often be subjected to the consequences of its own control even inside the process. This is definitely the case, for instance, for a user driving a car.

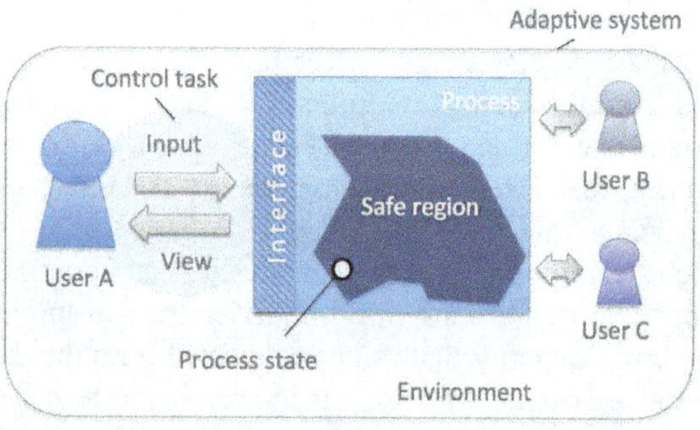

Figure 1. *Adaptive system (control situation) involving a control task for a human user: providing the right input for a process via an interface to steer the entire control situation (process and other users within an environment) such that the process state is kept within the safe region.*

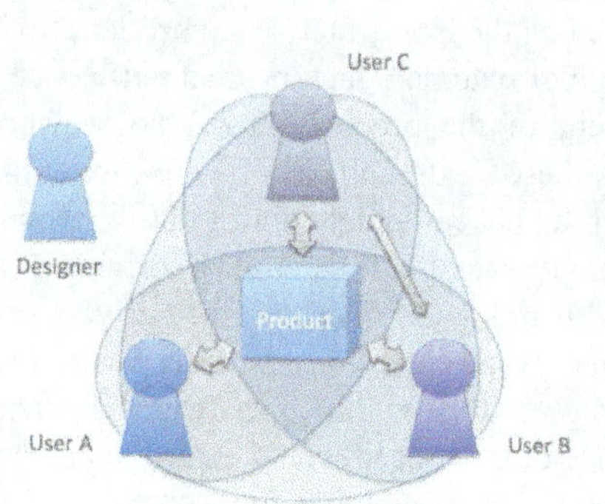

Figure 2. *Different points of view on a user-product system in a control situation; the designer determines how User A perceives other communicating Users B and C 'through' the product.*

3.2. A First Person View

When discussing control tasks related to a specific process we have to take different points of view of different users on the process into account: to one user, for instance, the partially autonomous reactions of the process may appear hostile, or confusing, while another user may experience the process' behavior as appropriate or exactly as expected [24]. Designers of such a process need to adopt the various relevant potential views of key stakeholders or users to design a process that makes sense to them.

In the following, a taxonomy of steering tasks is presented, where a first-person-view from a controller, i.e. a user, to the process and its environment including other users (in total: a control situation) is presupposed as the relevant perspective to classify the steering-task. This presupposition (that tasks are classified as to how they appear from a first-person perspective) is essential, because we want to base design conclusions on our (rather abstract) taxonomy. Figure 2 illustrates this presupposition in the context of a product such as a car in traffic. From the controller's (User A) point of view, not only the physical product is targeted by her steering task, but rather a whole control situation that combines the car (product), the road, signaling, weather (environment), and other users within the same environment. The same holds for all the other users in the control situation respectively. What might be experienced as product behavior can actually be a result of other users' actions rather than truly autonomous behavior of the product.

This is the control situation model, which shall be used in the following to outline important design challenges. The model describes a control situation structurally (cf. Figure 1), but also as a collection of first person views (cf. Figure 2). Understanding an adaptive interaction setting as a connected system of users, process, and environment is a prerequisite of defining the task of designing such a system as a meta-steering task which needs to take all users' steering tasks and experiences (actions and perceptions) into account to create a meaningful and holistic user experience. A designer should be able to temporarily adopt or assume all first person views that can be identified in the system of connected users and products, and design accordingly. Referring to the running example of controlling cars, this would be about how to design a crossing and associated traffic light system that optimizes traffic flow and still prevents accidents as much as possible. The designer has to understand the needs and temptations of various drivers approaching the crossing from different entry points. For innovative products and related processes, this requires the designer to predict and foresee potential future user actions. The taxonomy of steering and control tasks, which we propose in the following, highlights essential distinctions that help to guide the design process.

4. Taxonomy of User's Control Tasks

It is clear that control tasks that humans may encounter

can differ substantially in terms of complexity [25], and difficulty [26]. The *control situation model* described above helps us to examine and distinguish various relevant factors that may determine how challenging the control task is: apart from the complexity of the system state and its associated dynamics, we also need to consider the nature and extent of the external disturbances, together with the completeness of the view of the *controller*, and also the tolerances on the errors which are permitted. In this section we will provide arguments to distinguish different types of control tasks. Based on these distinctions we create a taxonomy that can serve as a useful characterization of the main types of control tasks that humans will encounter in future ambient and smart environments. The taxonomy is presented in Figure 3 and its details will be explained below.

When the user perceives an adaptive counterpart (the process), the first distinction we make is between *isolated* and *open* systems. This distinction is based on the nature of the external disturbances that may jolt the process, or even the entire system. If the external disturbances to a process are so weak and predictable that their effects can always still be corrected after they have occurred (the 'cared for' process state can always be brought back into the safe region after a disturbance), then it follows that the control task can always be performed based on a view that is limited to the process itself. The entire system can effectively be monitored in isolation, and control can be based on a process view, which is completely unaware of all other events taking place in the environment.

However, if some external disturbances may actually be fatal, in the sense that it cannot be avoided that the process state leaves the safe region (cf. Figure 1) after such a disturbance has occurred, it is necessary to prevent these disturbances from affecting the process in the first place.

For these kinds of systems, the user cannot rely on actions that are solely determined by the view on the process. Once a potentially fatal disturbance has compromised the process, it may be fatally derailed, and it may already be too late to avoid system failure. Instead, the user needs to predict the external disturbance and its effect on the process state. The user's view on the environment of the process is crucial here: He needs to be able to adopt a view that extends beyond the state of the process to be controlled, and he must consider other aspects of the environment, which therefore influence his actions. We can thus formulate the following distinction between *isolated* and *open* systems (see Table 1).

To illustrate the distinction between these two different types of control tasks through concrete examples, we look at a user driving and navigating a car through traffic scenario. One of the tasks the user is facing is to keep the car at the desired speed; we can call this cruise control. Cruise control is an isolated control task as it regulates the speed of the car by taking the desired speed given by the driver as the reference value, the current speed as a sensor input and using this input to adapt the speed by accelerating the engine or braking. The task can be executed based on a limited system view, therefore it can, at least in this case, also be easily automated (i.e., modern cars with cruise control).

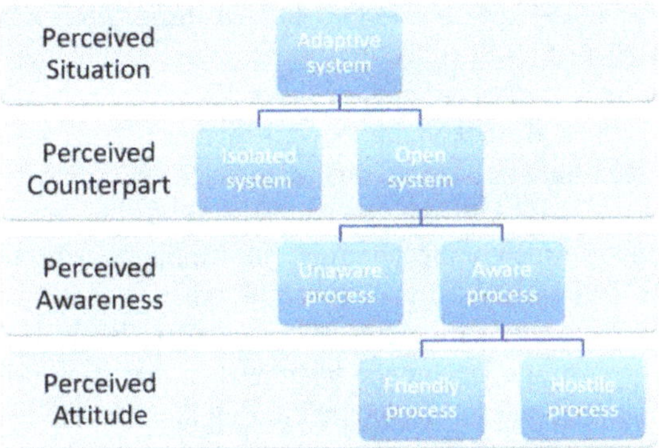

Figure 3. *A hierarchical taxonomy of product or process classes is shown with which a user can interact within a control situation. This hierarchy describes the encountered control situations from a first person view of a single user; not shown are other users or the surrounding environment, however, they might be represented indirectly by the product's reactions on their control processes.*

But, when navigating in traffic, there can also be external disturbances that are fatal; one needs to avoid collisions with obstacles and other cars. Avoiding the obstacles is a task that cannot be executed from an isolated system view. So, in this case we then deal with 'open' control tasks, where we need to monitor various other aspects of the environment to keep the system on track on the safe region.

Table 1. *Distinction between isolated and open systems*

'Isolated' or 'bounded' system	'Open' system
• Controllable from isolated system view; • Predictable behavior; • Disturbances non-fatal; • Remediation possible; • Disturbances compensated after they occurred.	• Environment view; • Unpredictable behavior; • External disturbances can be fatal; • Prevention of disturbances needed, i.e. developments in the system environment need to be anticipated and preventive action is needed.

Table 2. *Distinction between unaware and aware disrupting systems (counterparts)*

'Unaware' Disruptor	'Aware' Disruptor
• Disruptor is blind to user's input. • Prediction of disruptor's actions is independent of our own control actions.	• Disruptor is aware of environment and may even sense user's input. • Complex prediction, dependent on our own action, possibly Intentional Stance [15].

Table 3. *Distinction between friendly and hostile (aware and disrupting) systems*

'Friendly' system	'Hostile' system
• Cooperative, supportive process; • Signals reliable; • System wants to be predictable.	• Challenging, non-cooperative process; • Signals may be unreliable; • System may/want to be unpredictable.

4.1. Dealing with Open Control Tasks

The tasks dealing with isolated systems as studied by control theory are extensively covered in the literature [27] [28], and are less interesting from a design perspective. What is of interest to designers, are the complexities that can arise in the 'open' control tasks. These are common in real life and do not yield so well to formal analysis. We have seen that, in order to deal with an 'open' control task, a user needs knowledge which extends beyond the immediate process she controls; she needs at least to be able to make some estimations about other elements in the environment that may create disturbances in her task. This presupposed, among others, whether her first person view allows a meaningful decomposition of the environment in various interaction chunks, which is, of course an important design consideration in general [29]. What is crucial here is that the user needs to be able to predict the potential occurrence of certain external disturbances before they do indeed happen. Central questions are:

- What is the effect of her actions (including doing nothing)?
- What is the expected development of other elements in the environment, which may lead to a disturbance and a resulting fatal change of the system state?

If the user finds an element in the environment that may develop into a (fatal) disturbance of the process she may need to take appropriate action beforehand. The user needs to predict the future of elements of the environment that she does not control, in order to influence the process that she needs to control, and she needs to trust her prediction and act accordingly.

The situation resembles somewhat the way modern software engineering copes with unforeseen situations [30]. In early programming languages, the programmer was supposed to specify all situations and code the proper behavior. Contemporary languages support exception handling, a kind of rough classification of unforeseeable situations and what to do if they arise.

How well a user can deal with the presence of a possibly disruptive counterpart, an element of the environment, may now largely depend on the usability, but also reliability, with which she can predict its behavior, and on the complexity of the internal state that is attributed to this disruptor. Her ability to predict depends on her own ability to model this disruptive counterpart. Clearly, the models that she may employ can become very complex, and this is interesting from a design perspective as this model may even be supported by tools. For instance, nowadays, detailed short-term weather predictions are available, that can play an important role as planning aids for other applications. This poses no great problems as long as the behavior of such tools is largely straightforward. This, however, cannot always be guaranteed.

The disruptor that she tries to predict might be itself a process that adapts to the environment, and may produce various reactions. The prediction of such counterpart's behavior is fundamentally more difficult than in isolated systems. In particular, a threshold is crossed if the process to be predicted may somehow sense the users own controlling action and may react to that. In this case, calculating a favorable action depends also on the user's ability to predict how the other process may react to her action, and thus can become a significantly complex task. In such cases, it may even be useful to attribute intentions to the disruptive process, in order to model it.

This leads to the next dichotomy, which distinguishes two types of 'open' control tasks. A user may have to predict the course of another process not under her direct control. This prediction becomes fundamentally more difficult if this process itself is an adaptive process that can take aspects of the surrounding control situation such as the user's own action into account (see Table 2)

We already encountered concrete examples of control tasks involving 'unaware' and 'aware' disruptors: first, reacting to other 'unaware' obstacles in the environment, and second reacting to the other drivers that clearly are 'aware'. Thus, driving a car in an empty road can be seen as a control task with unaware disruptors as the environment such as the road, trees on its side, a nearby forest, and in general the weather are certainly not aware of the driver, but can influence directly the conditions the driver has to deal with: during a thunder storm, a tree struck by lightning might fall onto the road and the rain might reduce visibility, posing an easily life-threatening danger to the driver. Driving in rush hour traffic, on the other hand, forces the driver to deal with 'aware' disruptors.

4.2. Dealing with 'Aware' Disruptors

The case where a user needs to respond to an adaptive process that can show various reactions to the user's action is extremely interesting from a design perspective. In this case it is often impossible for the user to predict how this process will develop. On the other hand, however, this problem can now be alleviated because such an adaptive process may signal its planned actions or intentions [31].

The importance of signalling in this case is indisputable: for instance, signalling is often essential for synchronization, as the use of crude signals such as a car's brake lights and turn-signals in traffic already illustrates. Also, signals help to interpret the actions of others and may therefore support cooperation or even social learning [32].

But its design potential reaches much further, as is obvious from importance of phenomena like body language and intonation in cooperation and communication. The use of more intuitive and embodied types of communication through adaptive artefacts therefore seems a promising field of design research: These artefacts may for instance help to direct human attention [33], assist in interaction with autistic children [34], or support inter-human communication in virtual environments [35]. We can conclude that in the interaction between adaptive processes, signalling is extremely important and opens up a huge design space, which is still largely unexplored. While humans or adaptive processes may quite successfully communicate with others, the interpretation of their messages is unfortunately not always as reliable as one might hope, even if these messages have been carefully designed. The simple reason for this is that not all processes that we encounter need to be completely friendly or benign, or have interests that are aligned with our own. This means that controllers must, in certain cases, be aware that there may be adaptive processes in their environment that may not have their best interest at heart, and that might try to hinder our success or might even seek to do active harm (see Table 3) .

It is clear that this last distinction is quite crucial, and has enormous importance for designers: they can often shape part of a control task, and may, for instance, determine the extent and scope of the signals that are exchanged in a control task. Moreover, for such control tasks, the traffic example is demonstrative as actions of fellow drivers on the road can be both cooperative and uncooperative: cooperative behavior on the road is the usual, expected, *predictable* case, where accidents are avoided and actually prevented by applying a 'defensive' style of driving, e.g., leaving space, signalling clearly, and avoiding abrupt changes of lanes and directions. On the contrary, there might be bullies, people who drive carelessly or even provoke potentially dangerous situations with unpredictable, irrational moves, i.e. *unreliable* signals. When messages are not necessarily true, and views are not necessarily reliable, the designer may well be forced to take sides between divergent interests. The person's view on a process clearly plays a decisive role in her success to steer or control that process, and designers can determine that view to a certain extent. Furthermore, in a software and electronics age, the freedom in shaping this view has expanded in many ways. Designers should consider these issues because it may well be that their responsibility has increased lately [36]. It may even be possible, that, in order to take this responsibility, designers may need to cooperate more closely with lawyers, lawmakers, or legal professionals. As designers, we should try to avoid situations where users have to face the challenge of dealing with hostile disruptors.

An 'aware' disruptor is a cyclist who chooses to ignore a red traffic light and cross the road regardless of the traffic regulations (a situation not uncommon in many countries). Although the action is illegal and perhaps immoral, the cyclist still is observing the cars and trying to prevent a sure kill of herself. The car drivers are usually not pleased, but such a cyclist is still a fairly friendly disruptor. An example of a hostile disruptor is someone throwing a stone brick or a tile from a viaduct just to hit a car. Although this is illegal and immoral, it happens nevertheless (fortunately not very often, but it is hard for the police to trace and arrest the brick throwers).

5. From Theory to Application

In the last section, a taxonomy of different design challenges has been presented that implies a first-person view on adaptive systems [37]. The challenges that a process seems to pose to a user in a control situation may have a huge impact on the attitude of this user towards her interaction counterpart. As users are often involved in an increasingly more complex network of adaptive systems interfering with multiple users and their environments [38], the designer's challenge is not only to design a part of such a system that behaves according to a single user's expectations, but also to design elements that can function as part of a networked ecology of adaptive systems and still will be perceived by a majority of users as *actively supporting tasks under various circumstances in such an environment* [39].

While systems traditionally are understood as a composition of functional components and these elements are subordinate to the system, adaptive systems are a dynamic (transient) composition of users (human actors), processes, and the environment, in which interaction is performed as control tasks between the different 'elements' of the adaptive system. One of these elements is the process which users interact with, and, quite naturally, such an element of the ecology will be perceived differently by different users. These differences are not only due to different levels of understanding, expertise, intensity of use, or initial expectations, but first and foremost due to the respective roles of user towards this element. Additionally, these differences are due to the perspective that the different user have on this particular element. Mapping this to the taxonomy view, different users' views can see a particular process in different ways: for instance, a process that seems hostile to one user may seem friendly to another and even 'unaware' to a third. The responsibilities of designers that change the 'fabric' of the network are far from clear. But it seems obvious that they should shoulder some responsibilities: while designers may not be able to avoid that some of their creations will, to some users, appear as hostile disruptors, they might be able to strive towards minimizing these misinterpretations, or perhaps ascertain that their creations, even when perceived this way, will at least communicate via reliable signals.

Certainly, new questions are raised: On the one hand, we need a better understanding how the design can make use of signalling or refinement of the interface between user and process to ensure certain desired perceptions and expectations. On the other hand, we need to research how the design can allow for transparency within the interface and process to show other users' actions and allow them to transparently interact via the process.

6. Discussion and Conclusion

It is clear that humans are increasingly entwined in a complex network of (adaptive) systems (which is a complex system in itself) that is rapidly growing and covering many areas of human activity and interest. The speed, size, density and connectivity of this network are increasing rapidly.

This development offers many unprecedented possibilities to designers, but also poses new questions: We need ways to categorize different positions of humans (or artefacts) in this network, to cater for the various control tasks that humans need to perform, and to be aware of the basic assumptions that we make about these control situations.

In this paper we have proposed a hierarchical taxonomy of user challenges that can act as a starting point to help designers to group control tasks and determine at least some of the basic characteristics of a given control task that users will encounter. The taxonomy may, for instance, suggest in which control situations signaling may play a role, and to what extent controllers can allow themselves to rely on the content of the signals. The taxonomy also indicates that one of the promising areas of design research may well relate to control tasks where the crucial element is the synchronization and cooperation of a human controller with another adaptive controller. While these possibilities are exciting and leave much to be explored, our taxonomy also shows there is a case for concern. Although the characteristics of interactions with friendly or hostile processes are vastly different, it can happen that these types of interactions are extremely difficult to distinguish, both due to the support for long distance connections, and the almost unlimited flexibility and reproducibility of electronic interfaces and underlying processes (cf. challenges noted above).

It is therefore crucial that the control tasks and situations through which humans can interact with other humans or other systems guarantee or at least support interaction with a minimum of fairness. At present it is not at all obvious that this issue is being addressed [40]. If this does not happen, the position of individual humans in the network might become very vulnerable to say the least. If the current trends continue unabated, we have to face and address the following design challenges:

1. *Ambiguity*: Users will continuously interact (directly or indirectly) with process that can undertake an action they, as a user, did *not* initiate.

2. *Timeliness*: These processes may have (and often will have) fast or almost immediate access to information that informs or motivates their actions that users might *not* have access to.

3. *Lack of transparency*: Our knowledge about these processes, their existence, their presence, their structure, their true intent, their origin, and their functioning may be *nil* (or rather limited), even while we are interacting with them.

4. *Control*: Through these processes, we may participate in control situations, i.e. interacting with adaptive systems or other human beings, in ways that are *not* primarily determined by our intentions or attitudes towards these humans, but which are largely shaped by the way our interfaces to them are structured and organized.

It is fair to say that these trends are worrying. It is therefore imperative for us as designers and system engineers that we support and advocate the design of systems and the adoption of rules and practices that strengthen the position of individual humans in such adaptive environments.

Acknowledgements

This paper is the result of the discussions about 'intelligence', 'adaptivity' and other related topics among the authors and with Karl Tuyls and Razvan Cristescu. We are grateful for their valuable contributions.

References

[1] E. Aarts and B. de Ruyter, "New research perspectives on Ambient Intelligence," *Journal of Ambient Intelligence and Smart Environments,* vol. 1, pp. 5-14, 2009.

[2] J. H. Holland, "Complex adaptive systems," *Daedalus,* vol. 121, pp. 17-30, 1992.

[3] K. Kelly, *Out of Control: The Rise of Neo-biological Civilization*, Reading, Mass.: Addison-Wesley, 1994.

[4] D. Benyon, "Adaptive systems: A solution to usability problems," *User Modeling and User-Adapted Interaction,* vol. 3, pp. 65-87, 1993.

[5] H. Hagras, V. Callaghan, M. Colley, G. Clarke, A. Pounds-Cornish, and H. Duman, "Creating an ambient-intelligence environment using embedded agents," *Intelligent Systems, IEEE,* vol. 19, pp. 12-20, 2004.

[6] B. Latour, "On actor-network theory: A few clarifications plus more than a few complications," *Soziale Welt,* vol. 47, pp. 369-381, 1996.

[7] S. Evenson, J. Rheinfrank, and H. Dubberly, "Ability-centered design: From static to adaptive worlds," *interactions,* vol. 17, pp. 75-79, 2010.

[8] T. Rakow and B. R. Newell, "Degrees of uncertainty: An overview and framework for future research on experience-based choice," *Journal of Behavioral Decision Making,* vol. 23, pp. 1-14, 2010.

[9] E. Lopes, *Decision-Making under Uncertainty and Complexity: A Grounded Theory Approach*, Saarbrücken: VDM Verlag Dr. Müller, 2010.

[10] C. Hummels and J. Frens, "Designing disruptive innovative systems, products and services: RTD process," in *Industrial Design-New Frontiers*, D. Coelho, Ed., Shanghai: Intech Open Access Publisher, pp. 147-172, 2011.

[11] K.-H. Huang and Y.-S. Deng, "Social interaction design in cultural context: A case study of a traditional social activity," *International Journal of Design,* vol. 2, pp. 81-96, 2008.

[12] S. Kyffin and P. Gardien, "Navigating the innovation matrix: An approach to design-led innovation," *International Journal of Design,* vol. 3, pp. 57-69, 2009.

[13] S. A. Rijsdijk, E. J. Hultink, and A. Diamantopoulos, "Product intelligence: Its conceptualization, measurement and impact on consumer satisfaction," *Journal of the Academy of Marketing Science,* vol. 35, pp. 340-356, 2007.

[14] M. Davis, "Why do we need doctoral study in design?," *International Journal of Design,* vol. 2, pp. 71-79, 2008.

[15] D. C. Dennett, *The Intentional Stance*, MIT Press: A Bradford Book, 1989.

[16] F. Acker, "New findings on unconscious versus conscious thought in decision making: Additional empirical data and meta-analysis," *Judgment and Decision Making,* vol. 3, pp. 292-303, 2008.

[17] P. Dourish, *Where the Action is: The Foundations of Embodied Interaction*, MIT Press: Bradford Books, 2004.

[18] A. Flammer, *Erfahrung der eigenen Wirksamkeit: Einführung in die Psychologie der Kontrollmeinung*, Bern: Huber, 1990.

[19] A. Clark, *Being There: Putting Brain, Body, and World Together Again*, Cambridge, Mass.: MIT Press, 1997.

[20] B. Schmidt, "Human factors in complex systems: The modelling of human behaviour," in *Proceedings 19th European Conference on Modelling and Simulation*, Riga, Latvia, pp. 1-10 (CD), 2005.

[21] Z. Gao, "Active disturbance rejection control: A paradigm shift in feedback control system design," in *Proceedings of the 2006 American Control Conference*, Minneapolis, Minnesota, USA, June 14-16, pp. 2399-2405, 2006.

[22] P. Dourish, "What we talk about when we talk about context," *Personal and Ubiquitous Computing,* vol. 8, pp. 19-30, 2004.

[23] R. W. Ashby, *An Introduction to Cybernetics*, 2nd ed., London, UK: Chapman and Hall, 1957.

[24] P. Houston, M. Floyd, and S. Carnicero, *Spy the Lie: How to Spot Deception the CIA Way*, London: Icon Books, 2012.

[25] K. Byström and K. Järvelin, "Task complexity affects information seeking and use," *Information Processing & Management,* vol. 31, pp. 191-213, 1995.

[26] P. Robinson, "Task complexity, task difficulty, and task production: Exploring interactions in a componential framework," *Applied Linguistics,* vol. 22, pp. 27-57, 2001.

[27] R. E. Kalman, "A new approach to linear filtering and prediction problems," *Journal of Fluids Engineering,* vol. 82, pp. 35-45, 1960.

[28] T. Sugie and T. Ono, "An iterative learning control law for dynamical systems," *Automatica,* vol. 27, pp. 729-732, 1991.

[29] S. Wensveen, T. Djajadiningrat, and K. Overbeeke, "Interaction frogger: A design framework to couple action and function through feedback and feedforward," in *Proceedings of the 5th conference on Designing Interactive Systems: Processes, Practices, Methods, and Techniques*, Cambridge, MA, USA, pp. 177-184, 2004.

[30] K. Schwaber, *Agile Project Management with Scrum*, vol. 7, Redmond: Microsoft Press, 2004.

[31] J. Redish, "Expanding usability testing to evaluate complex systems," *Journal of Usability Studies,* vol. 2, pp. 102-111, 2007.

[32] N. E. Miller and J. Dollard, *Social Learning and Imitation*, New Haven, CT, US: Yale University Press, 1941.

[33] E. Deckers, S. Wensveen, R. Ahn, and K. Overbeeke, "Designing for perceptual crossing to improve user involvement," in *Proceedings of the ACM SIGCHI Conference on Human Factors in Computing Systems*, Vancouver, BC, Canada, pp. 1929-1938, 2011.

[34] E. Barakova, J. Gillessen, and L. Feijs, "Social training of autistic children with interactive intelligent agents," *Journal of Integrative Neuroscience,* vol. 08, pp. 23-34, 2009.

[35] J. Hu and L. Feijs, "A distributed multi-agent architecture in simulation based medical training," in *Transactions on Edutainment III.* vol. LNCS 5940, Z. Pan, A. Cheok, W. Müller, and M. Chang, Eds., Berlin Heidelberg: Springer, pp. 105-115, 2009.

[36] T. Swierstra and K. Waelbers, "Designing a good life: A matrix for the technological mediation of morality," *Science and Engineering Ethics,* vol. 18, pp. 157-72, 2012.

[37] G. Bateson and M. Mead, "For god's sake, Margaret," *Convolution Quarterly,* vol. 10, pp. 32-44, 1976.

[38] D. Preuveneers, J. Van den Bergh, D. Wagelaar, A. Georges, P. Rigole, T. Clerckx, *et al.*, "Towards an extensible context ontology for ambient intelligence," in *Ambient Intelligence.* vol. 3295 (LNCS), P. Markopoulos, B. Eggen, E. Aarts, and J. L. Crowley, Eds., Berlin Heidelberg: Springer, pp. 148-159, 2004.

[39] R. Holzer, H. de Meer, and C. Bettstetter, "On autonomy and emergence in self-organizing systems," in *IWSOS 2008.* vol. LNCS 5343, K. A. Humme and J. P. G. Sterbenz, Eds., Berlin Heidelberg: Springer, pp. 157-169, 2008.

[40] R. Gorbunov, E. Barakova, and M. Rauterberg, "Design of social agents foundations on natural and artificial computation," in *Foundations on Natural and Artificial Computation.* vol. 6686 (LNCS), J. Ferrández, J. Álvarez Sánchez, F. de la Paz, and F. Toledo, Eds., Berlin Heidelberg: Springer, pp. 192-201, 2011.

26

Differential flatness applications to industrial machine control

Ejike C. Anene[1], Ganesh K. Venayagamoorthy[2]

[1]Electrical Engineering Programme, Abubakar Tafawa Balewa University, PMB 0248, Bauchi, Nigeria
[2]Real-Time Power and Intelligent Systems Laboratory, Clemson University, Clemson, USA

Email address:
ejikeanene@yahoo.com (E. C. Anene), gkumar@ieee.org (G. K. Venayagamoorthy)

Abstract: In this article the applications of differential flatness to some industrial systems are presented. Computational methods of obtaining the flat output and the straight forward method of constructing the corresponding control law are given. Some theoretical and industrial systems are used as illustration including the third order synchronous machine model and the one degree of freedom magnetic levitation system model. Computations of the flat output are done using various approaches. The Levine's approach is presented in such detail as to facilitate quick understanding. Computations for the synchronous machine model yielded a flat output that is a function of the load angle while the magnetic levitation model yielded a flat output that is a function of the objects' position. Results showing the stabilization of the applied systems in fault and uncertain situations are discussed.

Keywords: Magnetic Levitation, Flatness, Feedback Linearization, Synchronous Machine

---

1. Introduction

THE concept of differential flatness proposed by Michel Fliess and co-workers [1],[2] about twenty years ago has evolved into a full-fledged field for the study of control systems in a practically new way. In this setting, controllability is linked with system flatness and controllable systems possess this flatness property [3],[4]. For such systems there is a solution set called flat output in the solution space consisting of a set of state variables that completely parameterize the system without the need for solving differential equations.

Once this output is shown to be flat, it in effect implies that the system possesses a well characterized dynamics[5] . This is because all system parameters and control becomes a function of the linearizing output that can enable the generation of reference trajectories a-priori. The construction of the feedback law is done by a simple inversion of system equations with respect to the control. The scheme in derivation is an extension from the input-output linearization scheme with zero internal dynamics.

Fliess et-al [1] proposed the notion of endogenous equivalence and defined a class of dynamic feedbacks for classification and linearization of systems in the form of Fliess' differential algebraic forms. Such classes of systems are the so-called differentially flat systems. One of the main consequences of their result is a constructive method of computing the feedback that exactly linearizes a flat system. Accordingly a control system M, F is differentially flat around p if and only if it is equivalent to a trivial system in a neighborhood of p. A trivial system can be defined as one which is without dynamics described by a collection of independent variables or R_s^{∞}, F_s where $F_s(y, y^{(1)}, y^{(2)}, ...) = (y, y^{(1)}, y^{(2)}, ...)$, with $y \subset R^s$y [6]. It is said to be differentially flat if it is differentially flat around every p of an open dense subset of M. The set $y = \{y_j | j = 1, ..., s\}$ is called a flat or linearizing output of M described by a collection of independent variables, the flat output having the same number of components as the number of control variables. The following deductions are shown with proofs in [1].

1. The number of components of a flat output is equal to number of input channels.
2. A classic linear system is flat if and only if it is controllable.
3. The controllability of differentially flat systems is related to the well known strong accessibility

property of nonlinear systems due to Sussmann and Jurdjevic.

4. If a classic nonlinear system is differentially flat around p, then it satisfies the strong accessibility at p.
5. Differential flatness means that the state and input may be completely recovered from the flat output without integrating the system differential equations.

After the introduction in Section 1, the paper discusses the basic theory of differential flatness in Section II. In Section III the procedure of computations of flat output is detailed. Section IV discusses the examples for computing flat outputs for some systems and the simulations done on the resulting controllers of some industrial systems on MATLAB. Conclusions are given in Section V while in Section VI the references are given.

2. Basic Theory of Flatness

A system variable is endogenous if it can be expressed as a linear combination of the input, the output and a finite number of their time derivatives. Otherwise it is exogenous. A single input single output (SISO) system is therefore flat or differentially flat if there exists an endogenous variable called the flat output, such that the input and the output can be expressed as a linear combination of the flat output and a finite number of its time derivatives [7]. Naturally any other endogenous variable of the system enjoys the same property with respect to the flat output. Thus the flat output differentially parameterizes all system variables.

Generally, the definition of system flatness can be cast in what follows:

The system

$$f(\dot{x},x,u) = 0 \tag{1}$$

with $x \in R^n$ and $u \in R^m$ is differentially flat if one can find a set of variables called flat output;

$$y = h(x,u,\dot{u},\ddot{u},.....,u^{(r)}) \tag{2}$$

$y \in R^m$ and system variables,

$$x = \alpha(y,\dot{y},\ddot{y},.....,y^{(q)}) \tag{3}$$

and control,

$$u = \beta(y,\dot{y},\ddot{y},.....,y^{(q+1)}) \tag{4}$$

with q a finite integer such that the system equation

$$\begin{gathered}0 = f(\frac{d\alpha}{dt}(y,\dot{y},\ddot{y},......,y^{(q)})),\\ \alpha(y,\dot{y},\ddot{y},......,y^{(q)}),\beta(y,\dot{y},\ddot{y},......,y^{(q+1)})\end{gathered} \tag{5}$$

are identically satisfied [8].

2.1. Equivalence and Feedback

The authors in [1] in their comprehensive paper unifying their theory of flatness and its associated dynamic feedback, formalized the concept that two systems are equivalent if there is an invertible transformation exchanging their trajectories, or if any variable of one system may be expressed as a function of the variables of the other system and of a finite number of their time derivatives. In a more general sense this transformation is said to be a Lie-Bäcklund isomorphism.

If two systems

$$\begin{gathered}\dot{x} = f(x,u),(x,u) \in X \times U \subset R^n \times R^m\\ \dot{y} = g(y,u),(y,v) \in Y \times V \subset R^r \times R^s\end{gathered} \tag{6}$$

and vector fields

$$\begin{gathered}F(x,u,u^{(1)},u^{(2)}....) = (f(x,u),u,u^{(1)},u^{(2)}....)\\ G(y,v,v^{(1)},v^{(2)}....) = (g(y,v),v,v^{(1)},v^{(2)}....)\end{gathered} \tag{7}$$

where,

$$\begin{gathered}u = \alpha(x,z,w)\\ \dot{z} = a(x,z,w), with, z \in Z \subset R^q\end{gathered} \tag{8}$$

are differentially equivalent, it becomes possible to go from one to another by a dynamic feedback as shown in Figure 1. That is by a diffeomorphism of the extended state space $X \times Z$. This dynamic feedback is endogenous if the original system is differentially equivalent to the closed loop system. It is called endogenous because the new z variables can be expressed as functions of the state and derivatives of the input. Thus from the work in [9] it can be stated that, if a system is differentially flat, there exists an endogenous dynamic feedback such that the closed loop system is diffeomorphic to a linear controllable system. Therefore for a nonlinear system equation (1),
where

$$f(0,0) = 0 \tag{9}$$

and rank

$$\frac{\partial f}{\partial u}(0,0) = m \tag{10}$$

its dynamic feedback linearizability means the existence of:

1) dynamic compensator;

$$\begin{gathered}\dot{z} = a(x,z,v), z \in R^q\\ u = b(x,z,v), v \in R^m\end{gathered}$$

where

$$a(0,0,0) = 0$$
$$b(0,0,0) = 0 \quad (11)$$

2) diffeomorphism;

$$\xi = \Xi(x,z), (\xi \in R^{n+q}) \quad (12)$$

such that the $(n+q)$ dimensional dynamics is given by

$$\dot{x} = f(x, b(x,z,v))$$
$$\dot{z} = \beta(x,z,v) \quad (13)$$
$$u = \alpha(x,z,v)$$

and becomes a constant linear controllable system

$$\dot{\xi} = F\xi + Gv \quad (14)$$

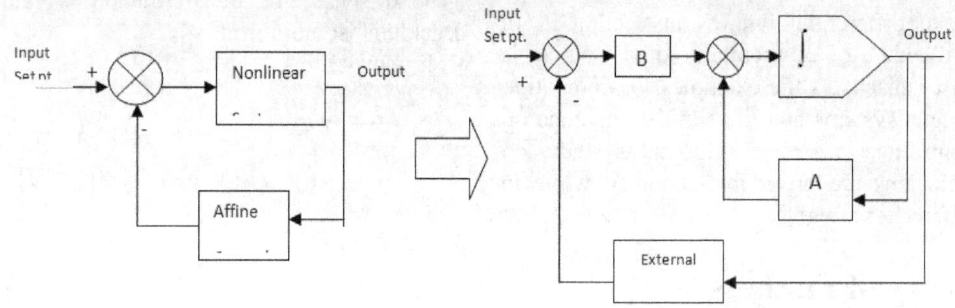

Figure 1. *Transformation of a Nonlinear System into a Linear Equivalent.*

The components of u and x can be expressed as real-analytic functions of the component of equation (2), and a finite number of their derivatives (equations (3), (4)).

The dynamic feedback is said to be endogenous if and only if the converse holds, that is, if and only if any component of y can be expressed as a real-analytic function of, x , u and a finite number of its derivatives. In a final remark in [1], the flat dynamics of a system whose output is given by equation (2) is square left and right input-output invertible system, where any component of u or x may, by definition be recovered from y without integrating any differential equation: It is said to possess a trivial zero-dynamics or a trivial residual dynamics. Figure 2 shows the endogenous dynamic feedback linearization process consisting of pole placement and linearization loops.

3. Generating Flat Outputs

Differential flatness is an idea that is naturally associated with underdetermined systems of differential equations where a system of n algebraic equations in $n+m$ unknowns [4] is written as:

$$Ax + Bf = 0, \ B \neq 0, \ rank[A, B] = n. \quad (15)$$

If A is invertible and B is full rank, then x solutions may be written in terms of f as

$$x = -A^{-1}Bf \quad (16)$$

and as such make all solutions parameterizable in terms of f. In this setting endogenous transformation φ in which the original variables of the system are transformed without creating new exogenous variables is realized [2].

3.1. Classical Methods

Following [4], consider a SISO system given by the transfer function

$$y(s) = \frac{n(s)}{d(s)}u(s) \quad (17)$$

the system is controllable if and only if the polynomials $n(s)$ and $d(s)$ are coprime, that is they have no non-trivial common factors. By Bezout's theorem, there exists polynomials $a(s)$ and $b(s)$ such that

$$a(s)n(s) + b(s)d(s) = 1 \quad (18)$$

for all $s \in C$. Define a new variable

$$f(s) = \frac{1}{d(s)}u(s), \quad (19)$$

we can write

$$y(s) = n(s)f(s), \ u(s) = d(s)f(s) \quad (20)$$

multiplying both sides of (18) by $f(s)$ we have,

$$a(s)n(s)f(s) + b(s)d(s)f(s) = f(s) \text{ or}$$

$$a(s)y(s) + b(s)u(s) = f(s) \quad (21)$$

which implies we have a variable f which is a differential function of the system input and output and a finite number of their time derivatives. Conversely all system variables and input are also differential functions of the new variable. This new variable qualifies as a flat output. Therefore given any controllable linear system in transfer function

form (17), the flat output can be chosen as any constant multiple of the variable $f(s)=\frac{1}{d(s)}u(s)$ or $f(s)=\frac{k}{d(s)}u(s)$ for any $k\neq 0$, for example consider the linear, coprime minimum phase function $y(s)=\frac{s+1}{s-1}u(s)$, from (18), $a(s)(s+1)+b(s)(s-1)=1$
$s\,a(s)+a(s)+s\,b(s)-b(s)=1$, $a(s)=\frac{1}{2}$ and $b(s)=-\frac{1}{2}$ satisfies the equation so that from (21), $f(s)=\frac{1}{2}y(s)-\frac{1}{2}u(s)$. f therefore parameterizes all system variables as given.

$$u(s)=d(s)f(s)=(s-1)f(s)=s\,f(s)-f(s)=\dot{f}-f$$

similarly

$$y(s)=n(s)f(s)=(s+1)f(s)=s\,f(s)+f(s)=\dot{f}+f$$

This treatment can be extended to the state space approach [4]: For a given linear time-invariant SISO system described by

$$y(s)=\frac{b_m s^m+b_{m-1}s^{m-1}+\cdots+b_0}{s^n+a_n s^{n-1}+\cdots+a_0}u(s),\quad m<n \tag{22}$$

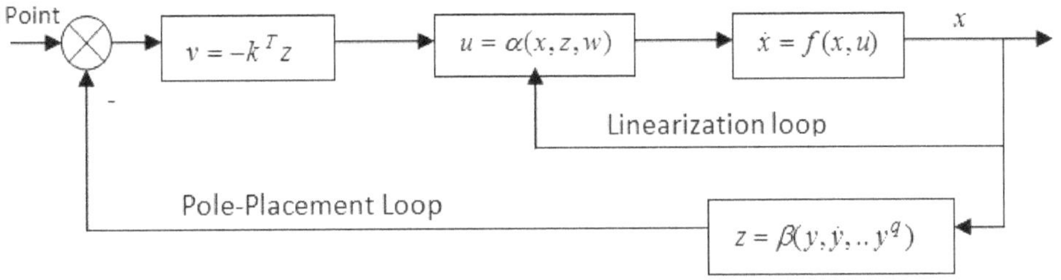

Figure 2. *Structure of Dynamic Feedback Linearization.*

with coprime polynomials in numerator and denominator admits a flat output $f(s)=\frac{\kappa}{s^n+a_n s^{n-1}+\cdots+a_0}u(s)$, which in terms of differential equation and scalar output equation gives: $\frac{d^n f}{dt^n}+a_{n-1}\frac{d^{n-1}f}{dt^{n-1}}+\cdots+a_0 f=\kappa u$ and

$$y=\frac{1}{\kappa}\left[b_m\frac{d^m f}{dt^m}+b_{m-1}\frac{d^{m-1}f}{dt^{m-1}}+\cdots+b_0 f\right]$$

if $x_1=f$, $x_2=\dot{f}$, ..., $x_n=f^{(n)}$

then $\frac{d}{dt}\begin{pmatrix}x_1\\ \vdots\\ x_{n-1}\\ x_n\end{pmatrix}=A\begin{pmatrix}x_1\\ \vdots\\ x_{n-1}\\ x_n\end{pmatrix}+bu$, $y=c\begin{pmatrix}x_1\\ \vdots\\ x_{n-1}\\ x_n\end{pmatrix}$

with $A=\begin{pmatrix}0 & 1 & \cdots & 0\\ \vdots & \vdots & \ddots & \vdots\\ 0 & 0 & \cdots & 1\\ -a_0 & -a_1 & \cdots & -a_{n-1}\end{pmatrix}$ $b=\kappa\begin{pmatrix}1\\ \vdots\\ 0\\ 1\end{pmatrix}$

$c=\frac{1}{\kappa}(b_0\ \cdots\ b_m\ 0\ \cdots\ 0)$

The flat output of such a system is given by $f=(0,0,\cdots 1)(C)^{-1}x$ where

$C=(b,Ab,\cdots\cdots,A^{n-1}b)$ is the Kalman controllability matrix.

Example: Given a DC motor dynamics [4]

$$\begin{aligned}L\dot{I}+RI&=v_a-k_e\omega\\ J\dot{\omega}+B\omega&=k_m I\end{aligned} \tag{23}$$

where I=Armature current, w=Angular VelocityL,R,K_e are electrical constants and J,B,k_m are mechanical constants. The state space representation is given by

$$C=\begin{pmatrix}\frac{1}{L} & -\frac{R}{L^2}\\ 0 & \frac{k_m}{JL}\end{pmatrix}\begin{pmatrix}\dot{x}_1\\ \dot{x}_2\end{pmatrix}=\begin{pmatrix}-\frac{R}{L} & \frac{-k_e}{L}\\ \frac{k_m}{J} & -\frac{B}{J}\end{pmatrix}\begin{pmatrix}x_1\\ x_2\end{pmatrix}+\begin{pmatrix}\frac{1}{L}\\ 0\end{pmatrix}u$$

$$F=\begin{pmatrix}0 & 1\end{pmatrix}C^{-1}x=\frac{JL^2}{k_m}\omega\ \ C^{-1}=\frac{JL^2}{k_m}\begin{pmatrix}\frac{k_m}{JL} & \frac{R}{L^2}\\ 0 & \frac{1}{L}\end{pmatrix} \tag{24}$$

Where C,C^{-1},F, are the controllability matrix, its inverse and flat output respectively. The control is computed using $\ddot{F}$ as follows:

$$\begin{aligned}&x_1=I=\frac{1}{k_m}\left(J\dot{F}+BF\right),\quad x_2=\omega=F\\ &u=v_a=\frac{JL}{k_m}\ddot{F}+\left(\frac{LB+RJ}{k_m}\right)\dot{F}+\left(\frac{RB}{k_m}+k_e\right)F\end{aligned} \tag{25}$$

3.2. The Implicit Representation (Lévine's Method)

Equation (1) can be locally transformed into an underdetermined implicit system

$$F(x,\dot{x})=0 \tag{26}$$

for $x\in X$, and $x,f(x,u)\in T_x X$ (Tangent space), for every u and rank $\frac{df}{du}=m$. This adopts a prolonged manifold of solutions to the implicit representation. The author in [3] extends the notion of endogenous transformation (Lie-Bäcklund Isomorphism) to the implicit system, stating that if two regular implicit systems of equation (6) are Lie-

Bäcklund equivalent then their linear cotangent approximation is locally Lie-Bäcklund equivalent. The Implicit system equation (26) is flat if and only if it is Lie-Bäcklund equivalent. The system is flat if there exists local mappings ϕ satisfying $\phi(\bar{y}_0)=\bar{x}_0$ such that $\Phi^* df_i = 0; i=1,\ldots\ldots, n-m$.

$$\Phi^* df = \frac{\partial F}{\partial x} + \frac{\partial F}{\partial \dot{x}}\frac{d}{dt} \tag{27}$$

where $\Phi^* df = P(F)$, which are actually polynomial matrices and the differential operator $\frac{d}{dt}$ is the indeterminate. The inverse of a polynomial is not a polynomial and the inverse of a square matrix is not a matrix. These polynomial matrices have the following characteristics [3]:

1. They require the use of special algebraic manipulations.
2. $P(F)\in M_{n-m,n}\left(\frac{d}{dt}\right)$ admits a Smith decomposition (or diagonal reduction) given by

$$VP(F)U = (\Delta.\theta_{n-m,m}) \tag{28}$$

3. A matrix is $M\in M_{p,q}\left(\frac{d}{dt}\right)$ is hyper-regular if and only if it's Smith decomposition leads to $I_p, 0_{p,q-p}$, if $p<q$; to I_p, if $p=q$; and to $\begin{pmatrix} I_p \\ 0_{p-q,q}\end{pmatrix}$ if $p>q$
4. A square matrix $M\in M_{p,q}\left(\frac{d}{dt}\right)$ is hyper-regular if and only if it is unimodular- denoted by $u_p\left(\frac{d}{dt}\right)$ a subgroup of invertible matrices $M_{p,q}\left(\frac{d}{dt}\right)$.
5. $P(F)$ is hyper-regular if and only if the linear cotangent approximation of the implicit system equation (26) is controllable implying that the system is flat.

These are the compact set of matrix manipulations that lead to the determination of the system's flat output.

4. Application to Synchronous Machine

4.1. Synchronous Machine Reduced Order Model

From the fourth order model of the synchronous machine , the direct axis $e_d^{'}$ can be assumed constant reducing it to a third order one-axis model [10] given by (29):

$$\tau_{d0}\dot{e}_q^{'} = e_{fd} - e_q^{'} - (x_d - x_d^{'})i_d$$

$$\frac{2H}{w_R}\frac{d^2\delta}{dt^2} = P_m - D(\omega-\omega_0) - e_d^{'}i_d - e_q^{'}i_q \tag{29}$$

$$\dot{\delta} = \omega - \omega_0$$

where

$$i_d = \frac{1}{(r_a+R_e)^2+(x_d^n+x_e)(x_q^n+x_e)}(-(r_a+R_e)(e_d^n-V_\infty\sin\delta) + (x_q^n+x_e)(e_q^n-V_\infty\cos\delta))$$

$$i_q = \frac{1}{(r_a+R_e)^2+(x_d^n+x_e)(x_q+x_e)}(-(x_d^n+x_e)(e_d^n-V_\infty\sin\delta) + (r_a+R_e)(e_q^n-V_\infty\cos\delta))$$

4.2. Implicit Method

Using Lévine's necessary and sufficient conditions for differential flatness [3] where for the system of equations (29) the system order $n=3$ and the number of system input $m=1$. The notion of linear cotangent approximation henceforth called cotangent approximation is defined thus.

Given a trajectory $t\mapsto x(t)$ of (6)of class C^∞ on an interval J of $\Re$, the linear time-varying implicit system

$$\left(\frac{\partial F}{\partial x}(x(t),\dot{x}(t))\right)\xi(t) + \left(\frac{\partial F}{\partial \dot{x}}(x(t),\dot{x}(t))\right)\dot{\xi}(t) = 0 \tag{30}$$

with $\bar{\xi}=(\xi,\dot{\xi},\ldots)\in TX$, is defined as the linear cotangent approximation of equation (6) around the trajectory x.

The system of equations (29) is first transformed to the implicit equivalent, obtained by eliminating the dynamics that contains the system input e_{fd}, and making $F(\delta,\omega,e_q^{'},\dot{\delta},\dot{\omega},\dot{e}_q^{'})$ equal 0, Such that

$$\frac{2H}{w_R}\frac{d^2\delta}{dt^2} - P_m + D(\omega-\omega_0) + e_d^{'}i_d + e_q^{'}i_q = 0; \tag{31}$$

$$\dot{\delta} - \omega + \omega_0 = 0 \tag{32}$$

The cotangent approximation to the implicit equations (31) and (32) is computed from:

$$P(F) = \left(\frac{\partial F}{\partial \delta} + \frac{\partial F}{\partial \dot{\delta}}\frac{d}{dt},\ \frac{\partial F}{\partial \omega} + \frac{\partial F}{\partial \dot{\omega}}\frac{d}{dt},\ \frac{\partial F}{\partial e_q^{'}} + \frac{\partial F}{\partial \dot{e}_q^{'}}\frac{d}{dt}\right) \tag{33}$$

It is noteworthy according to the characteristics above, that the cotangent approximation of system of equations (31) and (32) is hyper-regular if and only if it is controllable. And if it is locally flat around $\bar{x}_0$, its linear cotangent approximation around $\bar{x}_0$ is controllable. Therefore there must exist $V\in L-Smith\,(P(F))$ and $U\in R-Smith\ (P(F))$ such that

$$VP(F)U = (I_m, 0_{n-m,m}) \tag{34}$$

The cotangent approximation after applying equation (33) on equation (31) and (32) yields:

$$\begin{pmatrix} \frac{d}{dt} & -1 & 0 \\ a_{21} & a_{22} & a_{23} \end{pmatrix} \tag{35}$$

where:

$$a_{21} = \frac{\omega_0 V_\infty}{2H\det}\left(\dot{e}_d'(-R_e\cos\delta + x_{qt}\sin\delta)\dot{\delta} + e_q'(x_{dt}\cos\delta\dot{\delta} + R_e\cos\delta)\right);$$

$$a_{22} = \left(\frac{d}{dt} + \frac{\omega_0}{2H}D\right);$$

$$a_{23} = \left(\frac{\omega_0 e_d'}{2H\det}(x_{qt} - x_{dt}) + V_\infty(x_{dt}\sin\delta - R_e\cos\delta) + 2R_eV_\infty e_q'\dot{e}_q'\right);$$

and

$$\det = (r_a + R_e)^2 + (x_d' + x_e)(x_q' + x_e).$$

We now apply the Smith decomposition algorithm to equation (35) in successive polynomial matrix manipulations using unimodular matrices of rank n until $P(F)$ of rank $n-m$ reduces to lower or upper triangular polynomial matrix to prove its hyper-regularity. The unimodular matrices are constructed in such a way to shuffle the elements of the cotangent approximation matrix and achieve lower triangular form. Successive steps of the reduction are given as follows [11]:

Step a1: Multiplying equation (35) with the unimodular matrix-1 $\begin{pmatrix} 0 & 1 & 0 \\ 1 & 0 & 0 \\ 0 & 0 & 1 \end{pmatrix}$ gives

$$\begin{pmatrix} \frac{d}{dt} & -1 & 0 \\ a_{21} & a_{22} & a_{23} \end{pmatrix}\begin{pmatrix} 0 & 1 & 0 \\ 1 & 0 & 0 \\ 0 & 0 & 1 \end{pmatrix} = \begin{pmatrix} -1 & \frac{d}{dt} & 0 \\ \left(\frac{d}{dt} + \frac{\omega_0}{2H}D\right) & a_{21} & a_{23} \end{pmatrix} \tag{36}$$

Step a2: Multiplying equation (36) with unimodular matrix-2 $\begin{pmatrix} -1 & \frac{d}{dt} & 0 \\ 0 & 1 & 0 \\ 0 & 0 & 1 \end{pmatrix}$ reduces row 1 to $[1 \quad 0 \quad 0]$

$$\begin{pmatrix} -1 & \frac{d}{dt} & 0 \\ \left(\frac{d}{dt} + \frac{\omega_0}{2H}D\right) & a_{21} & a_{23} \end{pmatrix}\begin{pmatrix} -1 & \frac{d}{dt} & 0 \\ 0 & 1 & 0 \\ 0 & 0 & 1 \end{pmatrix} = \begin{pmatrix} 1 & 0 & 0 \\ -\left(\frac{d}{dt} + \frac{\omega_0}{2H}D\right) & \left(\frac{d}{dt} + \frac{\omega_0}{2H}D\right)\frac{d}{dt} + a_{21} & a_{23} \end{pmatrix} \tag{37}$$

Step a3: Multiplying equation (37) with unimodular matrix -3 $\begin{pmatrix} 1 & 0 & 0 \\ 0 & 0 & 1 \\ 0 & 1 & 0 \end{pmatrix}$ shuffles row 2 to make entry [2, 2] in (37) constant, yielding.

$$\begin{pmatrix} 1 & 0 & 0 \\ -\left(\frac{d}{dt} + \frac{\omega_0}{2H}D\right) & \left(\frac{d}{dt} + \frac{\omega_0}{2H}D\right)\frac{d}{dt} + a_{21} & a_{23} \end{pmatrix}\begin{pmatrix} 1 & 0 & 0 \\ 0 & 0 & 1 \\ 0 & 1 & 0 \end{pmatrix} = \begin{pmatrix} 1 & 0 & 0 \\ -\left(\frac{d}{dt} + \frac{\omega_0}{2H}D\right) & a_{23} & \left(\frac{d}{dt} + \frac{\omega_0}{2H}D\right)\frac{d}{dt} + a_{21} \end{pmatrix} \tag{38}$$

Step a4: Multiplying equation (38) with unimodular matrix-4 $\begin{pmatrix} 1 & 0 & 0 \\ 0 & \frac{1}{a_{23}} & -\frac{1}{a_{23}}\left(\left(\frac{d}{dt} + \frac{\omega_0}{2H}D\right)\frac{d}{dt} + a_{21}\right) \\ 0 & 0 & 1 \end{pmatrix}$ achieves the required lower triangular matrix $P(F)$.

$$\begin{pmatrix} 1 & 0 & 0 \\ -\left(\frac{d}{dt} + \frac{\omega_0}{2H}D\right) & a_{23} & \left(\frac{d}{dt} + \frac{\omega_0}{2H}D\right)\frac{d}{dt} + a_{21} \end{pmatrix}\begin{pmatrix} 1 & 0 & 0 \\ 0 & \frac{1}{a_{23}} & -\frac{1}{a_{23}}\left(\left(\frac{d}{dt} + \frac{\omega_0}{2H}D\right)\frac{d}{dt} + a_{21}\right) \\ 0 & 0 & 1 \end{pmatrix} = \begin{pmatrix} 1 & 0 & 0 \\ -\left(\frac{d}{dt} + \frac{\omega_0}{2H}D\right) & 1 & 0 \end{pmatrix} \tag{39}$$

Therefore $$P(F) = \begin{pmatrix} 1 & 0 & 0 \\ -\left(\frac{d}{dt} + \frac{\omega_0}{2H}D\right) & 1 & 0 \end{pmatrix} \tag{40}$$

Equation (40) which is a lower triangular polynomial matrix proves the hyper-regularity of equations (29). By right multiplying the unimodular matrices 1 to 4 used to generate $P(F)$, the U matrix is generated as given in equations 41 to 43:

Step b1: Unimodular matrix-1 by Unimodular matrix-2.

$$\begin{pmatrix} 0 & 1 & 0 \\ 1 & 0 & 0 \\ 0 & 0 & 1 \end{pmatrix}\begin{pmatrix} -1 & \frac{d}{dt} & 0 \\ 0 & 1 & 0 \\ 0 & 0 & 1 \end{pmatrix} = \begin{pmatrix} 0 & 1 & 0 \\ -1 & \frac{d}{dt} & 0 \\ 0 & 0 & 1 \end{pmatrix} \tag{41}$$

Step b2: Equation (41) by unimodular matrix-3

$$\begin{pmatrix} 0 & 1 & 0 \\ -1 & \frac{d}{dt} & 0 \\ 0 & 0 & 1 \end{pmatrix}\begin{pmatrix} 1 & 0 & 0 \\ 0 & 0 & 1 \\ 0 & 1 & 0 \end{pmatrix} = \begin{pmatrix} 0 & 0 & 1 \\ -1 & 0 & \frac{d}{dt} \\ 0 & 1 & 0 \end{pmatrix} \tag{42}$$

Step 3 Equation (42) by unimodular matrix-4

$$\begin{pmatrix} 0 & 0 & 1 \\ -1 & 0 & \frac{d}{dt} \\ 0 & 1 & 0 \end{pmatrix}\begin{pmatrix} 1 & 0 & 0 \\ 0 & \frac{1}{a_{23}} & -\frac{1}{a_{23}}\left(\left(\frac{d}{dt}+\frac{\omega_0}{2H}D\right)\frac{d}{dt}+a_{21}\right) \\ 0 & 0 & 1 \end{pmatrix} = \begin{pmatrix} 0 & 0 & 1 \\ -1 & 0 & \frac{d}{dt} \\ 0 & \frac{1}{a_{23}} & -\frac{1}{a_{23}}\left(\left(\frac{d}{dt}+\frac{\omega_0}{2H}D\right)\frac{d}{dt}+a_{21}\right) \end{pmatrix} \quad (43)$$

Equation (43) as U can be arranged compactly

$$U = \begin{pmatrix} 0 & 0 & 1 \\ -1 & 0 & \frac{d}{dt} \\ 0 & \frac{1}{a_{23}} & A_{33} \end{pmatrix} \quad \text{Where } A_{33} = -\frac{1}{a_{23}}\left(\frac{d}{dt}+\frac{\omega_0}{2H}D\right)\frac{d}{dt}+a_{21}.$$

Thus from

$$\hat{U} = U\begin{pmatrix} 0_{2,1} \\ I_1 \end{pmatrix} \quad (44)$$

$$\hat{U} = \begin{pmatrix} 1 \\ \frac{d}{dt} \\ -\frac{1}{a_{23}}\left(\frac{d}{dt}+\frac{\omega_0}{2H}D\right)\frac{d}{dt}+a_{21} \end{pmatrix} \quad (45)$$

Using the definition

$$Q\hat{U} = \begin{pmatrix} 1 \\ 0 \\ 0 \end{pmatrix} \quad (46)$$

it is possible to compute by further matrix manipulations $Q \in L-Smith(\hat{U})$ which yields

$$Q = \begin{pmatrix} 1 & 0 & 0 \\ -\frac{d}{dt} & 1 & 0 \\ -A_{33} & 0 & 1 \end{pmatrix} \quad (47)$$

where A_{33} is as defined above. Multiplying Q by the vector $(d\delta, d\omega, de_q^{'})^T$, the last two entries in the resulting vector are: $-\frac{d}{dt}d\delta + d\omega$ and $\left(\frac{1}{a_{23}}\left(\frac{d}{dt}+\frac{\omega_0}{2H}D\right)\frac{d}{dt}+a_{21}\right)d\delta + de_q^{'}$ which by (35) vanishes identically on $\overline{X}_0$. The first entry of the vector is therefore given by:

$$\begin{pmatrix} 1 & 0 & 0 \end{pmatrix}\left(d\delta, d\omega, de_q^{'}\right)^T = dy \quad (48)$$

Equation (48) is trivially strongly closed such that

$$d\delta = dy \quad (49)$$

and so gives the flat output

$$y = \delta$$

Verification of the Flat Output of the Third Order Single-Input (SMIBS) model is done by showing that all the system states and variables are a function of the flat output and its derivatives.

$$\omega = \dot{\delta} + \omega_0 \quad (50)$$

such that

$$\dot{\omega} = \ddot{\delta} \quad (51)$$

and thus

$$\frac{1}{\det}R_e e_q^{'2} - \frac{1}{\det}((x_{dt} - x_{qt})e_d^{'} - V_\infty(-x_{dt}\sin\delta - R_e\cos\delta))e_q^{'} - P_m + D(\omega - \omega_0) + \frac{e_d^{'}}{\det}(R_e e_d^{'} + V_\infty(-R_e\sin\delta - x_{qt}\cos\delta)) + \frac{2H}{\omega_0}\ddot{\delta} = 0 \quad (52)$$

Equation (52) is a quadratic function that can be evaluated for $e_q^{'}$.

Since the system states have been shown to be functions of the flat output and its derivatives, it follows that all other system variables which are functions of the states are also functions of the flat output and its derivatives.

Hence:

$$\zeta_i = f_i(\delta, \dot{\delta}, \ddot{\delta}) \ \forall \ \zeta_i \in [i_d, i_q, v_{dt}, v_{qt}, V_t] \quad (53)$$

4.3. Compensator Design and Simulation Results

It has been shown in the preceding section that the components of the system states and other system variables depending on the system states can be expressed as real-analytic functions of the component of δ and a finite number of its derivatives thus:

$$x = A(\delta, \dot{\delta}, \ddot{\delta}) \quad (54)$$

The dynamic feedback is shown to be endogenous since the converse holds, that is, the flat output y is expressed as a real-analytic function of δ one of the states of the system. Thus the state of the SMIBS is a function of the linearizing output δ and its derivatives up to order $\alpha = 2$. The endogenous feedback system to the following closed loop system is of order $\alpha + 1 = 3$, so that from the linear system

$$\dddot{\delta} = v \quad (55)$$

the compensator follows. Considering the systems' dynamical equations, perform the following state transformations:

$$\begin{aligned} \dot{z}_1 &= z_2 = \dot{y}_1 = \dot{\delta} = \omega - \omega_0 \\ \dot{z}_2 &= z_3 = \ddot{y}_1 = \ddot{\delta} = \dot{\omega} \\ \dot{z}_3 &= \dddot{y}_1 = \dddot{\delta} = \ddot{\omega} = v \end{aligned} \tag{56}$$

This yields the equivalent normal form for the system, from which we can compute the nonlinear controller by inverting the expressions from $\ddot{\omega}$ and e_{fd}. The state transformations are invertible and exist throughout the transient operating zone $0 < \delta < 180^o$. Using the network parameters of figure (3), the resulting excitation control is given by [11]:

$$e_{fd} = \frac{\tau_{d0}}{E}\left(\frac{2Hv}{\det\omega_0} + \frac{D\dot{\omega}}{\det} + A\dot{e}'_d + Be'_d - Ce'_q\right) + e'_q + (x_d - x'_d)i_d \tag{57}$$

where,

$$A = 2R_{eT}\dot{e}'_d - R_{eT}V_\infty \sin\delta - x_{qt}V_\infty \cos\delta;$$

$$B = x_{qt}V_\infty \sin(\delta)\dot{\delta} - R_{eT}V_\infty \cos(\delta)\dot{\delta};$$

$$C = (x_{dt} - x_{qt})\dot{e}'_d - x_{dt}V_\infty \cos(\delta)\dot{\delta} - R_{eT}V_\infty \sin(\delta)\dot{\delta};$$

$$E = (x_{dt} - x_{qt})e'_d - x_{dt}V_\infty \sin\delta - 2R_{eT}e'_q + R_{eT}V_\infty \cos\delta;$$

and

$$\dot{e}'_d = \frac{1}{\det}((x_q - x'_d) + x_{dt})(x_{dt}\cos(\delta) + R_{eT}V_\infty \sin(\delta)\dot{\delta}) + R_{eT}e'_q;$$

$$R_{eT} = (r_a + R_e);\; x_{dt} = (x'_d + x_e);\; x_{qt} = (x'_q + x_e)\cdot$$

e_{fd} is hereby proved also to be a function of the flat variable and its derivatives, that is

$$e_{fd} = \beta(\delta, \dot{\delta}, \ddot{\delta}) \tag{58}$$

The loop closure is then done to stabilize the system.

$$v = -k_1(\delta - \delta_1^*) - k_2(\dot{\delta} - \dot{\delta}_1^*) - k_3(\ddot{\delta} - \ddot{\delta}_1^*) \tag{59}$$

and choose k_i appropriately such that the linear time invariant error dynamics

$$e^{(3)} = k_1 e + k_2\dot{e} + k_3\ddot{e} \tag{60}$$

where $e^{(j)} = \delta^{(j)} - (\delta^*)^{(j)}$ are stable.

Equation (57) is the control law referred to as Field Voltage Dynamic Feedback Controller (FVDFC), [11] while (59) is the linear input that stabilizes the system to equilibrium. Simulation of the system was done by connecting the synchronous machine as a single machine infinite bus system (SMIBS) under a short circuit fault situation as shown in Figure 3.

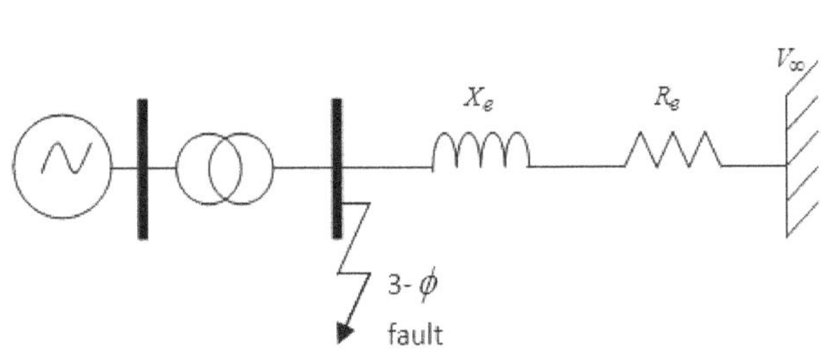

Figure 3. *Fault Location on the Single Machine Infinite Bus System (SMIBS)*

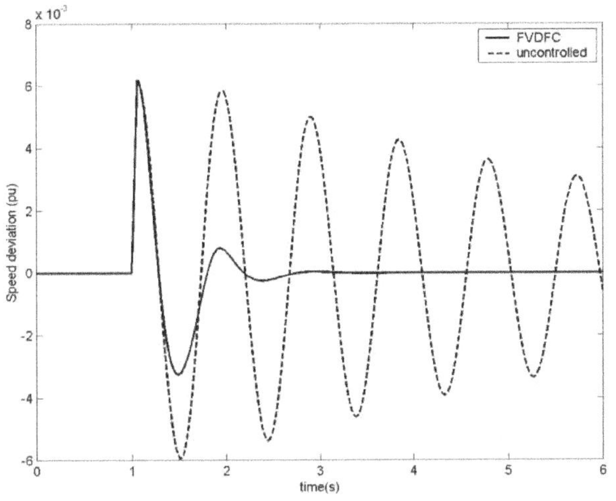

Figure 4. *Responses of Speed Deviation to 3-Cycle Fault with and without FVDFC.*

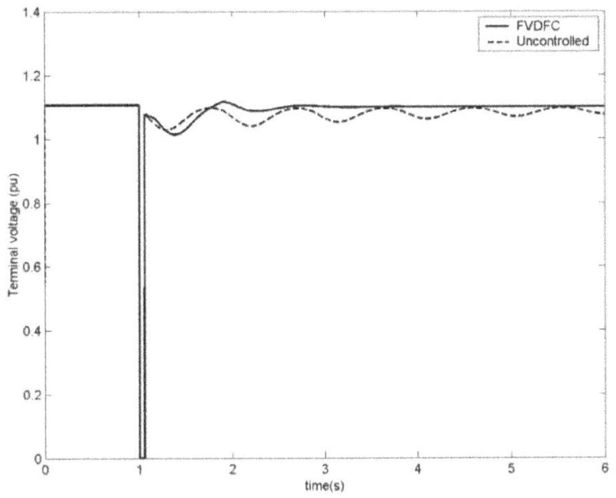

Figure 5. *Responses of Terminal Voltage to 3-Cycle Fault with and without FVDFC.*

Some simulation results with the system equipped with the designed controller are presented in Figures 4 and 5 which are representative of the system performance. These figures clearly show the responses of the controller to a three-phase short circuit fault of 3-cycles duration. The system was restored to steady state operating point as the controller damped the fault oscillations under three seconds as shown by the machine speed deviation and the corresponding terminal voltage. The oscillations in the uncontrolled system were not damped within the same time duration.

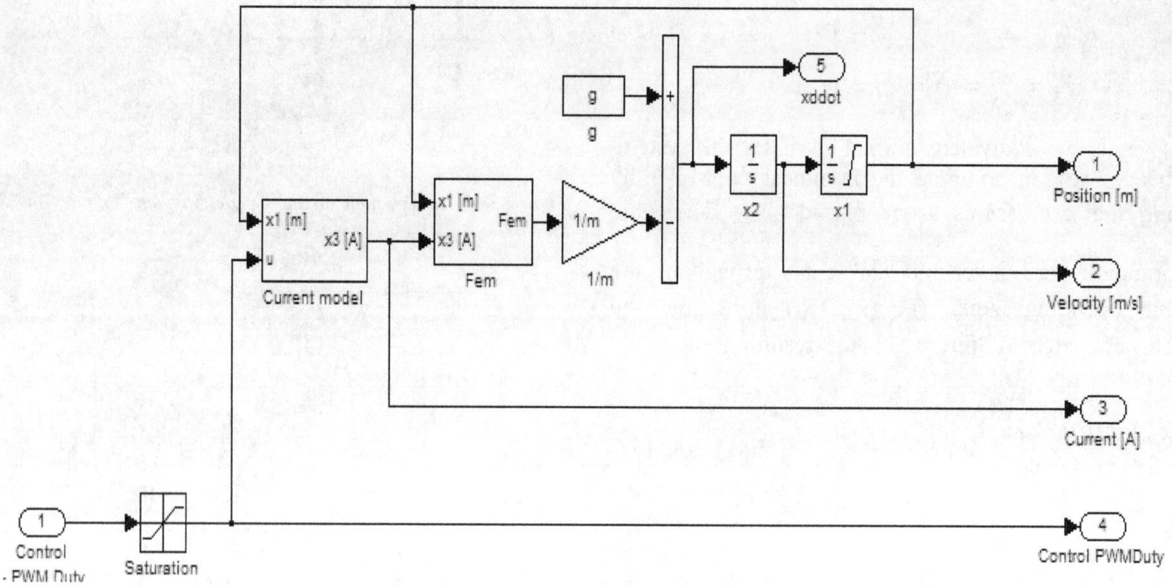

Figure 6. Block diagram of INTECO™ maglev model

5. Application to Magnetic Levitation

The model development of the magnetic levitation is based on the system developed by INTECO™ for the purpose of teaching. The system block diagram is shown in Figure 6. INTECO used empirical analysis to model the control of the current that goes to the electromagnet. The resulting linear relationship is found to be a straight line $i(u) = au + b$ with a dead zone. The constants a and b are determined from the experimental data. The system dynamics are described in (61) – (63).

$$\dot{x}_1 = x_2 \tag{61}$$

$$\dot{x}_2 = g - \frac{1}{m}x_3^2\left(\frac{f\_p_1}{f\_p_2}\right)e^{\left(\frac{-x_1}{f\_p_2}\right)} \tag{62}$$

$$\dot{x}_3 = (k_i u + c_1 - x_3)\frac{1}{\left(\frac{p_1}{p_2}\right)e^{\left(\frac{-x_1}{p_2}\right)}} \tag{63}$$

Where g is gravitational force, m is mass of object, $f\_p_1, f\_p_2, p_1, p_2, k_i, c_1$ are system constants.

Flat output

The flat output can be determined using Levine's method by applying the implicit function theory and eliminating the dynamics with control. The variational equation is given by[12]:

$$d\dot{x}_1 - ae^{\left(\frac{-x_1}{f-p_2}\right)}x_3^2\,dx_1 - ae^{\left(\frac{-x_1}{f-p_2}\right)}x_3^2\,dx_3 \tag{64}$$

Where $a = \frac{1}{m}\left(\frac{f\_p_1}{f\_p_2}\right)$

The polynomial matrix will therefore be

$$p(f) = \left[\frac{d^2}{dt^2} - ae^{\left(\frac{-x_1}{f\_p_2}\right)}x_3^2 \quad -ae^{\left(\frac{-x_1}{f\_p_2}\right)}x_3^2\right]\begin{bmatrix} dx_1 \\ dx_3 \end{bmatrix} \tag{65}$$

or compactly

$$p(f) = \begin{bmatrix} A & -b \end{bmatrix}\begin{bmatrix} dx_1 \\ dx_3 \end{bmatrix} \tag{66}$$

Where $A = \frac{d^2}{dt^2} - ae^{\left(\frac{-x_1}{f\_p_2}\right)}x_3^2$ - a polynomial and $b = -ae^{\left(\frac{-x_1}{f\_p_2}\right)}x_3^2$. Using Smith's algorithm for the manipulation of polynomial matrices, the following right Smith steps are performed.

$\begin{bmatrix} A & -b \end{bmatrix}\begin{bmatrix} 0 & 1 \\ -\frac{1}{b} & \frac{1}{b}A \end{bmatrix} = \begin{bmatrix} 1 & 0 \end{bmatrix}$, therefore $\hat{U} = \begin{bmatrix} 1 \\ \frac{1}{b}A \end{bmatrix}$, such that

$$Q\hat{U} = \begin{bmatrix} 1 & 0 \\ \frac{1}{b}A & -1 \end{bmatrix}\begin{bmatrix} 1 \\ 0 \end{bmatrix} \text{ as required.} \tag{67}$$

Therefore,

$$Q\,dx = \begin{bmatrix} 1 & 0 \\ \frac{1}{b}A & -1 \end{bmatrix}\begin{bmatrix} dx_1 \\ dx_3 \end{bmatrix} \tag{68}$$

Such that the first line reads $dy = dx_1$ which gives $y = x_1$ the flat output, while the second line is identically equal to zero from (66) showing the flatness of the system dynamics. This follows that the flat output of this maglev model is the ball position which is also a system variable.

5.1. Compensator Design and Simulation Results

From the computed flat output the control law follow from the following compensator

$$
\begin{aligned}
y &= x_1 \\
\dot{y} &= \dot{x}_1 = x_2 \\
\ddot{y} &= \ddot{x}_1 = \dot{x}_2 \\
\dddot{y} &= \dddot{x}_1 = \ddot{x}_2 = u_L
\end{aligned} \tag{69}
$$

From (62), we have

$$
x_3 = \left(\frac{m(g-\dot{x}_2)}{\frac{f\_p_1}{f\_p_2} e^{\frac{-x_1}{f\_P_2}}} \right)^{\frac{1}{2}} \tag{70}
$$

And from $\dot{x}_3$ and (69) the control law is computed as

$$
u = x_3 - c_1 + \frac{1}{2} \left(\frac{m\ddot{x}_2 + (m(g-\dot{x}_2))\frac{1}{f\_p_2}\dot{x}_1 M_p}{\left((m(g-\dot{x}_2))^{\frac{1}{2}} \left(\frac{f\_p_1}{f\_p_2} e^{\left(\frac{-x_1}{f\_P_2}\right)} \right)^{\frac{1}{2}} \right)} \right) \frac{1}{k_i} \tag{71}
$$

where $M_p = \frac{p_1}{p_2} e^{\frac{-x_1}{p_2}}$

And the linear control is given by

$$
u_L = -k_1(\delta - \delta^*) - k_2(\dot{\delta} - \dot{\delta}^*) - k_3(\ddot{\delta} - \ddot{\delta}^*) \tag{72}
$$

The gains k_i are chosen such that the linear time invariant error dynamics

$$
e^{(3)} = -k_1 e - k_2 \dot{e} - k_3 \ddot{e} \tag{73}
$$

where $e^{(j)} = \delta^{(j)} - (\delta^*)^{(j)}$ are stable. To compute the gains, (72) can be rewritten as a Hurwitz polynomial by

$$
s^3 + k_3 s^2 + k_2 s + k_1 = 0 \,. \tag{74}
$$

The closed loop characteristic polynomial of a third order equivalent system is given in terms of the natural frequency and damping ratio by

$$
(s^2 + 2\xi\omega_n s + \omega_n^2)(s + \beta) \tag{75}
$$

such that comparing (74) and (75) gives

$$
k_1 = \beta\omega_n, \; k_2 = 2\xi\omega_n\beta + \omega_n^2 \,, \; k_3 = \beta + 2\xi\omega_n
$$

Figures 7 and 8 shows the ball position and the Flatness control applied to stabilize it. The results are for a ten second simulation of the maglev system to levitate a ball to a set point of 0.006 m.

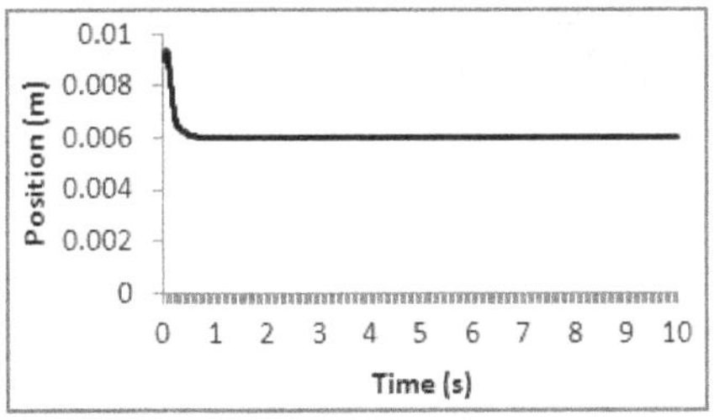

Figure 7. *Ball position for a ten second simulation*

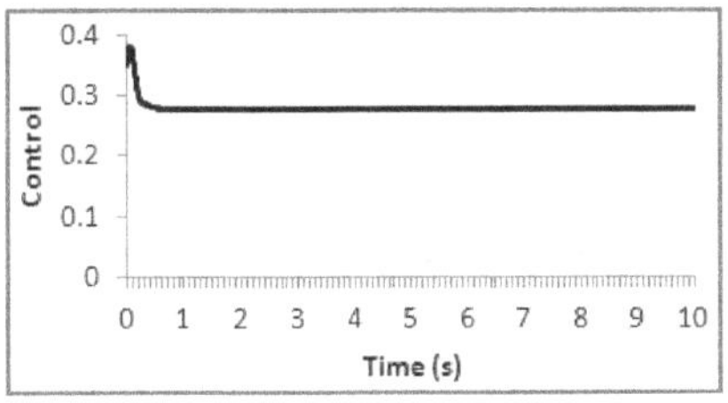

Figure 8. *Applied controls for a ten second simulation*

Figs. 9 and 10 show the response of the system to decreasing set point levels like in descending a staircase. This task seems to be a challenging control task as can be seen by the sloppy response of the PID controller used on the same system as seen in fig 10. The flatness based controller did not show the same sloppy behavior for the descending set point levels as seen in fig 9. The PID behaved like it is having difficulty coping with the sharp transitions of the ball position. Studies of other systems show that the flatness controller gives a strong first swing control and as well improves stability margin of the system.

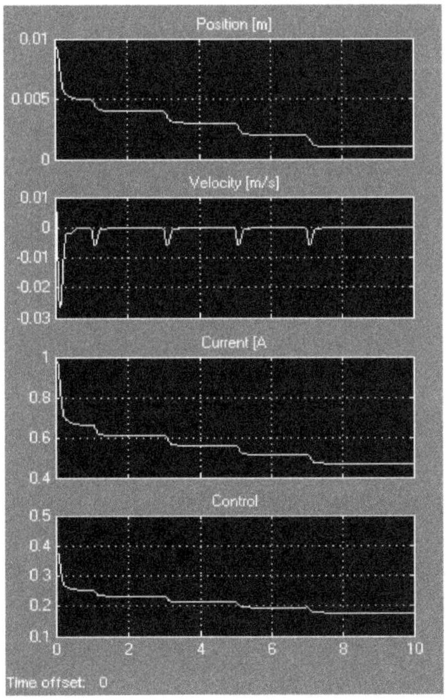

Figure 9. *Response to input [.005, .004, .003, .002, .001] mm using the Flatness based controller*

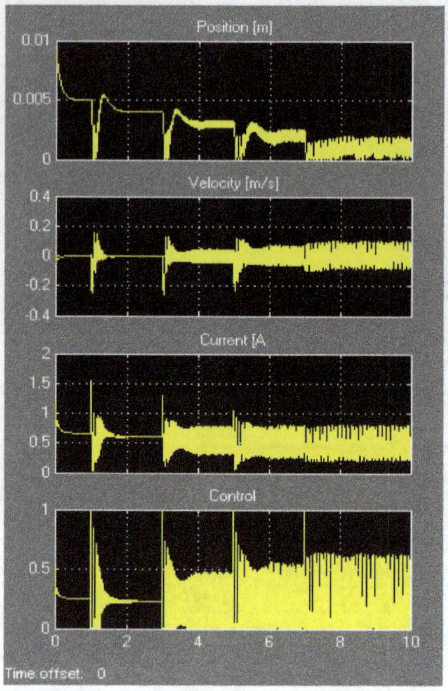

Figure 10. *Response to input [.005, .004, .003, .002, .001] mm using the PID controller*

6. Conclusion

This paper presented some basic theory of flatness-based feedback linearization, a variant of the well-known techniques of feedback linearization. Theoretical formulation and examples to enhance learning of the concept of flatness and how it is computed for certain industrial systems is given. A novel method of computation of the flat output developed by Jean Levine is introduced and two industrial systems used to illustrate its efficacy. An application to the synchronous machine and magnetic levitation system was achieved by constructing a control law around the flat output. The method requires the mathematical analysis of system models for flatness - a condition that describes how well characterized the model is with a view to determining its possession of a "virtual" (flat) output driven by contributions made by the system state variables. This output was determined for the given models and used to obtain corresponding feedback laws for the transformed linear systems and equipped with a linear controller used to stabilize the systems to steady state or damp system oscillations induced by fault. For the one-axis single input synchronous machine (SMIBS) model there exists a flat output the *rotor angle (delta)*– a system variable while for the magnetic levitation system the flat output computed is the *ball position* which is also a system variable. The simulation results obtained agreed with the expectations and performed well when compared with the PID schemes.

References

[1] Fliess M., Lévine J., Martin Ph., and Rouchon P. (1999) "A Lie-Bäcklund approach to equivalent and flatness of nonlinear systems", *IEEE Transactions on Automatic Control, 38: 700-716.*

[2] Fliess M., Lévine J., Martin Ph., and Rouchon P. (1993) "Flatness and defect of nonlinear systems: introductory theory and examples", *Int. J. of Control, 61(6): 1327-1361.*

[3] Lévine J. (2003) Revised (2006) "On the necessary and sufficient conditions for differential flatness" Electronic Print, Digital Library for Physics and Astronomy, Harvard-Smithsonian center for Astrophysics, (arXiv: math/0605405v).

[4] Hebertt Sira-Ramirez, and Sunil K. Agrawal (2004) Differentially flat systems, Marcel Dekker, Inc, New York

[5] Fliess M., Lévine J., Martin Ph., and Rouchon P. (1993) "Flatness and defect of nonlinear systems: introductory theory and examples", *Int. J. of Control, 61(6): 1327-1361.*

[6] Rouchon P., Fliess M, Lévine J., and Martin Ph. (1993) "Flatness and motion planning: the car with n-trailers". *In Proc. ECC'93, Groningen, Pages 1518-1522.*

[7] Lévine J. (1999) Are there new industrial perspectives in the control of mechanical systems? In Paul M. Frank, editor, Advances in Control, pages 195–226. Springer-Verlag, London.

[8] Kiss B., Lévine J., and Mullhaupt Ph. (2000) Control of a reduced size model of us navy crane using only motor position sensors. In A. Isidori, F. Lamnabhi-Lagarrigue, and W. Respondek, editors, Nonlinear Control in the Year 2000, volume 2, pages 1–12. Springer.

[9] Charlet B., Lévine J., and Marino R. (1991) "Sufficient conditions for dynamic state feedback linearization", SIAM *J. of Control and Optimization, 29(1):38-57.*

[10] Anderson P.M and Fouad A.A (1994) Power system control and stability, IEEE series on Power Systems.

[11] E. C. Anene, J. T. Agee, U. O. Aliyu and J. Levine, (2006) "A new technique for feedback linearisation and an application in power system stabilisation". IASTED International Conference on POWER, ENERGY and APPLICATIONS Gaborone Botswana, September.11-13, Pages 90-95.

[12] Ejike Anene, Ganesh K. Venayagamoorthy (2010), *Senior Member, IEEE* "PSO tuned flatness based control of a magnetic levitation system", 45th IEEE Industrial Automation and Control Annual Conference, 3rd – 7th October, Houston Texas.

Permissions

All chapters in this book were first published in ACIS, by Science Publishing Group; hereby published with permission under the Creative Commons Attribution License or equivalent. Every chapter published in this book has been scrutinized by our experts. Their significance has been extensively debated. The topics covered herein carry significant findings which will fuel the growth of the discipline. They may even be implemented as practical applications or may be referred to as a beginning point for another development.

The contributors of this book come from diverse backgrounds, making this book a truly international effort. This book will bring forth new frontiers with its revolutionizing research information and detailed analysis of the nascent developments around the world.

We would like to thank all the contributing authors for lending their expertise to make the book truly unique. They have played a crucial role in the development of this book. Without their invaluable contributions this book wouldn't have been possible. They have made vital efforts to compile up to date information on the varied aspects of this subject to make this book a valuable addition to the collection of many professionals and students.

This book was conceptualized with the vision of imparting up-to-date information and advanced data in this field. To ensure the same, a matchless editorial board was set up. Every individual on the board went through rigorous rounds of assessment to prove their worth. After which they invested a large part of their time researching and compiling the most relevant data for our readers.

The editorial board has been involved in producing this book since its inception. They have spent rigorous hours researching and exploring the diverse topics which have resulted in the successful publishing of this book. They have passed on their knowledge of decades through this book. To expedite this challenging task, the publisher supported the team at every step. A small team of assistant editors was also appointed to further simplify the editing procedure and attain best results for the readers.

Apart from the editorial board, the designing team has also invested a significant amount of their time in understanding the subject and creating the most relevant covers. They scrutinized every image to scout for the most suitable representation of the subject and create an appropriate cover for the book.

The publishing team has been an ardent support to the editorial, designing and production team. Their endless efforts to recruit the best for this project, has resulted in the accomplishment of this book. They are a veteran in the field of academics and their pool of knowledge is as vast as their experience in printing. Their expertise and guidance has proved useful at every step. Their uncompromising quality standards have made this book an exceptional effort. Their encouragement from time to time has been an inspiration for everyone.

The publisher and the editorial board hope that this book will prove to be a valuable piece of knowledge for researchers, students, practitioners and scholars across the globe.

List of Contributors

H. Souilem and N. Derbel
National School of Engineers of Sfax BP.W, 3038, Sfax-Tunisia

H. Mekki
National School of Engineers of Sousse

A. A. Ojugo and D. Oyemade
Department of Mathematics/Computer Sci, Federal University of Petroleum Resources Effurun, Delta State

R. E. Yoro
Department of Computer Science, Delta State Polytechnic Ogwashi-Uku, Delta State, Nigeria

A. O. Eboka, M. O. Yerokun and E. Ugboh
Department of Computer Sci. Education, Federal College of Education (Technical), Asaba, Delta State

Matej Ciba and Ivan Sekaj
Institute of Control and Industrial Informatics, Bratislava, Slovakia

A. Fratu
Transilvania University of Brasov, 500036 Brasov, Romania

M. Dambrine, L. Vermeiren and A. Dequidt
Univ. Lille Nord de France, F-59000 Lille, France
UVHC, LAMIH, F-59313 Valenciennes, France
CNRS, UMR 8530, F-59313 Valenciennes, France

Sonja Pravilović
Montenegro Business School, "Mediterranean" University, Podgorica, Montenegro

Dipartimento di Informatica, Universita degli Studi di Bari Aldo Moro, Bari, Italy

Annalisa Appice
Dipartimento di Informatica, Universita degli Studi di Bari Aldo Moro, Bari, Italy

Suparna Roy, Dhrubojyoti Banerjee, Chiranjib Guha Majumder and Amit Konar
ETCE Department, Jadavpur University, Kolkata-700032

R. Janarthanan
Jaya College of Engineering, Chennai, Tamil Nadu

Shinichi Tamura
NBL Technovator Co., Ltd, 631 Shindachimakino, Sennan City, 590-0522 Japan
Graduate School of Medicine, Osaka University, Suita, 565-0871 Japan

Yoshi Nishitani, Tomomitsu Miyoshi and Hajime Sawai
Graduate School of Medicine, Osaka University, Suita, 565-0871 Japan

Takuya Kamimura and Yasushi Yagi
ISIR, Osaka University, 8-1 Mihogaoka, Ibaraki City, Osaka, 567-0047 Japan

Chie Hosokawa
AIST Kansai, 1-8-31 Midorigaoka, Ikeda 563-8577 Japan

Yuko Mizuno-Matsumoto
Graduate School of Applied Informatics, University of Hyogo, Kobe, 650-0047 Japan

Yen-Wei Chen
Graduate School of Sci. and Eng., Ritsumeikan University, Kusatsu, 525-8577 Japan

Atul Ikhe and Anant Kulkarni
P. G. Department, College of Engineering Ambajogai, Dist. Beed, Maharashtra, India

Neeraj Srivastava, Deoraj Kumar Tanti and Md Akram Ahmad
Electrical Engineering Department, BIT Sindri, Dhanbad, India

Matej Ciba and Ivan Sekaj
Institute of Control and Industrial Informatics, Bratislava, Slovakia

L. Guenfaf
LSEI Laboratory, USTHB University BP 32 El Alia 16111, Bab Ezzouar, Algiers, Algeria

M. Djebiri, M. S. Boucherit and F. Boudjema
LCP, Laboratory, ENP, Hassan Badi El harrach, Algiers, Algeria

Nguyen Trong Cac and Nguyen Van Khang
School of Electronics and Telecommunications, Hanoi University of Science and Technology, Hanoi, Vietnam

Yasabie Abatneh and Omprakash Sahu
Department of Chemical Engineering, KIOT, Wollo University, Kombolcha, Ethiopia

Koji Abe
Kinki University, Osaka, Japan

Takeshi Tahori
Contec EMS, Co. Ltd, Japan

Masahide Minami
The University of Tokyo, Japan

Munehiro Nakamura
Kanazawa University, Kanazawa, Japan

Haiyan Tian
Chongqing University, Chongqing, China

YUE Xiangyu
School of Management and Engineering, Nanjing University; Nanjing Jiangsu; 210093; PR China

Faraj El Dabee, Romeo Marian and Yousef Amer
School of Engineering, University of South Australia, South Australia, Australia

Kingsley Monday Udofia
Department of Elect/Elect/Computer Engineering, University of Uyo, Uyo, Nigeria

Joy Omoavowere Emagbetere and Frederick Obataimen Edeko
Department of Electrical/Electronic Engineering, University of Benin, Benin City, Nigeria

Reza Narimani
Department of Financial Engineering, University of Economic Sciences, Tehran, Iran

Ahmad Narimani
Department of Economics, University of AllamehTabatabae'i, Tehran, Iran

Yuan Wang and Keming Tang
College of Information Science and Technology, Yancheng Teachers University, Yancheng 224002, People's Republic of China

Zhudeng Wang
School of Mathematical Sciences, Yancheng Teachers University, Jiangsu 224002, People's Republic of China

Hiroki Shibasaki, Takehito Fujio, Ryo Tanaka and Hiromitsu Ogawa
Graduate School of Science and Technology, Meiji University, Kawasaki, JAPAN

Yoshihisa Ishida
Graduate School of Science and Technology, Meiji University, Kawasaki, JAPAN

School of Science and Technology, Meiji University, Kawasaki, JAPAN.

Yahya Hassanzadeh-Nazarabadi and Abolfazl Saravani
Ferdowsi University, Park Sq, Mashhad, IRAN

Bahareh Alizadeh
Khayam University, Ghasem Abad, Mashhad, IRAN

Mustafa Demetgul
Marmara University, Technology Faculty, Department of Mechatronics Engineering, Istanbul, TURKEY

Osman Ulkir and Tayyab Waqar
Marmara University, Institute of Pure and Applied Sciences, Department of Mechatronics, Istanbul, TURKEY

Luo Yiping, Xu Biao, Ren Hongjuan and Chen Fuzhi
College of Automobile Engineering, Shanghai University of Engineering Science, Shanghai, China

Khalid Abd
School of Engineering, University of South Australia, Mawson Lakes 5095, South Australia

School of Industrial Engineering, University of Technology, Baghdad, Iraq

Kazem Abhary and Romeo Marian
School of Engineering, University of South Australia, Mawson Lakes 5095, South Australia

René Ahn, Emilia Barakova, Loe Feijs, Mathias Funk, Jun Hu and Matthias Rauterberg
Designed Intelligence Group, Industrial Design Department, Eindhoven University of Technology, Eindhoven, the Netherlands

Ejike C. Anene
Electrical Engineering Programme, Abubakar Tafawa Balewa University, PMB 0248, Bauchi, Nigeria

Ganesh K. Venayagamoorthy
Real-Time Power and Intelligent Systems Laboratory, Clemson University, Clemson, USA

Index

CPSIA information can be obtained
at www.ICGtesting.com
Printed in the USA
BVHW01*1208250718
522613BV00005B/24/P

9 781682 853481